Preface

The study of the environmental changes of the last three million years, which forms the subject of this book, can be approached from a wide variety of viewpoints. The viewpoint adopted here is that of the geographer—the aim being to show how the physical environment and landscape of the Earth have changed during the time that man has been living on the Earth, and to suggest, by example, some ways in which the great environmental changes may have influenced his development.

These environmental changes include not only those of climate, but also changes of sea-level, of vegetation associations, desert limits, lake levels, river discharges, hurricane frequencies, sea-ice cover, numbers of mammals, and many others. Particular attention is paid to the degree and frequency of change, for it is often insufficiently appreciated how frequent and substantial the changes have been, even within historical time. To understand the nature and origin of present-day soils, landforms, and floral and faunal distributions, it is essential to be aware of their history and evolution. Many features of the environment and of the landscape are not necessarily in equilibrium with present processes, and thus it is frequently inappropriate to examine them purely in terms of currently functioning systems.

The changes which man himself has made to his environment and landscape, however, only form an incidental part of this book. This is not because the influence of man is considered to be unimportant—the reverse is the case—but because it could in itself form the basis of a lengthy volume. However, particularly in the last century or so, man has become an increasingly potent agent of environmental change, and to put 'natural' environmental changes into context, some reference has had to be made to these developments.

Another feature of this volume is that, inevitably, the intensity of treatment of the various phases of the last three million years increases as one moves towards the present. This reflects two underlying facts: our knowledge becomes less uncertain, and our chronology more exact, as the present is approached; and the relationships of environmental changes to human affairs is more evident, not least because of the exponential growth of human population. Nevertheless, this book is not meant to be crudely deterministic. All it seeks to do is to show that changes have taken place in man's environment to a remarkable degree, and to indicate some of the ways in which such changes may have been related to the development of man and the landscape.

During the preparation of this work I have been conscious of the debt which I owe to numerous teachers from my undergraduate days. At Cambridge the degree of dedication to the Quaternary is considerable, and this once led an Oxford visitor to make the quip that 'If the Ice Ages had not existed Cambridge would have had to invent them so as to have something to lecture on.' Soil stripes in the Breckland, the ice wedges of Fen margin gravel pits, the fossil frost mounds of the Lark valley, the organic deposits of the West Runton section, and the inland dunes at Lakenheath Warren were part of one's undergraduate diet. The post-graduate diet was scarcely more varied, but the flavours were hotter, for I was offered the opportunity to study environmental changes in the Kalahari sandveld, the Namib desert, the lakes region of the southern Ethiopian Rift Valley, and the fringes of the Thar. At an even earlier stage I owe much to P. G. Foster, E. S. Hoare, M. A. Girling, and C. Kenyon. Without their stimulus, flair, and dedication I would not have become a geographer.

Amongst those I would like to thank for giving me field experience are Dick and Jean Grove, Bridget and Raymond Allchin, and K. T. M. Hegde. I should also like to thank David Stoddart for encouraging me to write this book, Councillor John Patten for tolerating my early efforts to produce this volume in a Woodstock cottage, the Librarians of the School of Geography and the Radcliffe Science Library for their resourcefulness and assistance, and Peter Masters and Chris Jackson for helping in a multitude of tasks. The Figures were kindly drawn by Mrs Ursula Miles and Miss Margaret Loveless, and I received some helpful comments from C. G. Smith, Fellow of Keble College, Professor Gordon Manley, and Miss Alayne Street. Some of the material in this book was tried out on a remarkable trio of Hertford undergraduates, Mary Francis, Ken Pye and John Johnson, to whom go my best wishes.

A. S. GOUDIE

Oxford.

Environmental Change

This Book
belongs to the Library of
King Edward VI's
Grammar School,
Guildford, Surrey.

Environmental

Change

THIRD EDITION

Andrew Goudie

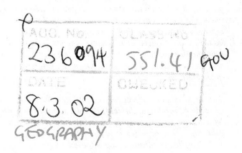
CLARENDON PRESS · OXFORD

Oxford University Press, Great Clarendon Street, Oxford OX2 6DP

Oxford New York
Athens Auckland Bangkok Bogota Bombay
Buenos Aires Calcutta Cape Town Dar es Salaam
Delhi Florence Hong Kong Istanbul Karachi
Kuala Lumpur Madras Madrid Melbourne
Mexico City Nairobi Paris Singapore
Taipei Tokyo Toronto
and associated companies in
Berlin Ibadan

Oxford is a trade mark of Oxford University Press

Published in the United States by
Oxford University Press Inc., New York

Hardback and paperback editions first published 1992
Paperback edition reprinted 1993, 1994, 1995, 1996

British Library Cataloguing in Publication Data
Data available

Library of Congress Cataloging in Publication Data
Goudie, Andrew.
Environmental change/Andrew Goudie.—3rd ed.
Includes bibliographical references and index.
1. Geology, Stratigraphic—Quaternary.   2. Glacial epoch.
3. Paleoclimatology.   I. Title.
QE696.G68   1992   551.7'92—dc20   91–45544
ISBN 0–19–874166–9
ISBN 0–19–874167–7 (Pbk.)

Printed in Great Britain by Biddles Ltd.,
Guildford and King's Lynn

Preface to the third edition

Environmental change has become a matter of great academic, public, and political concern over the last few years. As a consequence the quantity of work that is being done to ascertain the pattern and causes of change is enormous. A consequence of this is that I have had to undertake a very substantial revision of almost all parts of this book. The structure is essentially the same as in previous editions, but whole areas have been rewritten. Although human influences are being regarded as of ever greater significance (particularly in Holocene and modern times) I have deliberately refrained from giving them full treatment here. To do otherwise would be to make this volume excessively long and to repeat material and ideas that have appeared elsewhere.

ASG

1992

Contents

List of Figures

List of Tables

1

Introduction

This book is concerned with the history of the time during which humans have inhabited the Earth, a period which the geologist calls the Quaternary. The Russians, mindful that this period was unique because of man's presence, have sometimes called it the *Anthropogene*. Man, the 'tool-making animal', has only been evident on our planet for 2 to 3 million years, which is a mere fraction of the Earth's 4500 million years. None the less, the environmental changes that have taken place during the time that man the toolmaker has been an inhabitant of the Earth have not been slight. Changes, whether in climate, sea-levels, vegetation belts, animal populations, or soils and landforms, have been both many and massive. Some changes are still taking place, and even if one rejects some of the more extreme proposals as to the effects that they have had on man and his affairs, the changes have been of such an order as to influence markedly both man and landscape. In the last 20 000 years alone, and man, as already noted, has been in existence one hundred times longer than that, the area of the Earth covered by glaciers has been reduced to one-third of what it was at the glacial maximum; the waters thereby released have raised ocean levels by over one hundred metres; the land, unburdened from the weight of overlying ice, has locally risen by several hundred metres; vegetation belts have swung through the equivalent of tens of degrees of latitude; permanently frozen ground has retreated from extensive areas of Europe; the rainforest has expanded; desert sand fields have advanced and retreated; inland lakes have flooded and shrunk; and many of the finest mammals have perished in the catastrophe called 'Pleistocene overkill'. Even at the present time, smaller fluctuations of climate are leading to changes in fish distributions in northern waters, marked fluctuations in valley glaciers, extensive flooding of African lakes, and difficulties for agricultural schemes in marginal areas.

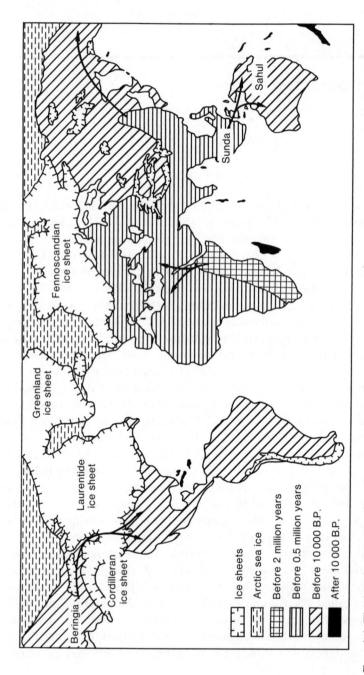

FIG. 1.1 The human colonization of ice-age Earth (after Roberts 1989, fig. 3.7). Note the contrast between the Old World and the New, and the late colonization of Oceania.

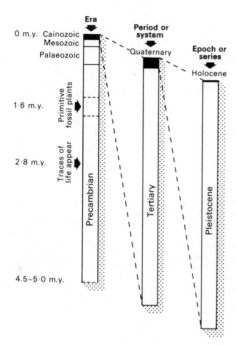

Fig. 1.2 The Quaternary and its subdivisions in relation to the geological time-scale (modified after Vita-Finzi 1973, fig. 1); m.y. refers to the left column and is an abbreviation for millions of years. The numbers are $\times 10^3$.

During the past few million years man has gradually diffused over the face of the Earth, and has now become a powerful agent of environmental change in his own right (Goudie 1990). The oldest records of human activity, crude stone tools which consist of a pebble with one end chipped into a rough cutting edge, have been found in conjunction with bone remains from various parts of Africa, notably in the vicinity of Lake Turkana, the Omo valley, and Olduvai Gorge in East Africa. Elsewhere (see Fig. 1.1) the arrival of man is considerably later, and the colonization of Australia and the Americas has taken place only in the last few tens of thousands of years: that of New Zealand, Madagascar, and Oceania even later.

The terminology used to describe some of the natural events and changes which provide the backcloth of human history needs to be stated at this point. As Fig. 1.2 shows, the major geological divisions currently employed are as follows: the most recent *era* is called the Cainozoic, and this has been divided into two *periods*: the Tertiary and the Quaternary.

4 Environmental Change

Table 1.1 Orders of climatic variation

Time-scale unit		
(1) Minor fluctuations within the instrumental record	10 years	Minor fluctuations which give the impression of operating over intervals of the order of 25–100 years, with somewhat irregular length and amplitude
(2) Post-Glacial and historic	10^2 years	Variations over intervals of the order of 250–1000 years, e.g. the sub-Atlantic recession and others affecting vegetation in Europe and N. America
(3) Glacial	10^4 years	The phases within an ice age, e.g. the duration of the Würm was of the order of 50×100^3 years
(4) Major geological	10^6 years	Duration of ice ages as a whole, periods of evolution of species
(5) Minor geological	10^8 years	e.g. ice ages at intervals of 2.5×10^8 years

Principal bases of evidence

(1)   Instrumental; behaviour of glaciers; records of river-flow and lake levels; non-instrumental diaries: crop yields, tree-rings (also for dating).

(2)   Earlier records of extremes: fossil tree-rings; archaeological finds; lake levels; varves and lake sediments; oceanic core-samples; pollen analysis; radio-carbon dating; ice-cores.

(3)   Fauna and flora characteristic of interglacial deposits; pollen analysis; variation in height of snowline and extent of frozen ground; oceanic core-samples (dating through latter).

(4, 5) Geological evidence: Character of deposits; fossil fauna and flora; dating largely through radioactivity of rocks.

(After Manley, 1953)

These periods are in turn divided into various *series* or epochs, with the Pliocene, for example, being the final series of the Tertiary. The Pleistocene and Holocene series together constitute the Quaternary.

It is exceedingly difficult to draw boundaries between some of these units, for they represent terminological and classificatory devices rather than reality. As a consequence, some authorities currently consider the Tertiary and the Quaternary as constituting the Late Cainozoic. Likewise, because the boundary between the Pleistocene and Holocene is no different in character from the boundaries between the various preceding glacials and interglacials, some authorities would consider the Holocene as the most recent stage of the Pleistocene rather than as a separate series. Other arguments against retaining the term Holocene for a series are that the Pleistocene is still in progress, and that the Holocene has been too brief to merit the status of a series or epoch. In spite of these difficulties, in this book the terms Pleistocene and Quaternary are used in the conventional way because of their familiarity.

Whatever terms we may use, it is still clear that changes in environment have occurred over a wide range of time-scales. Table 1.1 summarizes the main orders of change which can be identified, from the minor fluctuations within the period of instrumental record (with durations of the order of a decade or decades) to the major geological periods, with durations of many millions of years. The shorter-term changes (described in Chapter 5) include such events as the period of warming that took place in the first decades of the twentieth century. The changes with durations of hundreds of years were characteristic of the Holocene (see Chapter 4) and include various phases of glacial advance and retreat such as the Little Ice Age between about 1500 and 1850. The fluctuations *within* the Pleistocene, consisting of major glacials and interglacials, lasted for the order of 10 000 to 100 000 years, while the Pleistocene itself, termed a minor geological variation, consisted of a group of glacial and interglacial events that lasted in total about two million or so years. Such major phases of ice age activity appear to have been separated by about 250 million years.

Various curves which show the nature of these different orders of change are illustrated in Fig. 1.3. They serve to show the frequency and magnitude of changes during the Quaternary, and to place such changes in the context of the longer span of Earth history.

The evolution of ideas

Appreciation of the fact that the world had such a history of environmental change emerged at much the same time that it was appreciated that the world had a history of some length beyond the constricting 6000 years of Archbishop Ussher's biblical time span. This concept, which saw the Earth as having been created in 4004 BC, was relatively little contested until the end of the eighteenth century, and it was accompanied by the belief that much of the denudation and deposition evident on the face of the Earth could be explained through the agency of Noah's Flood and other catastrophic events. Gradually, however, these ideas were shown to be erroneous through the evidence collected by geologists and natural historians like Guettard, Count Buffon, and George Poullet Scrope. In particular, James Hutton and his friend John Playfair, the Edinburgh scientists, are often regarded as the men who most effectively propagated the new ideas, for in the geological record they saw 'no vestige of a beginning, no prospect of an end'. They realized that the complexity of the sedimentary record could be explained through the operation of processes akin to those of the present day extending over a long time-span.

The idea that climate and other aspects of the environment had fluctuated or changed during this enlarged span of time resulted initially from the discovery that Norwegian and Alpine glaciers had formerly extended further than their current limits. Some suggestions as to this

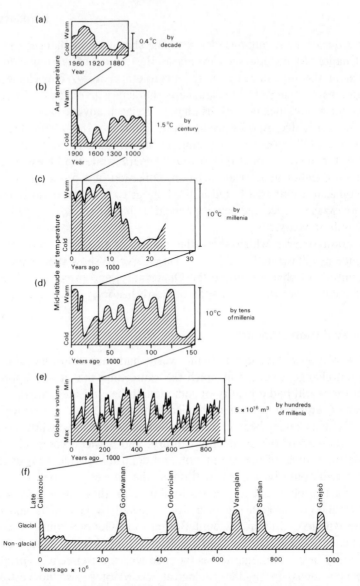

Fig. 1.3 The different scales of climatic change:

A. Changes in the five-year average surface temperatures over the region 0–80° N;

B. Winter severity index for eastern Europe;

C. Generalized northern hemisphere air-temperature trends, based on fluctuations in alpine glaciers, changes in tree-lines, marginal fluctuations in continental glaciers, and shifts in vegetation patterns recorded in pollen spectra;

D. Generalized northern hemisphere air temperature trends, based on mid-latitude sea-surface temperature, pollen records, and on world-wide sea-level records;

E. Fluctuations in global ice-volume recorded by changes in isotopic composition of fossil plankton in a deep-sea core;

F. The spacing of ice ages in geological time.

had originally been made by scientists at the end of the eighteenth century. In 1787 de Saussure recognized erratic boulders of palpably Alpine rocks on the slopes of the Jura Ranges, and Hutton reasoned that such far-travelled boulders must have been glacier-borne to their anomalous positions. Playfair extended these ideas in 1802, but it was in the 1820s that the Glacial Theory, as it came to be known, really became widely postulated. Venetz, a Swiss engineer, proposed the former expansion of the Swiss glaciers in 1821, and his ideas were supported and strengthened by Charpentier in 1834. The poet Goethe expressed the idea of 'an epoch of great cold' in 1830. However, the ideas of both Venetz and Charpentier were extended and widely publicized by their fellow countryman, Louis Agassiz, who was one of the originators of the term *Eiszeit* or Ice Age.

Agassiz wrote with vigorous prose, and with statements like the following both attracted attention and succeeded in spelling out, albeit inaccurately, some of the implications of the great new discoveries:

The development of these huge ice-sheets must have led to the destruction of all organic life at the Earth's surface. The ground of Europe, previously covered with tropical vegetation and inhabited by herds of great elephants, enormous hippopotami, and gigantic carnivora became suddenly buried under a vast expanse of ice covering plains, lakes, seas and plateaux alike. The silence of death followed . . . springs dried up, streams ceased to flow, and sunrays rising over that frozen shore . . . were met only by the whistling of northern winds and the rumbling of the crevasses as they opened across the surface of that huge ocean of ice. (Agassiz, cited in Imbrie and Imbrie, 1979: 33.)

In spite of this convergence of opinion from numerous sources, the ideas of these original minds were not easily accepted or assimilated into prevailing dogma, and for many years it was still believed that glacial till, called drift, and isolated boulders, called erratics, were the result of marine submergence, much of the debris, it was thought, having been carried on floating icebergs. Sir Charles Lyell noted debris-laden icebergs on a sea-crossing to America, and found that such a source of the drift was more in line with his belief in the power of current processes—the uniformitarian belief—than a direct glacial origin. For years even glacial or subglacial depositional features like eskers were thought to be of a marine origin, and were classified into fringe eskers, bar or barrier eskers, and shoal eskers. Moreover, in Britain, some drift deposits contained marine shells, a fact that gave prima-facie support to marine ideas.

The study of environmental change progressed somewhat further in the 1860s when Sir Archibald Geikie, A. C. Ramsay, and T. F. Jamieson showed that the spread of the glaciers had not been a single event, and that there had been various stages of glaciation, represented by moraines, which had been separated by warmer phases, called interglacials. These interglacials can be defined as a particular type of non-glacial climatic condition with a climatic optimum at least as warm as that experienced during the Holocene. Another type of interruption to full glacial conditions was the interstadial, a period which was either too cold or too protracted to allow the development of temperate deciduous forest of the full interglacial type. The term stadial refers to an ice advance.

The concept of multiple glacials, interglacials, stadials, and interstadials, during the Ice Age was further confirmed in the first years of the twentieth century by Penck and Brückner (1909), working in the Alps (see pp. 41–2). They developed much of the terminology and interpretation of sequences used to this day, and their work is one of the great landmarks in the study of environmental change.

In areas outside those which were subjected to Pleistocene glaciation other types of fundamental environmental change were recognized as having taken place, though Louis Agassiz, after a journey to Brazil, postulated that even equatorial regions had undergone glaciation. He believed that the Amazon Basin had been overwhelmed, but he seems, like certain others at the time, to have allowed his enthusiasm to convince him that what was in reality deeply weathered soil and boulders produced by intense chemical activity under tropical conditions, was glacial boulder clay. A truer appreciation of the effects of climatic change in non-glaciated regions of the tropics and subtropics came from Jamieson, Lartet, Israel Russell, and Grove Karl Gilbert, all of whom studied the fluctuations of Pleistocene lakes in semi-arid areas, as represented by old strandlines, deltas, and algal limestones. Their work in effect established the supposed general relation between high-latitude glacials and mid- and low-latitude pluvials, and Russell was able to tie in the moraines of the former Sierra Nevada glaciers with old strandlines around Lake Mono in California. Gilbert demonstrated that Lake Bonneville, in much the same way as the glaciers, had fluctuated several times, with alternations of high water and phases of desiccation. Other geologists investigating the west of the United States also believed that a diminution of rainfall might account for some of the anomalous drainage features they encountered on their travels, and the term 'pluvial' was itself coined by Alfred Taylor in 1868. It means basically a period of comparatively abundant moisture in regions beyond the limits of the ice-caps. Whether or not such pluvial periods were in phase or not in phase with the high-latitude glacial periods is a controversial problem referred to in Chapter 3.

Comparable in importance to the climate changes of the Pleistocene, and associated in large measure with them, were the great changes in the relative levels of land and sea. Once again the name of Playfair is an important one, for in his *Illustrations of the Huttonian theory of the Earth* (1802) he made detailed descriptions of the emerging shorelines of Fennoscandia, and assessed correctly that the cause in that particular case was crustal uplift, though this was then supposed to result from progressive cooling of the crust being accompanied by shrinkage. However, it was Jamieson working in Scotland in the 1860s who first proposed that the weight of glacial ice-sheets might cause subsidence of the crust beneath them, and that the release of this weight by melting in post-glacial times would lead to uplift. Gilbert extended this idea, and attributed the observed warping of the pluvial Lake Bonneville shorelines to removal of the weight of water on desiccation, whilst another North American worker, Dutton, was perhaps the most successful person in indicating the importance of such processes, to which the name 'isostasy' has been given (see Chapter 6).

Many years earlier, however, in the 1830s and 1840s, Lyell and Maclaren had argued that more extensive glaciers and ice-sheets would have stored up considerable bodies of water, and that, as a consequence, the level of the sea must have been many metres lower than at present. This was the eustatic theory of 'swinging sea-levels'. The causes and effects of sea-level changes produced by this and other mechanisms are discussed in the penultimate chapter of this book.

Traditional techniques

Since the Second World War the number and type of techniques used to determine the nature and chronology of Quaternary environmental changes have been vastly expanded, but the new techniques have tended to supplement rather than to replace the more traditional ones. Of these older techniques, the study of varves, tree rings, geomorphological features, plant macro-fossils, pollen, and faunal remains, have been especially significant.

The aim of the various techniques falls into one or both of two categories: either to date or to give environmental information. Varves were used to provide dating evidence. They are regular alternations in a sediment of layers of different composition or texture, forming pairs or couplets attributable in general to an annual seasonal rhythm. Such layers were produced by the great summer inputs of meltwater from glaciers into lake basins. The coarser material was deposited in the summer, but

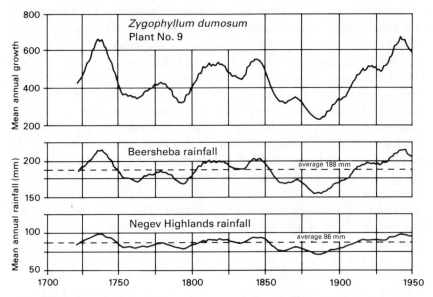

FIG. 1.4 Tree-rings and derived annual rainfall in the central Negev, Israel, 1720–1950. Annual growth of the woody shrub, *Zygophyllum dumosum*, in the central Negev, Israel, has been shown to give reliable estimates of the average rainfall of Beersheba and thus also for the Negev highlands. Existing 45-year rainfall records of Beersheba were used to estimate the average annual rainfall for the Central Negev highlands. The 25-year rainfall records were then derived, as shown in the above curves (21-year moving weighted means for the period 1720–1950). (From Shannan *et al*. 1967.)

the fine suspended clays in the meltwater did not settle out until the autumn or following winter. It is this which produces the banding. Some years would tend to produce particularly thick bands, and varve chronology depends on the correlation of such distinctive bands in different localities, and on the assumption that the couplets are annual. This technique was pioneered with great effect by de Geer in Sweden, and results have sometimes been sustained by radio-carbon dating (Tauber 1970). Where material datable by C14 is scarce or unavailable this technique still has a degree of utility. In a sense it utilizes the same principle as tree-ring analysis (dendrochronology), whereby attempts are made to correlate the annual growth rings of trees from area to area. In favourable situations tree-rings may be related in their growth to precipitation levels and can thus be used for climatic reconstruction (see Fig. 1.4). The time range covered can extend back some three or four thousand years where trees of such ages survive, as is the case with the Bristlecone Pine of the south-west USA. The nature of the rings also affords the possibility of dating structures in which tree remains are included and has provided an independent check on the validity of C14 dating. A magnificent compendium on tree-rings and climate has been published by Fritts (1976).

Some estimates of dates were obtained by the relative dating of landforms, for when geomorphological features such as moraines are formed their surfaces have certain characteristics which change through time. For example, slopes become degraded, boulders become pitted, weathering rinds form, soils develop, vegetation colonizes, and loessic silts may accumulate on the surface. Thus in the absence of methods of absolute dating, comparison of the degree of slope change, weathering, soil development, vegetation colonization, and loess accumulation may give an indication of the relative ages of features in an area. The changes that occur through time depend on a variety of factors other than time (geomorphological situation, lithology, etc.), and the changes may not be linear in type (see, for example, Colman and Pierce 1981). None the less, these methods have been widely employed, especially for the dating of moraines of uncertain age (Birkeland 1984, Porter 1975).

Pollen analysis, or palynology, is one type of micro-fossil analysis. It makes use of the fact that some sediments contain pollen grains and spores which mostly come from the air by fallout (pollen-rain), and they are thus derived from the regional and local vegetation. Vegetational changes, which may be caused by climatic, edaphic, or biotic factors, can be recorded by the preservation of pollen in a section. Pollen grains may be counted and recorded by dispersing sediments with appropriate agents and then looking at them under a high-power binocular microscope. Results of such analysis give a picture of the vegetation at a given point in time, and also allow the sequence of vegetational change to be examined over a period. In recent years the length of record obtained from pollen analysis has been increased by the taking of long cores from lakes and swamps. Among the longest known pollen records are those from Tenaghi-Philippon in Macedonia, Greece (Van der Hammen *et al.* 1972), where a core 120 m long did not reach the bottom of the deposit, Colombia in South America, where a 357 m core has been obtained from the Sabana de Bogota (Hooghiemstra 1989), and Lake Biwa in Japan where a 900 m long pollen core has been retrieved (Fuji 1988). Horowitz (1989) reports a continuous pollen core from Israel that goes back 3.5 million years.

An equally laborious technique, but one which has produced good results, is that which utilizes non-marine molluscs, remains of which are found very commonly in Pleistocene deposits. Molluscan assemblages have been found to be indicative of particular types of climate. Cold faunas show a dominance of a few species in great numbers, and temperate faunas a larger number of species, many of which occur frequently.

Similarly, especially at the University of Birmingham, techniques have been developed to use remains of beetles (Coope *et al.* 1971). Their wing cases are found in suitable sediments, and as distributions of living species are known quite well, especially in Scandinavia, it has proved possible to interpret palaeoenvironments on the basis of insect faunas. Results have tended to correlate well with results obtained by pollen analysis, and by the study of non-marine mollusca, though certain discrepancies have been noted. At Lea Marston, Warwickshire, for example, a deposit dated to about 9500 BP (before present) shows a relatively warm beetle fauna but a cool vegetation assemblage dominated by *Betula*, *Salix*, and some *Pinus*. This appears to be because the very mobile insect fauna was able to react to the very rapid climatic amelioration around that time in advance of the more slowly migrating trees (Osborne 1974).

Assemblages of diatoms, microscopic unicellular algae, which are often preserved in lake and mire sediments, can give indications of past lake-water quality, salinity, and nutrient status. Diatom studies can also be used to study the interface of fresh and saline environments, thereby enabling statements to be made about such phenomena as the locations of past shorelines (Battarbee 1986). Microscopic animals may also be preserved, including Cladocera (Frey 1986) and Ostracods (Loffler 1986). These too give an indication of past hydrological conditions. Another biological means of reconstructing past environments has been the analysis of plant remains preserved in pack-rat middens. In the southwest of the USA, for example, the middens of *Neotoma* are found in many caves, and these middens contain plant material foraged by the rats from the local area. Cemented by dried rat urine, these middens may remain preserved for tens of thousands of years, and provide an inventory of past vegetation conditions around a site (Wells 1976, Betancourt *et al.* 1990).

Plant macro-fossils may also be preserved in the form of charcoal. Such charcoal can, through microscopic analysis, give an indication of past vegetation conditions, and of the incidence of fires (Tolonen 1986).

Human artifacts, which may occur abundantly in some Quaternary deposits, have often been used for dating. While pottery, coins, and other metal objects may give relatively precise dates for deposits laid down in the last few thousands of years, the use of stone tools for dating older deposits offers much less precision. Such tools may often be derived from even earlier deposits, their characteristics tend not to have changed very rapidly through time, and changes have rarely been concurrent in different areas. In general it is normally preferable to date stone artifacts using other dating methods than to expect the artifacts to provide precise dates themselves.

New chronological techniques

In the last couple of decades the study of environmental change has been transformed. There are various reasons for this. First, the geographical spread of studies has been expanded so that scientific investigations have been made of areas hitherto neglected, thereby helping to erect a global picture. Second, there has been a rapid development in our knowledge of the oceans, and in particular of our knowledge of what has been going on on the sea floor. Many of the recent developments have depended on the recovery of a continuous core of deep-sea sediment, a possibility that was created with the development of the Kullenberg piston corer in 1947. The importance of the oceans is that their record is less disrupted than the record of continental sediments. As Ewing expressed it (1971: 572):

the problem of estimating the number of glaciations from evidence on continents is contrasted with the problem of estimating the number of glaciations from evidence in deep-sea sediments. The first may be compared in complexity to estimating the number of times a blackboard has been cleaned; while the second may be compared to finding the number of times the wall has been painted.

The third major advance has been in the available methods of dating. New geochronometric techniques, sometimes favoured with the rather optimistic label of 'absolute dating', have enabled sequences to be established and correlations to be made with a degree of confidence that was previously unattainable.

Of especial interest have been the new isotopic dating techniques, especially radio-carbon, uranium series, and potassium-argon. Some of the features of these methods are shown in Table 1.2.

These three isotopic dating techniques all depend on the measurement of amounts of elements which through time are either formed by, or are subject to, radioactive decay. The rate of decay being known for a particular element, the time-interval may be assessed between the present, and the time when the particular parent material was fixed and its decay began. Thus, for example, a growing organism incorporates radio-carbon, and on its death the radio-carbon is trapped and then begins to decay. As half the radioactivity will be lost after an interval calculated to be about 5730 years, by measuring the radioactivity of fossil material containing

Table 1.2 Some isotopic methods of dating Quaternary deposits

Name	Isotope	Half-life (years)	Range (years)	Materials
Radio-carbon	C14	5730 ± 40	0–75 000	Peat, wood, shell, charcoal, organic muds, algae, tufa, soil carbonates
Uranium Series	U^{234}	250 000	50 000–1 000 000	Marine carbonate, coral, molluscs
	Th^{230}	75 000	0–400 000	Deep-sea cores, coral and molluscs
	Pa^{231}	32 000	5 000–120 000	Coral and molluscs
Potassium-argon	K^{40}	1.3×10^9	greater than 20 000	Volcanic rocks, granites, etc.

carbon, the date at which death took place can be determined. Radio-carbon or C14 dating, formerly used mainly for organic carbonate, in the form of peat and wood, is now being extended to a wider range of Late Pleistocene materials, especially soil carbonates and mollusca. This technique has evolved steadily since its first application in 1949, and it provides a chronology for approximately the last 50 000 years, though practical problems become severe beyond about 40 000 years BP. Some laboratories can now bring material 75 000 years old within dating range (Grootes 1978).

Useful though it may be, radio-carbon dating still has many problems which need to be considered in assessing the reliability of the very large number of dates which are now available. In particular, contamination of samples may take place. Humic acids, organic decay products, and fresh calcium carbonate may be carried downward to contaminate underlying sediments. In the case of inorganic carbonates, 'young' carbonate may be precipitated in, or replace, the carbonate which one is interested in dating, and removal of the contaminant from pore spaces and fissures is almost impossible. Additional to problems such as these are miscellaneous other constraints, including the fact that different laboratories may have used different half-lives, and the discovery that fluctuations in cosmic radiation with time may produce slight differences in the C14 equilibrium of the atmosphere, biosphere, and hydrosphere.

Since the early 1960s potassium-argon (K/Ar) dating has been applied to Pleistocene and Pliocene chronology and has greatly changed our views on the length of the Pleistocene, and on the time when glaciation was initiated. Whilst radio-carbon dating utilizes organic and inorganic carbonates, potassium-argon dating, which can cover a theoretically unlimited time-span, utilizes unaltered, potassium-rich minerals of volcanic origin in basalts, obsidians, and the like. It is, however, only usable in practice for materials older than about 50 000 years.

Also since the 1960s the Thorium-Uranium and other Uranium Series dating methods have been applied to such materials as molluscs and coral. Although still subject to certain deficiencies, particularly for molluscs, these techniques are extremely valuable when applied to coral in bridging the gap between radio-carbon and potassium-argon techniques. These methods have been used for materials up to about 200 000 years in age with some success, and Uranium Series dates obtained for coral terraces have caused a change in ideas on the fluctuations of sea-level before the Last Glaciation (see Chapter 6).

In addition to the isotopic techniques, great use has recently been made of a palaeomagnetic calendar of magnetic events. Currently the Earth has what is termed a 'normal' magnetic field so that at the north magnetic pole a compass dips vertically towards the Earth's surface. However, the magnetic field, for reasons not fully understood, can switch to become 'reverse'. As some rocks and sediments may preserve the characteristic signal of the magnetic field during the time the unit was deposited, it has been possible to produce a calendar of magnetic events marked by switches from 'normal' to 'reverse'. As many of these switches have now been dated by independent means, in a conformable sequence of sediments these magnetic switches enable a particular section to be dated against a master system (Glass *et al.* 1967). Thus sediments from deep-sea cores can be given an age-scale of considerable length.

A two-level system of names has been introduced to describe the observed sequence of polarity reversals. At the lower end are polarity events—short intervals of normal or reversed polarity lasting in the order of 150 000 years or less. At the higher level are the polarity epochs— longer intervals during which the magnetic field was predominantly of one polarity, and which may contain one or more events (Cox *et al.* 1968). The dates of these epochs and events are shown in Fig. 1.5.

Volcanic eruptions can also provide important stratigraphic markers for the Quaternary. Different ash falls may be recognizable on the basis of petrology and chemical composition. The falls of ash are placed in chrono-stratigraphic position by C14 dating of associated sediments, or by K/Ar dating of the source volcanic unit. Once the age has been established, an ash can be used as a marker horizon in otherwise undated sections. This technique is termed tephrochronology: Self and Sparks (1981) provide numerous examples of its application.

Another technique, which like K/Ar dating can be applied to volcanic materials, is fission-track dating. This is based on the principal that traces

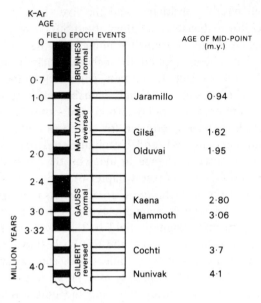

FIG. 1.5 Time-scale for geomagnetic reversals. Times when the field was normal are indicated by shading (after Cox *et al.* 1968, fig. 1.1). Reproduced from the *Quarterly Journal of the Geological Society of London*, 124, by permission of the Geological Society.

of an isotope, ^{238}U, occur in minerals and glasses of volcanic rocks, and that this isotope decays by spontaneous fission over time causing intense trails of damage, called tracks. These narrow tracks, between 5 and 20 microns in length, vary in their number according to the age of the sample. Thus by measuring the numbers of tracks an estimate can be gained of the age of the volcanic minerals. The method has been applied to the dating of tephrochronological events and provides cross-checks on other methods such as K/Ar dating.

The application of fission-track dating to relatively young geological events is summarized by Naeser and Naeser (1988). At the moment the time frame over which the technique can be used is limited by the time required to develop a statistically significant number of tracks. This is generally in excess of 100 000 years.

Recently, another dating technique has been developed which is based on the fact that protein preserved in the skeletal remains of animals undergoes a series of chemical reactions, many of which are time-dependent. After death of the organism the protein slowly degrades, as does the nature of the amino acids which form the basis of the protein.

Thus an examination of the amino acid composition of bone and of carbonate fossils has potential for dating purposes (Miller *et al.* 1979). The method has the advantage that only very small amounts of sample are required. It also covers a wide time range. However, the degree of change in amino acid composition depends on factors other than time, and therefore involves making certain assumptions, particularly about temperature conditions, that may create substantial errors.

Thermoluminescence is another technique that has come into use for the dating of sediments in the last two decades, and it has been applied, particularly by Soviet workers, to time spans of the order of 10^3 to 10^6 years. It employs quartz grains in material like loess (see pp. 67–9) and dune sand and is based on the principal that if a sample has been irradiated and subsequently heated, light is emitted as a function of temperature. This is called a 'glow curve', the intensity of which depends in part on the age of the sample. Further information is provided by Dreimanis *et al.* (1978), and Aitken (1989).

Closely related to thermoluminescence dating is optically stimulated luminescence dating (OSL), an approach that was first demonstrated by Huntley *et al.* (1985). This has great potential for the dating of quartz grains of wind-borne and water-borne origin. The luminescence emission caused by optical excitation has the advantage that it is only derived from the light-sensitive sources of luminescence from within the grains, i.e. those sources that are most likely to undergo resetting—'bleaching'—in the process of being transported and deposited, as compared to the spectrum of both light-sensitive and light-stable traps that are sampled during a TL analysis (Rhodes 1988).

Electron spin resonance has emerged as an important dating technique in recent years (Ikeya 1985). It is a method of measuring the paramagnetic defects in minerals or the skeletal hard parts of organisms. These defects are created by penetrating radiation from radioactive elements within, or in the sediment surrounding, the sample. It shares many fundamental principles with thermoluminescence dating. It has been applied successfully to the dating of materials such as corals, molluscan fossils, and tooth enamel, and can be used to date carbonate fossils back to *c.*400 000 years, with an uncertainty of ±10–20 per cent. Obsidian hydration dating can be applied to obsidian, a type of rhyolitic volcanic glass. This material absorbs water to form a hydrated rind that increases in thickness with time. Although many factors affect the rate of obsidian hydration, rind thickness can be used to develop relative chronostratigraphies (Trembour and Friedman 1984).

One other dating technique needs mention: lichenometry. This has

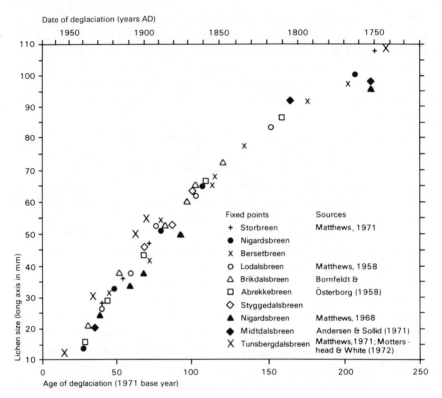

Fig. 1.6 Compilation of the results of measuring the longest diameter of the crustose lichen *Rhizocarpon geographicum* on moraine ridges in southern Norway, where the age of ridge construction is known to a high probability (first hundred years) and reasonable possibility (older dates). Each symbol represents a mean of the five largest lichens, although field techniques do differ to some extent (after Matthews 1974).

become increasingly important since it was developed in the 1950s, and is especially useful for dating glacial events over the last 5000 or so years. It is believed that most glacial deposits are largely free of lichens when they are formed, but that once they become stable, lichens colonize their surfaces. The lichens become progressively larger through time. Thus by measurement of the largest lichen thallus of one or more common species, such as *Rhizocarpon geographicum*, an indication of the date when the deposit became stable can be attained. An illustration of the relation between lichen size and the age of moraine ridges in Norway is shown in Fig. 1.6. Good reviews of lichenometry, which stress the

Table 1.3 Categories of dating methods

(1) METHODS THAT PROVIDE AGE ESTIMATES
    (a) Radiometric dating techniques
        Uranium-series dating
        Potassium-Argon dating
        Fission Track dating
        Thermoluminescence dating
        Optically stimulated luminescence dating
        Radiocarbon dating
        Electron spin resonance
    (b) Incremental methods
        Dendrochronology
        Varves
        Lichenometry
(2) METHODS THAT ESTABLISH AGE EQUIVALENCE
    Palaeomagnetism
    Tephrochronology
(3) RELATIVE AGE METHODS
    Amino acid stratigraphy
    Obsidian hydration
    Weathering development and
    geomorphological evolution

inherent problems in the method, are provided by Innes (1985) and Worsley (1990).

Thus a wide range of dating techniques is now available, and these can be put into three main categories (Lowe and Walker 1984). The first of these is methods that provide age estimates in years BP. These include radiometric methods (which rely on radioactive decay and related changes) and incremental methods (which are based on the accumulation through time in a regular manner of biological or lithological materials) (see Table 1.3). The second category is methods that establish age equivalence. This is based on the use of certain distinctive stratigraphic markers that are contemporaneous over wide areas, and which if they can be dated in one location can then be used to date other locations where they occur. The third category, relative age methods, establish the relative order of antiquity of materials. Relative age may, for example, be discerned from the degree of weathering produced through time.

Developments in stratigraphy

Of comparable importance to the new dating techniques has been the development of deep-sea coring procedures, for although by no means completely stable (burrowing organisms, solution, and currents creating problems) the sea floor does offer a more continuous and lengthy stratigraphic record than do most terrestrial sections. It is from the evidence

Table 1.4 Information to be gained from deep-sea cores

Indicator	Environmental information
O^{18}/O^{16}	Temperature
	Ice Volume
Coarse debris	Iceberg rafting
Aeolian dust	Aridity
Sensitive species Foraminifera	Temperature
Total species Foraminifera	Temperature
Clay minerals	Weathering regimes on land
Fluvial sediments	River inputs
Coccolith carbonates	Temperature
Aeolian sand turbidities	Aridity

of the deep-sea cores that a series of cold and warm episodes can be dated, identified, and perhaps related to glacials and interglacials. The cores, normally obtained by piston corers, have also helped to establish the age of the Plio-Pleistocene boundary—formerly a matter of great dispute. The deep-sea cores can be used and interpreted in a variety of ways (Table 1.4). Core materials can be dated by radiometric means, palaeomagnetic epochs, whether normal or reversed, can be identified in the sediment layers, micro-fossils (especially Foraminifera and Radiolaria) can be examined, and the lithological characteristics of the sediments within the cores can be determined with a view to finding out about changes in terrigenous sediment sources.

One of the most productive ways of examining the cores has been the study of changes in the frequency of particular 'sensitive species' Foraminifera. These are thought to reflect changes in the temperature of ocean waters (Kennett 1970). *Globorotalia menardii* tests, for example, may be counted, and the ratio of their number to the total population of Foraminifera tests can be worked out. The ratio can range from as high as 10 or 12, to as low as nearly zero. A high ratio appears to be associated with the warm water of interglacial conditions, while a low ratio appears to be associated with the colder water of glacial stages. Thus, by taking samples along the length of a core, the alternations of warmth and cold can be established. Similarly, another of the Foraminifera, *Globorotalia truncatulinoides*, can be used to the same end. In any portion of a core-sample some of its tests will show a left-hand direction of coiling, and others will show a right-hand direction of coiling. It has been found by some workers that right-coiling tests are associated with warmer conditions, and left-coiling tests with cooler. Thus ratios of left and right coiling may enable an assessment to be made of palaeo-climates. Some workers have attempted to use more sophisticated methods of utilizing

Foraminifera, and instead of approaching the problem through the study of 'sensitive species', they have tried to establish climatic sequences based on 'total fauna' (see Shackleton 1975 for a discussion).

Another way in which the Foraminifera can be utilized is by the measurement of the O^{18}/O^{16} ratios in the calcitic tests. This oxygen isotope method was developed in the 1950s, by Emiliani and others (Emiliani 1961). He supposed that the O^{18}/O^{16} ratio depended substantially on the temperature of the water in which the Foraminifera lived. While there is now some controversy as to the value of this technique in giving quantitative data on palaeotemperature changes (Shackleton 1967), the method does appear to give a fairly clear picture of the periodicity of the major glacial and interglacial episodes, and it has helped to show that the Pleistocene was characterized by more glacial cycles than has been suspected on the basis of evidence from the terrestrial record. However, besides the direct effects of temperature, the changing state of the ice sheets has played an important role in determining the oxygen-isotope record. During the glacial episodes immense ice sheets of isotopically light ice (depleted in O^{18}) accumulated in northern America and Europe. When this occurred, the oceans diminished in volume, became slightly more saline, and became isotopically more positive (i.e. enriched in O^{18}). This enrichment is recorded in the isotopic composition of calcareous Foraminifera preserved in deep-sea sediments (Shackleton 1975). Also obtainable from the deep-sea core evidence is the extent to which iceberg-rafted debris is present. In middle latitudes this is an indirect indicator of cold climate, though in high latitudes ice-rafted debris maxima may be associated with interglacial periods (Keany *et al.* 1976). This technique was extensively applied in the 1960s, notably to the North Pacific (Kent *et al.* 1971), the Southern Ocean (Opdyke *et al.* 1966), and the Arctic (Herman 1970).

When one compares the results obtained from these different approaches to gaining palaeoclimatic information from deep-sea cores, one often finds a remarkable degree of similarity in the pattern of the curves, especially for the upper portions of cores. This is illustrated by Fig. 1.7 which shows curves derived from oxygen isotope studies, from the proportion of glacial material, from the amount of carbonate, and from the frequencies of polar Foraminifera.

Comparable to such sedimentological evidence is that provided by the presence of aeolian debris on the ocean floors (Parmenter and Folger 1974). This, together with the presence of large quantities of unweathered minerals, including feldspars, has been used to assess whether tropical climates were dominantly arid and semi-arid, or whether they were

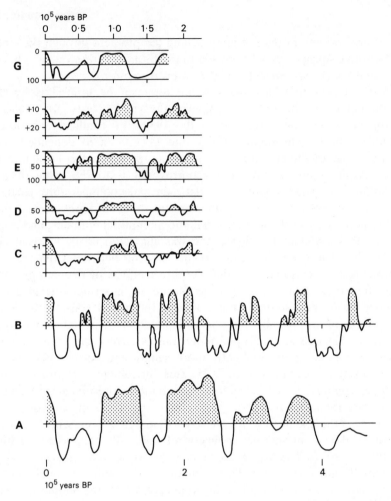

FIG. 1.7 Fluctuations in climate over the last 400 000 years as revealed by miscellaneous indicators from ocean cores. Shaded areas = warmth, non-shaded = cold.
A. Generalized curve for the Equatorial Atlantic based on ratios of cold to warm Foraminifera (after Bowles 1976, fig. 4).
B. Generalized curve for the Polar Front in the N.E. Atlantic based on ratios of cold to warm Foraminifera (after Bowles 1976, fig. 4).
C. Oxygen-isotope curve for a Caribbean core at 15° N (P6304-9) (after McIntyre *et al*. 1972, fig. 6).
D. Curve of percentage coccolith carbonate for an Atlantic core (V-23-83) at 51° N (after McIntyre *et al*. 1972, fig. 6).
E. Curve of percentage polar fauna (Foraminifera) for Atlantic core at 50° N (V-23-83) (after McIntyre *et al*. 1972, fig. 6).
F. Curve based on oxygen-isotope variation at core 280A (40° N) in the Atlantic (after McIntyre *et al*. 1972, fig. 6).
G. Curve based on the weight (per cent) of glacial detritus in the sand fraction of core RE5-36 at 50° N in the Atlantic (after McIntyre *et al*. 1972, fig. 6).

dominantly humid during particular phases. During arid phases rivers would tend to carry unweathered feldspars, while under more humid conditions the quantity of feldspars relative to the more stable quartz would be less. Similarly, in oceanic sediments off west Africa, opal phytoliths and freshwater diatoms are common in sediments deposited when waters were warm, but less common in sediments deposited when waters were cold (Parmenter and Folger 1974).

Cores have also been put down into the floors of lakes both in temperate and in tropical areas. These can indicate changes in the nature of sediments deposited over a long time-sequence. In some tropical lakes, for example, evaporite layers may be identified, and regarded as being the product of dry conditions (see, for instance, Kendall 1969). Alternatively the core-samples can be subjected to sophisticated chemical analysis (Degens and Hecky 1974). In Lake Kivu, East Africa, for instance, Fe-Ni sulphide contents were thought to indicate strongly reducing conditions, and thus high water stands, while high Al and Mn contents were thought to indicate low stands. As some lake cores may be several hundred metres thick, these types of investigations can extend back a considerable time.

Some success has been attained through the isotopic study of carbonate-rich cave sediments. Their history, and the temperature conditions associated with them, have been studied, for example, in France (Duplessy *et al.* 1970), in New Zealand (Hendy and Wilson 1968) and in England (Atkinson *et al.* 1978) using Th^{230}/U^{234} techniques.

The lengthy record provided by the deep-sea record has been supplemented by deep cores obtained from the ice-caps, at Byrd Station in Antarctica, at Devon Island in Arctic Canada, at Camp Century and Crete in Greenland, and elsewhere. The Camp Century Core (Epstein *et al.* 1970), put down in north-west Greenland, attained a length of no less than 1390 m. It represents a seemingly continuous sequence of annual layers of former snow. The core was sampled at regular intervals by Dansgaard and his co-workers (1971) for its O^{18}/O^{16} ratio. This ratio depends mainly on the temperature of condensation at the time of ice formation. Thus a plot of the O^{18}/O^{16} ratio down the length of the core should provide a sequence of temperature changes of varying amplitudes (see Fig. 2.5).

The main problem involved in this technique is that of time-calibration (Mörner 1972). The annual accumulation layers become progressively less visible at depth, in that they are squeezed out by compaction. Thus certain theoretical assumptions have to be made in the interpretation of the lower parts of these cores. In general, however, the results both from

Table 1.5 Potential palaeoclimatic information from ice-cores

Information	Analysis
Palaeotemperatures	
annual	$\delta 18_0$
summer	Melt layers
Palaeoaccumulation (net)	Seasonal signals
Ice sheet stability/flowlines	Gas content
Volcanic activity	Acidity, trace elements
Tropospheric turbidity (e.g. as result of dust)	Microparticle content, trace elements
Atmospheric composition	CO_2 content, trace elements
Solar activity/geomagnetic field strength	NO_3 content

(After Bradley 1985, Table 5.1)

Byrd Station, and from Camp Century, have tied in well both with each other, and with other lines of evidence.

A great deal of information can be gained from ice cores (Table 1.5) including precipitation amounts, air temperatures, atmospheric composition (including particulate matter and gaseous composition), past variations in solar activity, and the occurrence of explosive volcanic eruptions. Most cores have been retrieved from the polar ice-caps (see Bradley 1985, ch. 5), but tropical high altitude ice-caps have also proved productive (see, for example, Thompson *et al*. 1986). In addition to obtaining important palaeoenvironmental information, studies of the composition of gas bubbles trapped in the ice enable conclusions to be drawn about the role of past carbon dioxide concentrations in influencing climatic changes (see Barnola *et al*. 1987). Oescher and Langway (1989) provide a useful and detailed assessment of the many aspects of environmental reconstruction associated with cores from ice-sheets and glaciers.

Much of the data that have been obtained by these methods have been used to try and reconstruct the nature of world environmental conditions at certain selected points in the Pleistocene (Climap Project Members 1976), and attempts have been made to simulate conditions on computers with a global atmospheric model (Gates 1976). Such attempts at reconstruction and simulation on such a broad scale are in their infancy, but may do much to increase our understanding of the past, and, it is hoped, enable some prediction of the future.

Geomorphic and pedologic evidence for environmental change

Although the data provided by ocean floor stratigraphy and by palaeo-ecological examination of terrestrial sequences have proved them to be

Table 1.6 Some quantitative and semi-quantitative geomorphic indicators of environmental change

Landform	Indicative of	Example
Pingos, palsen, ice-wedge casts, giant polygons	Permafrost, and through that, negative mean annual temperatures	Williams (1975), Great Britain
Cirques	Temperatures though their relationship to snowlines	Kaiser (1969), European mountains
Closed lake basins	Precipitation levels associated with ancient shoreline formation	Dury (1973), New South Wales
Fossil dunes of continental interiors	Former wind directions and precipitation levels	Grove and Warren (1968), Saharan margins
Tufa mounds	Higher groundwater levels wetter conditions	Butzer and Hansen (1968), Kurkur Oasis
Caves	Alternations of solution (humidity) and aeolian deposition, etc. (aridity)	NW Botswana (Cooke, 1975)
Angular screes	Frost action with the presence of some moisture	McBurney and Hey (1955) Libya
Misfit valley meanders	Higher discharge levels which can be determined from meander geometry	(Dury, 1965) Worldwide
Aeolian-fluted bedrock	Aridity and wind direction	Massachusetts, Rhode Island, Wyoming (Flint, 1971)
Oriented and shaped deflation basins	Deflation under a limited vegetation cover	De Ploey (1965), Congo Basin
Dune breaching by rivers	Increased humidity	Daveau (1965), Mauritania
Lunette type	Hydrological status of lake basin	Bowler (1976), Australia
Old drainage lines	Higher humidity	Graaf et al. (1977), Australia
Fluvial siltation	Desiccation	Mabbutt (1977), Australia
Colluvial deposition	Reduced vegetation	Price Williams et al. (1982), Swaziland

two of the most profitable ways of reconstructing Pleistocene conditions, the importance of the evidence provided by fossil landforms and soils should not be forgotten. It is not possible here to go into detail about the relations between landforms and climate, and fossil landforms and fossil climates, but there are certain landforms which can give relatively precise information about past environments (Table 1.6). Under cold conditions with permafrost, for example, various types of patterned ground and ice-cored mounds, such as pingos, will develop. As permafrost distribution is closely related to mean annual temperatures former mean annual temperatures can be inferred from the distribution of fossil patterned ground and pingos. Likewise, the presence of glacial cirques can be used to infer the positions of former snowlines, which are themselves climatically controlled. The median level of a cirque floor tends to be at, or just above, the local snowline, so that the lowest cirque floor of a group of contemporaneous cirques will give a close approximation of the local snowline. Thus the height of Pleistocene snow-lines can be compared with present-day snowlines, and, by a knowledge of lapse rates, some estimate of temperature change can be obtained. In warmer areas landforms can also be used to reconstruct past climatic conditions. For example, as we shall see in Chapter 3, large continental sand dunes only develop over wide areas where precipitation levels are below about 100–300 mm. Above that figure sand movement is drastically reduced by the development of an extensive vegetation cover. Thus if fossil sand dunes are currently found in areas of high rainfall, it tends to suggest that rainfall levels have increased since the dunes were formed. Conversely, the presence of extensive fossil lake shorelines may be used to infer hydrological changes from wet to dry, and attempts have been made on the basis of old lake volumes to estimate former precipitation levels.

Soils too may be used profitably in Pleistocene studies. The development of a soil, whose duration depends on the nature and chemistry of the sediments, the climate, the character of the fauna and flora, and the balance between erosion and deposition, requires considerable time. Times of soil formation tend to be times of relative geomorphic stability. Thus thick palaeosols in a sequence of loess, dune sand, or alluvium may give important evidence for a halt to deposition, and a change to a period of stability. In the case of sand dunes, to take one example, the stability may result from a period of increased vegetation cover brought about by a phase of greater precipitation. Moreover, within a complex depositional sequence the character of the palaeosols themselves may change, and features in the soils such as gleying, carbonate accumulation, the snail fauna, degree of leaching, and frost structures may be used to assess environmental changes (see Chaline 1972: 44 *et seq.*, and Kukla 1975).

The loess deposits of China, Tajikistan, and Central Europe have provided a great deal of useful environmental information (Kukla 1977). The great advantage of loess stratigraphy is that it tends to be relatively, though not completely, continuous, with dust deposits having accumulated at different phases in the periglacial zones beyond the glacial margins. The loess contains a record of palaeosols and soil fauna, which in the case of the loess of Central Asia in Tajikistan shows something like 45 palaeosols over a period of about 2.5 million years. The loess is also susceptible to radio-carbon dating, palaeomagnetic dating, and thermoluminescent dating (Pye 1987). It is generally believed that phases of active dust deposition correspond to dry, cold phases (glacials?) whereas the palaeosols represent periods of relative stability associated with less dust input and a greater vegetation cover (interglacials?).

However, it needs to be appreciated that the formation of soils and most landforms results from a wide range of factors, of which climate forms but one group. Climate itself is also extremely complex. The problems of interpretation that these considerations present can be examined through a study of river terraces. They may sometimes form as a result of non-climatic causes, such as tectonic change, sea-level change, glacial invasion of catchment areas, and so forth. However, even if one can eliminate non-climatic causes, it is difficult to draw precise inferences as to climate from alluvial stratigraphy within terraces because of the variety of possible climatic influences: amount of precipitation, distribution of precipitation throughout the year, mean and seasonal temperature, and other climatic variables. Moreover, the response of a stream, in terms of load and discharge, to changes in such climatic variables, will be influenced by vegetation cover, slope angle, the range of the altitude of the basin, and other circumstances. Thus a change in any one climatic factor might, within one area, lead to different responses in different streams, and even in different segments of a single stream. Consequently, extreme care needs to be exercised in the utilization of landforms such as terraces for the reconstruction of past climates and environments.

There are, therefore, many ways in which past environmental conditions can be determined. These are summarized in Table 1.7. One needs to remember, however, that there are various problems with the use of these various methods. For instance, one must be aware that erroneous environmental reconstructions could result from the use of modern climate-proxy data relationships when past conditions have no analogue in the modern world. The resolution of some of the techniques is poor, and some climate-dependent phenomena may lag in their response to a climatic change. Moreover, not all data sources provide a continuous record. Many provide discontinuous or episodic information.

The era before the Pleistocene ice: the Cainozoic climatic decline

To appreciate the role which Pleistocene environmental changes, reconstructed on the basis of such techniques as those already described, have played in the history of the Earth and of man, and to grasp the magnitude of the changes involved, it is necessary to have some regard for the environmental conditions pertaining in the preceding period—the Tertiary (see Table 1.8).

It is, it must be pointed out, extremely difficult to make any division which is both logical and rigid between the Pleistocene and the last series of the preceding Tertiary period, the Pliocene. Indeed, the Villafranchian

Table 1.7 Principal sources of proxy data for palaeoclimatic reconstructions

(1) GLACIOLOGICAL (ICE-CORES)
    (a) oxygen isotopes
    (b) physical properties (e.g. ice fabric)
    (c) trace element and microparticle concentrations
(2) GEOLOGICAL
    (A) Marine (ocean sediment cores)
        (i) Organic sediments (planktonic and benthic fossils)
            (a) oxygen isotopic composition
            (b) faunal and floral abundance
            (c) morphological variations
        (ii) Inorganic sediments
            (a) mineralogical composition and surface texture
            (b) accumulation rates, and distribution of terrestrial dust and ice-rafted debris
            (c) geochemistry
    (B) Terrestrial
        (a) glacial deposits and features of glacial erosion
        (b) periglacial features
        (c) glacio-eustatic features (shorelines)
        (d) aeolian deposits (loess and sand dunes)
        (e) lacustrine deposits and erosional features (lacustrine sediments and shorelines)
        (f) pedological features (relict soils)
        (g) speleothems (age and stable isotope composition)
(3) BIOLOGICAL
    (a) tree rings (width, density, stable isotope composition)
    (b) pollen (type, relative abundance and/or absolute concentration)
    (c) plant macrofossils (age and distribution)
    (d) insects (type and assemblage abundance)
    (e) modern population distribution (refuges and relict populations of plants and animals)
(4) HISTORICAL
    (a) written records of environmental indicators (parameteorological phenomena)
    (b) phenological records
(5) ARCHAEOLOGICAL
    (a) Stone tools
    (b) Pottery, coins, metal objects

Source: Bradley (1985), Table 1.1, with modifications.

Table 1.8 Subdivisions of the Cainozoic Era

	Date of beginning in millions of years
Pleistocene	1.8
Pliocene	5.5
Miocene	22.5
Oligocene	36
Eocene	53.5
Palaeocene	65

(After Berggren, 1969)

(the earliest unit of the European Pleistocene), together with its marine equivalent, the Calabrian, have only been assigned to the Pleistocene rather than to the Pliocene since 1948.

On faunal grounds the Pleistocene has come to be regarded as the time when many of the modern genera, including elephant, camel, horse, and wild cattle, first appeared. Attempts have also been made to place the boundary between the Upper Pliocene and the Villafranchian of the Lower Pleistocene by using tectonic breaks in the stratigraphic succession, though this provides a generally inadequate and unusable correlation basis, except on a very local scale. In Britain the division between the Pliocene and the Pleistocene is placed at the boundary between the Coralline Crag and the Red Crag of East Anglia. At this point there is a relatively clear stratigraphic break, a marked increase in the proportion of modern forms of marine Mollusca, and of Mollusca of northern aspect, and the first arrival (in the Red Crag) of elephant and horse.

In Europe the base of the Pleistocene Series and of the Quaternary System has been set by the appearance in Late Cainozoic sediments occurring in various parts of Italy of a cold-water marine fauna (the Calabrian) differing from the underlying Pliocene fauna. The newer fauna is characterized by the appearance of a dozen species of North Atlantic molluscs, and by certain Foraminifera. In northern Italy the marine strata of the Calabrian grade into Villafranchian and Upper Villafranchian continental sediments containing a distinctive mammal fauna (Emiliani and Flint 1963).

An alternative way of placing the boundary is to do so on climatic grounds. In essence some people would place the line according to the first point where there is evidence of glaciation, and of rapid and relatively sudden temperature depression. Recent work involving some of the new techniques outlined above, including potassium-argon dating, Foraminiferal studies, the examination of volcanic lava structures, and the interpretation of deep-sea cores, suggest, however, that the former belief that glaciation was confined to the Pleistocene, and was not characteristic of the Tertiary, must now be discarded. It seems clear that glaciation was initiated in some areas in the middle Tertiary, and this has led R. F. Flint (1972: 2) to remark that 'perhaps the most stirring impression produced by recent great advances in the study of the Quaternary period is that the Quaternary itself is losing its classical identity'. Previously it had been widely believed that the period from the Triassic to the Tertiary was a lengthy span when ice-sheets and glaciers were absent, and when climatic fluctuations were less frequent and less severe than they were to be in the Pleistocene. Bandy (1968), however, has written that 'the magnitude of the planktonic faunal changes indicates that the paleo-oceanographic changes of the later Miocene and the middle Pleistocene are almost as great as those of the classic Quaternary'. By a study of *Globigerina* types in ocean cores he found evidence for a major expansion of polar faunas no less than between 10 and 11 million years

Table 1.9 Tertiary mean annual temperatures (°C)

	NW Europe	WUSA	Pacific Coast of North America
Recent	—	—	10
Pliocene	14–10	8–5	12
Miocene	19–16	14–9	18–11
Oligocene	20–18	18–14	20–18.5
Eocene	22–20	25–18	25–18.5

(From data in Butzer, 1972)

ago. This expansion was followed, he suggested, by another mid-Pliocene expansion between 5 to 7 million years ago, and the classic Pleistocene expansion 3 million years ago.

Equally, some of the lithified glacial tills (non-sorted and non-stratified sediments carried or deposited by a glacier) interbedded with volcanic lavas in the White River Valley area of Alaska, have been dated at 9 to 10 million years old, and similar methods, combined with a study of the iceberg-rafted debris in Southern Ocean deep-sea cores, suggest that the east Antarctic glaciers reached a full-bodied stage somewhat before 5 million years ago. Another core from the area has led to the even more striking conclusion that glacial conditions may have been present in the Eocene (Geitzenauer *et al.* 1968), since a coexistence of micro-fossils of Eocene age, and of sediments of glacial type, has been proved. In eastern Antarctica ice-field initiation dates back to the Lower Miocene (Drewry 1975).

Confirmatory evidence of Eocene glaciation in Antarctica, with all that this implies for world sea-levels and world climate, is provided by the nature of dated volcanic materials on that continent. Volcanoes that have erupted beneath an ice-sheet display a suite of textural and structural characteristics that are especially distinctive in volcanoes composed of basaltic lava, and such materials have been found in Antarctica back to the Eocene. Similar evidence implies also that glaciation may well have been uninterrupted until the present in this South Polar region (Le Masurier 1972), though doubts about the reliability of this evidence have been expressed (Frakes 1978: 53).

All these dates for the onset of ice-caps and glaciation are very considerably earlier than the classic dates of around one million to one and a half million years formerly given for the start of a 'climatic' Pleistocene. Some authorities gave even shorter duration to the Pleistocene. Zeuner, for example, in 1959 gave an estimate of only 600 000 years.

Even though this new evidence has greatly altered our view of the

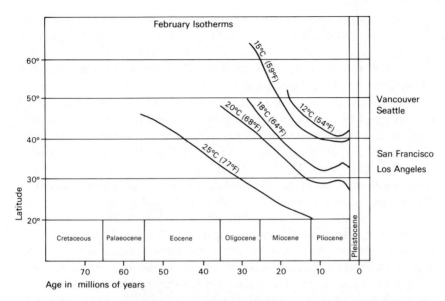

FIG. 1.8 Tertiary isotherms for Pacific Ocean water. During Tertiary times the isotherms for the water of the Pacific Ocean gradually shifted to the south as the climate became colder. About 35 000 000 years ago, for instance, the 20 °C isotherm was at the latitude of Seattle, but it moved to Baja California as the cold advanced from the north (after Kurten 1972: 28).

Tertiary era there is probably a considerable amount of truth in the oft-stated idea that in general temperatures did show a tendency to decline in many parts of the world during the course of the Tertiary. On the basis of deep-sea core studies, Emiliani (1961), a pioneer of the technique, has suggested that there was a broad temperature decrease amounting to about 8 °C for the middle latitudes during the Tertiary. A similar temperature depression is indicated in the Equatorial Pacific, but on the basis of land flora studies somewhat greater changes than these have been proposed for the western United States between 40° and 50 °N and the picture for other areas is the same (Table 1.9). The pattern of temperature decline for Pacific Ocean water, with a progressive southward shift in isotherms as the climate became cooler, is illustrated in Fig. 1.8.

In Australia a comparable sequence of decline of Tertiary temperatures has been proposed, based on the former great extent of the plants *Araucaria* and *Agathis* into Tasmania. At the present time they are limited to Queensland and the warmer parts of Australia. However, rainfall decline may have been equally important in such changes of floral distributions, and Gentilli (1961) has said that 'areas that now receive

12.5 cm of rain a year must then have received at least 125 cm with no rainy season. If there was even a short dry season the annual rainfall must have reached some 200 or 250 cm for these laurisilvae to grow as they did.' Large trees then existed in the Lake Eyre Basin and other stretches of the 'dead heart' of Australia (Gentilli 1961, Gill 1961).

The North Atlantic region in the early Tertiary may have been characterized by a widespread, tropical moist forest type of vegetation (Pennington 1969, ch. 1). Indeed, the classic interpretation of the fossil plant evidence from the Palaeocene London Clay of south-east England, made by Reid and Chandler (1933), was that it represented a flora comparable to that of the Indo-Malayan region of today. Malaysian *Nypa* palm, mangrove swamp with *Avicennia*, and *Pandanus* (a tropical plant with stilt roots) were present. Whether there really was a tropical rainforest environment in southern Britain in the Eocene is, however, a matter for debate. As Daley (1972: 187) has pointed out, 'Such a climate, with its markedly uniform character throughout the year, could not have existed at the latitude of southern Britain during the Eocene (40 °N), since seasonal climatic variations would undoubtedly have occurred.' In addition there is a small proportion of extratropical types within the London Clay flora. One possibility is that the mixture of tropical and extratropical plants may have resulted from a type of climate not represented at the present day. Daley envisages that given frost-free conditions and high rainfall levels tropical plants may have become established in low-lying damp areas even though temperatures were not truly tropical.

Nevertheless the picture of early and middle Eocene warmth is confirmed by more recent studies. For example, Collinson and Hooker (1987) report that beds of that age contain the highest proportion of potential tropical flora (up to 82 per cent) and the smallest proportion of potential temperate nearest living relatives (up to 42 per cent). Floras from the late Palaeocene and latest Eocene and early Oligocene contain a smaller proportion of potential tropical elements (e.g. 70–82 per cent). Collinson and Hooker (1987: 267) consider that: 'A great many of the near living relatives of fossils co-exist today in paratropical rainforest in eastern Asia. Many are large forest trees, others lianas. The greatest diversity of these elements in the Early and Middle Eocene suggests the presence of a dense forest vegetation.' Tree rings were also particularly large, indicating rapid growth in response to high temperatures (Creber and Chaloner 1985).

A further source of information on climates is provided by fossil soils in the Palaeocene Reading Beds. At Alum Bay, on the Isle of Wight, 'a warm climate with a marked dry season' is suggested (Buuman 1980). Seasonality is also to be inferred from fluvial sedimentary structures in the Middle Eocene beds of the Hampshire Basin (Plint 1983), which indicate 'a fluctuating, perhaps seasonal, discharge regime'.

The climatic deterioration at the end of the Eocene, which may have been abrupt (Buchardt 1978) or gentle (Collinson *et al.* 1981), according to the type of evidence used, created a different climatic regime in the Oligocene. In the Oligocene the climate of Britain was more comparable to that of a region like the south-eastern United States, though there may have been very dry intervals, as indicated by the presence of gypsum evaporite crusts and ancient continental dunes in the Ludien and Stampian sediments from the Paris Basin.

The termite fauna of the early Oligocene Bembridge Marls from the Isle of Wight confirm that, although it was cooler than the Eocene, it was none the less warmer than today (Jarzembowski 1980).

The causes of a warm Palaeogene in Britain are both local and global. At a local scale, Britain was at a lower latitude than today, being 10–12° further south in the Palaeocene. Movements of the continents may also have affected global climates by creating a unique configuration of the continental masses. Parrish (1987: 52–3) has suggested that:

By the beginning of the late Cretaceous, global palaeogeography had entered a mode of maximum continental dispersion, and the monsoonal circulation that had dominated palaeoclimatic history during the previous 250 million years had disappeared. From this time until the collision of India and Asia re-established strong monsoonal circulation during the Miocene, continental palaeoclimates were characterized by the zonal circulation that is the fundamental component of atmospheric circulation on Earth.

Other possible contributing factors to global Palaeogene warmth include: elevated atmospheric carbon dioxide levels associated with a phase of rapid sea floor spreading and plate subduction (Barron 1985, Creber and Chaloner 1985) or a marked reduction in the angle of tilt of the Earth's axis, possibly by as much as 15° in the Middle Eocene (Wolfe 1978).

However, by Pliocene times, the 'tropical' vegetation of the North Atlantic region was being replaced by a largely deciduous warm temperate flora including *Sciadopitys*, *Tsuga*, *Sequoia*, *Taxodium*, and *Carya*, some of which were to be eradicated by the intense cold of the Pleistocene itself (Montford 1970). The last appearance of palms north of the Alps in Europe is in the Miocene flora at Lake Constance.

The widespread nature of warm, wet conditions in certain parts of the world in Tertiary times had diverse environmental effects. Deep-

Table 1.10 Differences in temperature and annual precipitation between Reuverian B (the end of the Pliocene) and the present, estimated from climatic values based upon the current distribution of certain plants

Region and genus	1	2	3	4
NW Germany and Netherlands				
Liquidambar	+3 to 4		= or + 1 to 2	
Nyssa near coastline edge of Mittlegebirge	+3			
Tsuga (*canadensis*)	+2			+300
Carya and *Liriodendron* near coastline edge of Mittlegebirge	+3			
Ilex				
Fagus				
Central and east Poland, Lithuania, and White Russia				
Liquidambar	+5	+5	+3 to 4	+350
Nyssa	+4 to 5	+1	+3 to 4	+250
Tsuga (*canadensis*)	+2 to 3	+3	+1	+400
Carya	+3			
Ilex		+5 to 6		+150
Fagus		+4	+3	
Central reaches of the Volga				
Tsuga (*canadensis*)	+2 to 3	+3	+1	400
Carya (and *Liriodendron*?)	+3	+6	+5	350
Ilex		+15		250
Fagus		+12	+5 to 6	200
Lower reaches of the Don				
Tsuga (*canadensis*)				+350
Fagus		+7	+3	+200
Bulgaria—central lowlands				
Liquidambar	+3	+1	+1 to 2	+300
Tsuga (*canadensis*)				+150
Eastern Siberia				
Tsuga heterophylla and *mertensiana*		+10 to 15	+10	+200

Notes: 1 = July mean temperature (°C). 2 = January mean temperature (°C). 3 = Mean annual temperature (°C). 4 = annual precipitation (mm).
(From Frenzel, 1973, p. 89).

weathering profiles were produced over wide areas in mid-latitudes, in the form of laterites and silicified layers called silcretes. Limestone areas were subjected to intense solutional processes, and rocks were rotted so that they were to be particularly susceptible to glacial erosion in the Pleistocene.

Immediately before the Ice Age, however, in a period which the Dutch call Reuverian B, the climate of the present temperate latitudes of the northern hemisphere seems to have favoured the development of woodland. At that time varied woodlands extended from the Atlantic coasts to the Sea of Japan, a picture that has not been repeated since (Frenzel 1973). Over wide reaches of the present warm and cool temperate latitudes, temperature and rainfall conditions were particularly favourable, and resembled those of a subtropical climate. Some tentative quantitative estimates based on the analysis of present and past distributions of certain species or genera, are shown in Table 1.10. In central and eastern Europe, temperatures were probably 3° to 5 °C higher than they are now, and precipitation levels were several hundred millimetres higher.

The causes of the 'Cainozoic climate decline' (Fig. 1.9) are still not fully understood, but the trend seems to be associated with the break-up of the ancient supercontinent of Pangaea into the individual continents we know today. Round about 50 million years ago Antarctica separated from Australia and gradually shifted southwards into its present position centred over the South Pole. At the same time the continents of Eurasia and North America moved towards the North Pole. As more and more land became concentrated in high latitudes ice-caps could develop, surface reflectivity increased, and as a consequence climate probably cooled all over the world.

Late Cainozoic cooling may have promoted the expansion and development of some of the world's great deserts.

In late Cainozoic times (van Zinderen Bakker 1984) aridity became a prominent feature of the Saharan environment, probably because of the occurrence of several independent but roughly synchronous geological events (Williams 1985):

(i) As the African plate moved northwards there was a migration of northern Africa from wet equatorial latitudes (where the Sahara had been at the end of the Jurassic) into drier subtropical latitudes.

(ii) During the late Tertiary and Quaternary, uplift of the Tibetan plateau had a dramatic effect on world climates, helping to create the easterly jet stream which now brings dry subsiding air to the Ethiopian and Somali deserts.

(iii) The progressive build-up of polar ice-caps during the Cainozoic climatic decline created a steeper temperature gradient between the equator and the poles, and this in turn led to an increase in trade wind velocities and their ability to mobilize sand into dunes.

(iv) Cooling of the ocean surface may have reduced the amount of evaporation and convection in low latitudes, thus reducing the amount of tropical and subtropical precipitation.

Thus although the analysis of deep-sea cores in the Atlantic offshore from the Sahara indicates that some aeolian activity dates back to the early Cretaceous (Lever and McCave 1983), it was probably around 2 to 3

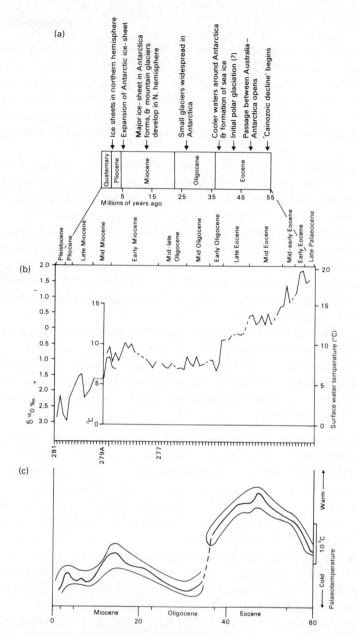

Fig. 1.9 The Cainozoic climate decline:
A. A generalized outline of significant events in the Cainozoic climate decline.
B. Oxygen isotopic data and palaeotemperatures indicated for planktonic foraminifera at three subantarctic sites (277, 279, and 281) (after Kennett and Shackleton 1975).
C. Temperature changes calculated from oxygen isotope values of shells in the North Sea (after Buchard 1978).

million years ago that a high level of aridity became established. From about 2.5 million years ago the great tropical inland lakes of the Sahara began to dry out, and this is more or less contemporaneous with the time of onset of mid-latitude glaciation. Aeolian sands become evident in the Chad basin at this time, and such palynological work as there is indicates substantial changes in vegetation characteristics (Servant-Vildary 1973, Street and Gasse 1981).

If we consider the deserts of southern Africa, we find that Siesser's investigation of offshore sediments (Siesser 1978, 1980) has indicated that upwelling of cold waters intensified significantly from the late Miocene (7–10 million years BP) and that the Benguela Current developed progressively thereafter. Pollen analysis of such sediments indicates that hyper-aridity occurred throughout the Pliocene, and that the accumulation of the main Namib *erg* (sand sea) started at that time.

Likewise, the Atacama of South America has shown hyper-aridity since the middle to late Miocene (Alpers and Brimhall 1988). The uplift of the Central Andes cordillera during the Oligocene and early Miocene was a critical palaeoclimatic factor, providing a rain-shadow effect and also stabilizing the south-eastern Pacific anticyclone. However, also of great significance (and analogous to the situation in the Namib) was the development between 15 and 13 million years ago of cold Antarctic bottom waters and the cold Humboldt Current as a result of the formation of the Antarctic ice-sheet.

Selected reading

The literature on environmental change is massive and this list merely gives a short guide to some of the more accessible modern literature. A most readable account of the development of ideas, especially on the causes of the Ice Age, is provided by J. and K. P. Imbrie (1979), *Ice Ages*. D. Q. Bowen's (1978) *Quaternary Geology* provides a good basic review. Methods for studying environmental change are given in J. J. Lowe and M. J. C. Walker (1984), *Reconstructing Quaternary Environments*, R. S. Bradley (1985), *Quaternary Paleoclimatology*, and A. S. Goudie (1990) (ed.), *Geomorphological Techniques* (2nd. edn., especially Part 5). Two books by J. A. Catt provide a useful geological, pedological, and geomorphological perspective: *Soils and Quaternary Geology* (1986) and *Quaternary Geology for Scientists and Engineers* (1988). Some useful specialist journals are *Quaternary Research*, *Climate Change*, *Boreas*, *Quaternaria*, *Palaeoecology of Africa*, *Palaeogeography, Palaeoclimatology and Palaeoecology*, *Quaternary Science Reviews*, and *Journal of Quaternary Science*.

2

The Chronology and Nature of the Pleistocene

Introduction

The Pleistocene did not consist of just one great ice age, but was composed of alternations of great cold (glacials, stadials), with stages of relatively greater warmth (interglacials, interstadials).

The expansion and thickening of the ice-sheets and glaciers in the glacial stages led to the erosion of underlying rock and to the transport of large quantities of debris over long distances. This debris, which is given a large variety of names, including till and boulder clay, is a highly characteristic, normally ill-sorted combination of boulders, sands, and clays. The debris frequently contains rocks derived from some hundreds of kilometres away. These are called erratics and may attain a considerable size. In eastern England, near Ely and also on the Norfolk coast, some of the erratics produced by glacial erosion of the chalk attain colossal dimensions, some being 400–600 m long and up to 50 m thick. In interglacial periods, when conditions became warmer, like those of the present time, the ice retreated and left moraines and other related glacial and fluvio-glacial landforms and deposits. These were then weathered. Other sediments might also accumulate on top of the glacial deposits, and these contain characteristic faunal and floral remains. In another glacial period such sediments might themselves become covered by boulder clay. Classic interpretations of the history of the Ice Ages or the Pleistocene have been based on a study of the extent and character of these alternations of glacial and interglacial deposits on land.

Although Geikie and Ramsay established over a century ago that the Pleistocene Ice Age was composed of multiple glaciations, and in spite of the great deal of work that is being devoted to establishing the duration and divisions of the Pleistocene epoch at the present time, there is still a marked degree of controversy over the number of glaciations, stadials, interglacials, and interstadials. This exists partly because of the problem of definition of these events, a matter discussed further on p. 8. There is also a lack of agreement with regard to correlations of events between different areas, and there is still no universally agreed idea, as mentioned in the previous chapter, on the date of the Pliocene–Pleistocene boundary. Nevertheless, the great increase in the use of new dating techniques, and of deep-sea core evidence, have enabled some statements to be made with a greater degree of confidence than hitherto. Indeed, there are those who would claim that Pleistocene studies have now been transformed by the new techniques. Bowen (1978: 193), for example, has argued, 'There is no doubt whatsoever that Quaternary systematics have been subject to a revolution comparable to that of plate tectonic theory on geology as a whole.' In particular he points to the fact that whereas traditionally the Ice Age was regarded as having comprised four, five, or at most six major glacials, now there are indications from the ocean cores that there have been no less than seventeen glacial cycles in the last 1.65 million years that make up the Quaternary *sensu stricto*.

The length of the Pleistocene

The slicing up of geological time, and the attribution of names to particular phases, has already been noted (pp. 3–4). It is a cause of perennial argument. In attempting to define the boundary between the Pliocene and the Pleistocene there has been a considerable range of views. On faunal grounds the base of the Quaternary is officially placed at about 1.65 million years ago (Bowen *et al.* 1986), and this conveniently coincides with a major geomagnetic reversal (the top of the Olduvai Event), which can be recognized on a global basis. On the other hand other major faunal changes took place earlier than this (see Table 2.1), including the appearance of certain Foraminifera and the extinction of a distinctive group of planktonic organisms called *Discoasteridae*, which secreted six-rayed star-shaped skeletal elements called *discoasters*.

There are also those who would place the boundary on the basis of some major climatic deterioration, namely the marked appearance of mid-latitude, as opposed to polar, glaciers. In some parts of the world (Table 2.1) this event seems to have taken place more than 3 million years ago. The dating of such glaciations is provided either by the presence of iceberg-rafted debris in ocean cores of known age, or by using the stratigraphic relations of tillites and volcanic rocks (such as basalt) which can be dated by means of Potassium Argon. Kukla (1989: 3)

Table 2.1 Faunal and climatic dates for the start of the Pleistocene

Source	Location	Evidence	Date (m yrs. BP)
Faunal dates			
Leakey (1965)	Olduvai (Tanzania)	K/AR dating of upper Villafranchian fauna	More than 1.75
Glass et al. (1967)	Deep-sea	Coiling reversal in Foraminifera	2.1
Glass et al. (1967)	Deep-sea	Discoasteridae die out	2.0–1.8
Zagwijn (1974)	North Sea Basin	Faunal extinctions	2.5
Climatic dates (mid-latitude glaciation)			
McDougall and Wensink (1966)	Iceland	Tillite/basalt	3.0
Mathews and Curtis (1966)	New Zealand	Tillite/basalt	2.47
McDougall and Stipp (1968)	Sierra Nevada, USA	Tillite/basalt	2.7–3.1
McDougall and Stipp (1968)	Taylor Valley, Antarctica	Tillite/basalt	2.7
Opdyke et al. (1966)	Southern Ocean cores	Ice-rafted debris	2.5
Mercer (1969)	Argentina	Tillite/basalt	2
Clapperton (1979)	Bolivia	Tillite	3.27
Shackleton and Opdyke (1977)	Northern Hemisphere	Oxygen isotopes in deep sea core	3.2
Kvasov and Blazhchishin (1978)	Barents Sea	Glacial	3.5
Shackleton et al. (1984)	N. Atlantic	Ice-rafted debris	2.5
Horowitz (1989)	Israel	Pollen core	2.6–2.4

summarizes the problems of demarcation that exist: '. . . the transition to cooler climates between 3 and 1.5 Ma ago took place in several pulses, tens and hundreds of thousands of years apart, which did not necessarily show up with the same relative amplitudes everywhere. The Plio-Pleistocene boundary will always have to be chosen arbitrarily.'

Analysis of deep-sea core sediments indicates that the initiation of moderate-sized ice-sheets in the northern hemisphere occurred at about 2.40 million years BP and that this represents the culmination of longer-term high-latitude cooling that began about 750 000 years earlier (Ruddiman and Raymo 1988).

The divisions of the Pleistocene: the classic terrestrial story

Between 1901 and 1909, Albrecht Penck and E. Brückner produced a major three-volume work, *Die Alpen im Eiszeitalter*. For decades this formed the basis for much glacial chronology, and in particular they proposed in it the classic four-fold glaciation model of the Pleistocene. By studying what was left of old moraines, they were convinced that there had been four great glaciations of different intensities in the Alpine

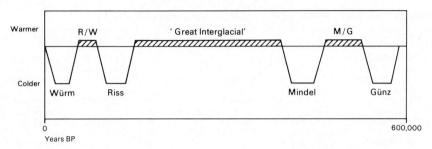

FIG. 2.1 The simple Penck and Brückner model of Pleistocene glaciations, showing the four main glacials. Note the extent of the 'Great Interglacial'.

region of Europe. They also proved, from the preservation of plant remains at some sites such as Hötting, that some of the intervening periods were fairly mild. They further demonstrated the correspondence between these four glaciations and the successive gravel terraces of the Rhine and other rivers. They named the four glacial periods after valleys in which evidence of their existence occurred: the Günz, Mindel, Riss, and Würm.

Penck and Brückner also provided an estimate of the relative duration of the glacial-interglacial cycles. Of particular note was their belief that the interglacial between the Mindel and the Riss lasted a longer time than any other. In the classic model this event was therefore termed *The Great Interglacial*.

Their model is summarized in Fig. 2.1. It was immensely influential, was widely accepted and almost attained the status of a law. Evidence for additional glaciations obtained by other workers was generally forced into their scheme, and assigned the secondary role of substages or stadia within the four major glaciations. In many respects, however, the Alpine sequence was an unfortunate base for long-distance correlation, in that the Alpine area is not ideal for Quaternary stratigraphic studies. In addition to the universal problem that earlier glaciations may have had their effects obliterated by subsequent erosional events, there is also a lack of organic deposits, there is some difficulty in relating such deposits to the glacial moraines, there are considerable possibilities for confusion resulting from earth movements, and complications are introduced by the separation of key areas by mountain chains. As Sparks and West (1972) proclaim, 'The eradication of the Alpine nomenclature, which should never have been applied widely in the first place, has proved and indeed is still proving a Herculean task.' Also proving to be a Herculean task is the establishment of an acceptable stratigraphic scheme to replace the

Table 2.2 Sequences of Pleistocene phases in the northern hemisphere

	Rhine Estuary[1]	Britain[2]	Alpine foreland[3]	European Russia[4]	North America[5]
	Holocene[6]———				
Glacial Pleistocene	WEICHSELLAN	DEVENSIAN	WÜRM	VALDAI	WISCONSIN
	Eemian	Ipswichian	Riss–Würm	Mikulino	Sangamon
	SAALIAN	WOLSTONIAN	RISS	MIDDLE RUSSIAN	ILLINOIAN
	Holsteinian	Hoxnian	Great Interglacial	Likhvin	Yarmouth
	ELSTERIAN	ANGLIAN	MINDEL	WHITE RUSSIAN	KANSAN
	Cromerian	Cromerian	Günz–Mindel	Morozov	Aftonian
	MENAPIAN	BEESTONIAN	GÜNZ	ODESSA	NEBRASKAN
Pliocene	Waalian	Pastonian	Donau–Gunz	Kryshanov	
	EBURONIAN	BAVENTIAN	DONAU		
	Tiglian	Antian			
	PRETIGLIAN	THURNIAN			
		Ludhamian			
		WALTONIAN			
	'Pre-Glacial'				

[1] From Zagwijn (1975).
[2] From Sparks and West (1972).
[3] From Penck and Brückner (1909) and others.
[4] From Flint (1971).
[5] From Flint (1971). Richmond & Fullerton (1986) suggest some of these names should now be abandoned.
[6] Interglacials are in lower case. Glacials in capitals.

Penck and Brückner model. Pilbeam (1975: 819) has remarked that: 'A wide variety of stratigraphic schemes have been proposed for the last 2 to 3 million years. . . . None of these . . . agrees precisely with any of the others; most are probably about equally acceptable: and, no doubt, none is absolutely correct. However, all are useful advances beyond the four glacial-pluvial schemes still utilized so widely in many anthropological textbooks.' Correlation of glacial and interglacial phases in different regions is a hazardous procedure given the inadequacies even of modern dating methods, and the fragmentary nature of the stratigraphic evidence. However, in Table 2.2 a list of major events is given for different regions from the northern hemisphere, but deliberately no direct correlations are attempted. Its purpose is to enable the reader to obtain some meaning from the local terminology which is utilized in various parts of this book. Nevertheless, radio-carbon dates are now available which allow the last glaciations (Weichselian, Devensian, Würm, Valdai, Wisconsin) to be correlated, and many workers have suggested that the last four cold events (equivalent to the four classic glacials of the Alpine sequence) may also be correlated, though the means whereby this can be done with any degree of certainty are sparse. It is certain that correlations become increasingly hazardous as one moves back through the classic 'glacial' Pleistocene into the 'pre-glacial' Pleistocene. The main problem is one of dating, and as Vita-Finzi (1973) has remarked, 'When correlation precedes dating it is difficult to argue about contemporaneity, let alone about age differences.'

In some areas cold phases may not have led to actual glaciation. This is, for example, the case in the Early Pleistocene of eastern England, where sediments in a borehole at Ludham indicate fluctuations from forest to a type of oceanic heath, and back to forest. This oceanic heath is interpreted as a result of climatic change for the worse, which probably produced glacial conditions in higher latitudes. Evidence for similar Lower Pleistocene climatic oscillations has also been found from boreholes in the Netherlands.

The Netherlands continental sequence (after Zagwijn) is illustrated in Fig. 2.2. While it cannot yet be directly correlated with the East Anglian sequence it shows a similar pattern of fluctuations with miscellaneous Early Pleistocene oscillations following on from the rather warm conditions of the Pliocene Reuverian. The palaeobotanical evidence for the early cold phases is good, and permafrost structures have also been indentified, but actual glaciation in the Netherlands would only appear to have been widespread in the penultimate cold phase, the Saalian. In all there were at least six major cold phases revealed in the non-marine Pleistocene sediments of the Netherlands. Zagwijn (1975) believes that the climatic curve shows two important trends during the course of the Pleistocene. The first is that the amplitudes of the oscillations seem to increase. This was caused mainly by the temperatures of the cold phases becoming cooler, for temperatures in the warm phases remained fairly similar throughout the Pleistocene. The second trend that can be identified is that the frequency of the oscillations shows a distinct increase upwards, especially so after the Jaramillo palaeomagnetic event about 900 000 years ago.

Recent studies in the USA (summarized in Fig. 2.3) indicate that the classic picture given in Table 2.2 is now inadequate. The picture that emerges is one of early initiation of glaciation in the Pliocene with repeated advances and retreats after that time. Full details of the complex stratigraphy of glacial events in the different parts of the USA are given in Richmond and Fullerton (1986).

The sequence of glacial events in the erstwhile USSR is summarized in Fig. 2.4, and shows a series of glacials and interglacials since the Brunhes–

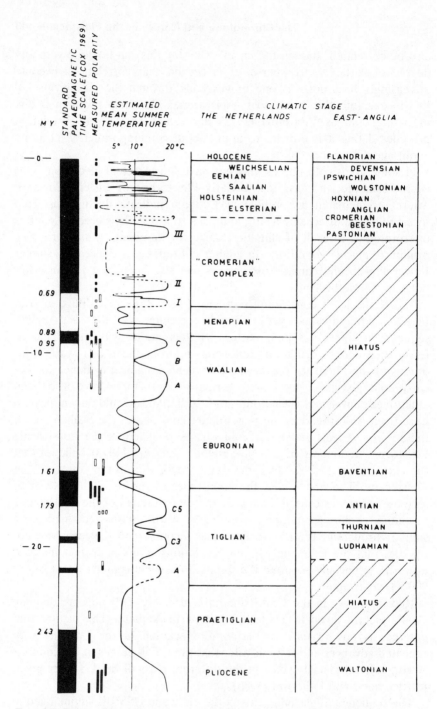

FIG. 2.2 Palaeoclimatic measurements, climatic curve, and stratigraphic climatic division of the Quaternary of the Netherlands (after Zagwijn 1975: 8).

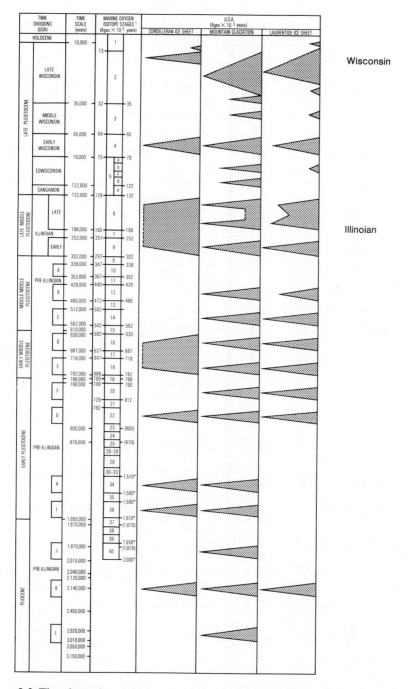

Fig. 2.3 The chronology of glaciations in the USA. The main glacial advances are shaded (modified after Bowen *et al.* 1986, Chart 1). Note that glaciation took place as early as the Pliocene.

FIG. 2.4 Glacial chronology for the erstwhile USSR in relation to palaeomagnetic divisions (modified after Velichko and Faustova 1986, Table 1).

Matuyama boundary (c.0.7 million years ago). The last of these was called the Valdai glaciation, and it corresponds with the Wisconsin of the USA and the Weichsel and Würm of Europe.

The Pleistocene record in the oceans

The various terrestrial sequences that have been laboriously compiled by classic stratigraphic techniques in Britain and Europe have demonstrated beyond doubt the fact both that major glacial and interglacial fluctuations have taken place and that the Pleistocene was a period of remarkable geomorphological and ecological instability. However, because of the incomplete nature of the evidence it is not possible, using this inform-ation, to construct a correct, long-term model of environmental changes in the Pleistocene. For this to be achieved it is necessary to look at the more complete record of deposition preserved in the ocean core sediments, and, to a lesser extent, in the loessic sequences of the Old World.

Many cores have now been extracted from the world's oceans, and as recounted in Chapter 1 many techniques have been used to extract palaeo-environmental information and to obtain a reliable dating frame-work. The most important information has been gained from the oxygen isotope composition of foraminiferal tests, and piston core V28–238 has been used as the type locality for the O^{18}/O^{16} record (Fig. 2.5). It was taken from the Pacific near the Equator (01°01′ N 160°29′ E), and was raised from a water depth of 3120 m (Kukla 1977). The upper 14 metres of the core have been subdivided into 22 oxygen-isotopic stages numbered in order of increasing age (Table 2.3). Odd-numbered units are relatively deficient in O^{18} (and thus represent phases when temperatures were relatively high and the ice-caps small), while the even-numbered units are relatively rich in O^{18} (and represent the colder phases when the ice-caps were more extensive). Boundaries separating especially pronounced isotopic maxima from exceptionally pronounced minima have been called 'terminations'. They are in effect rapid deglaciations, and are conventionally numbered by Roman numerals in order of in-creasing age. The segments bounded by two terminations are called glacial cycles. The youngest glacial cycle, which starts with isotopic stage 1, and follows Termination I, is not yet completed.

The results of the analysis of this major type core, illustrated in Fig. 2.6, can be summarized thus. There have been eight completed glacial cycles (named b to i), and nine terminations (I–IX) in the last epoch of normal polarity (the Brunhes) which lasted just over 0.7 million years.

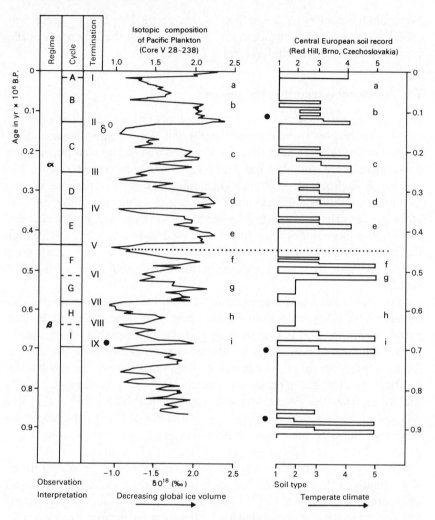

FIG. 2.5 The history of the last 900 000 years as revealed in the isotopic composition of a Pacific core (V 28–238) and in the soil record of a loess profile from Brno, Czechoslovakia. Compare this figure with Fig. 2.1.

There are ten completed glacial cycles and eleven terminations in the one million years following the Jaramillo normal polarity event. Oscillations of the ice-cap volume within each glacial cycle seem to follow a saw-tooth pattern, progressing from an early minimum to a late maximum. The number of secondary fluctuations in isotopic composition within each glacial cycle varies. Commonly it varies between 4 and 6. Over the last

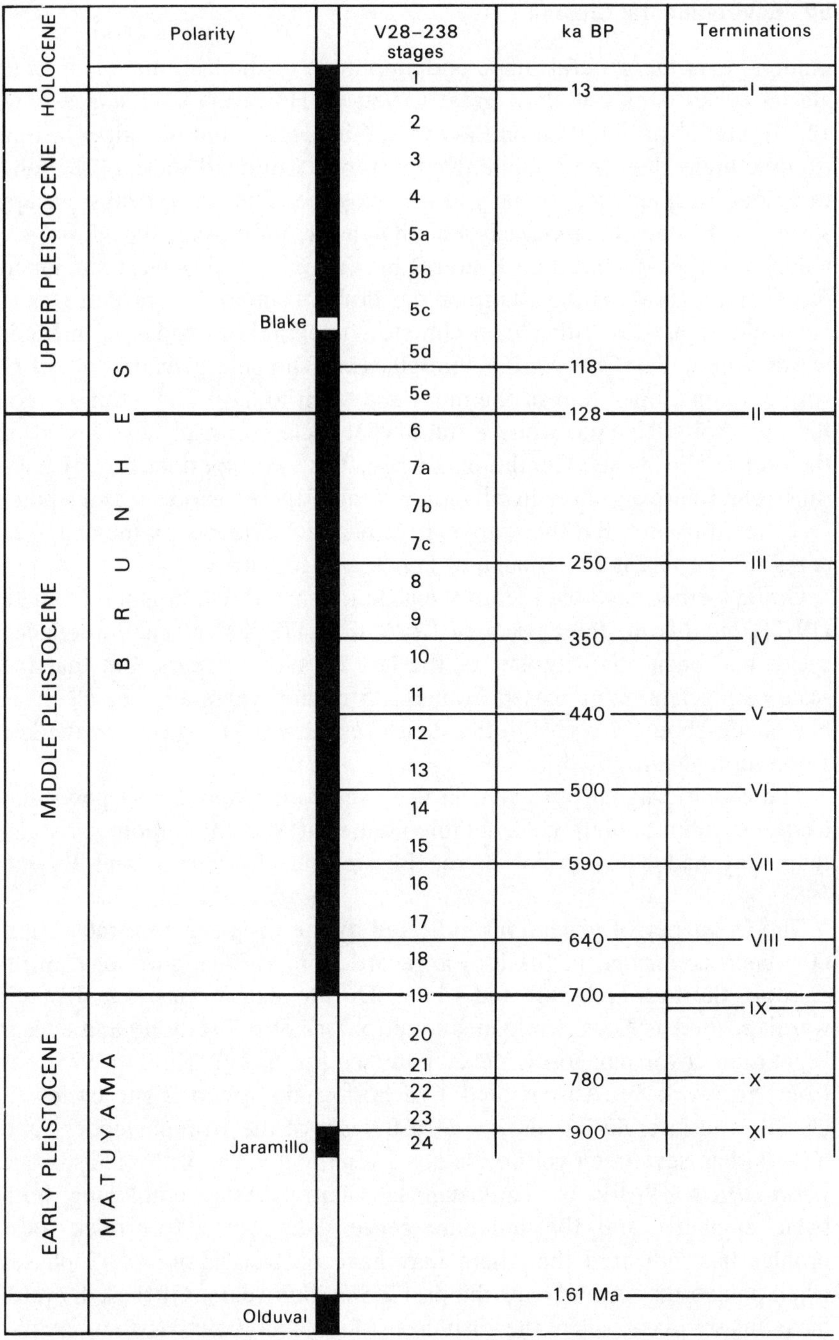

	Polarity		V28–238 stages	ka BP	Terminations
HOLOCENE			1	13	I
UPPER PLEISTOCENE	B R U N H E S	Blake	2		
			3		
			4		
			5a		
			5b		
			5c		
			5d	118	
			5e	128	II
MIDDLE PLEISTOCENE			6		
			7a		
			7b		
			7c	250	III
			8		
			9	350	IV
			10		
			11	440	V
			12		
			13	500	VI
			14		
			15	590	VII
			16		
			17	640	VIII
			18		
EARLY PLEISTOCENE	M A T U Y A M A	Jaramillo	19	700	IX
			20		
			21	780	X
			22		
			23	900	XI
			24		
		Olduvai		1.61 Ma	

FIG. 2.6 The subdivisions of Pacific core V28–238. An explanation is given in the text.

million years the general shape and amplitude of the last nine completed glacial cycles does not show great variation. However, cold stages 2, 6, 12, 16, and 22 are marked on the average by deeper and/or longer lasting isotopic highs than the remaining cold stages, with cold stage 14 seeming to be one with an exceptionally low ice volume. The warm peaks (except stage 3) all seem to approach a similar level, indicating that global ice volume and sea-surface temperatures in peak interglacials were similar to those of the present day. During the last 0.9 million years there have been nine episodes with global climate comparable to today's. In other words, there have been nine interglacials. The interglacials only constitute about 10 per cent of the time, and seem to have had a duration of the order of 10^4 years, while a full glacial cycle seems to have lasted of the order of 10^5 years. Conditions such as those we experience today have thus been relatively short-lived and atypical of the Pleistocene as a whole. One can also note that there appears to be scant evidence in the deep-sea cores for the Great Interglacial of Penck and Brückner.

Going further back the record seems less clear. Shackleton and Opdyke (1977) found from their study of Core V28–179 that glacial/interglacial cycles had been characteristic of the last 3.2 million years, but that the scale of glaciations increased around 2.5 million years ago. In all there have been about 17 cycles in the Pleistocene *sensu stricto* (i.e. in the last 1.6 million years).

The climatic signal preserved in deep-sea core sediments suggests that ice-sheet volumes were considerably greater after *c*.0.9 million years ago than they had been earlier in the Pleistocene (Ruddiman and Raymo 1988).

The frequency of glaciations indicated by the deep-sea core record has now been confirmed in the loessic record from various parts of Central Europe, the former USSR and China. During glacial phases aeolian silt was deposited as loess, while in warmer phases of soil stability and denser vegetation cover palaeosols developed (see pp. 67–9). Kukla (1975) has from his work in Austria and Czechoslovakia found eight cycles of glacials and interglacials in the 700000 years of the Brunhes epoch, and no less than seventeen within the last 1.6 million years of the Pleistocene *sensu stricto*. Work in Tajikistan in Central Asia, employing both palaeomagnetic and thermoluminescence dating of 200 m thick loess profiles has indicated that there may have been as many as 45 phases when palaeosols formed over the past 2.5 million years. Of these 8 cycles have taken place since the Brunhes–Matuyama boundary. A similar picture has emerged from the classic loess terrains of China, where 12 major cycles have been recognized over the same span of time. Thus

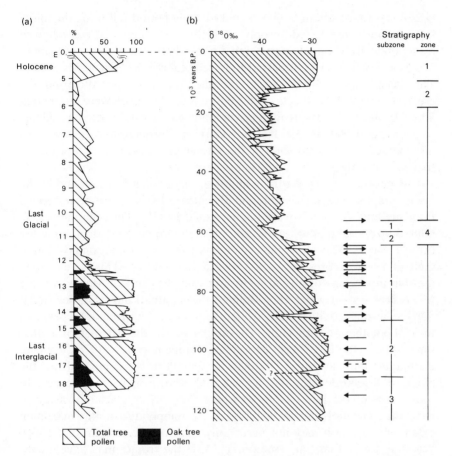

FIG. 2.7 A. The pollen record from Grande Pile, France (modified after
            Woillard 1978).
         B. Oxygen isotope variations in the Camp Century Core, Greenland
            (after Johnson *et al.* 1970).

although there may be slight differences in the record from these three
loessic areas one is none the less impressed by the way in which the
frequency of change broadly mirrors that found in the oceanic record
(Goudie *et al.* 1983).

A fairly precise comprehension of the nature of one glacial cycle, the
last, has been gained from other relatively continuous depositional
sequences (Fig. 2.7) including the polar ice-caps and some large bogs like
Grand Pile in France. From about 116000 years ago until 60000–70000
years ago there was a relatively warm interval. The onset of the last

glacial was round about 60 000 BP, but the cold period following the initial fall in temperature lasted for only a few thousand years and was followed by a relatively clement period (an interstadial) that lasted until about 30 000 BP. A further cold period followed, called the Glacial Maximum. This lasted about 10 000 years and the lowest temperature and the greatest extent of ice sheets of this entire glacial episode were attained at about 18 000 BP. Widespread deglaciation began abruptly about 14 000 BP, so that by 10 000 BP the North American Cordilleran Ice Sheet had disappeared. At around 9500 BP the Greenland Ice-Cap had accelerated to a rate of retreat of around 3 km per century.

This period of rapid warming and deglaciation can be identified in the oceanographic record (Kennett and Shackleton 1975). Freshwater derived from melting ice is relatively impoverished in O^{18}. Thus marine waters diluted by glacial meltwater are isotopically more negative than undiluted oceanic waters. There is a massive anomaly in the isotopic composition of Gulf of Mexico cores between 15 000 and 12 000 BP. This is related to massive inpourings of glacial meltwater into the Gulf of Mexico via the Mississippi system which then drained the melting Laurentide ice-sheet. Surface water salinities were evidently reduced by around 10 per cent throughout the western Gulf. The discharges of the Mississippi at that time appear to have been enormous. Emiliani *et al.* (1978) estimate discharges that averaged 100 000 to 230 000 m^3/s, compared to the 57 000 m^3/s recorded for the highest peak known in recent decades. In those parts of the lower latitudes where the state of glaciers is as much controlled by moisture availability as by temperatures, the maximum extent of glaciation may not necessarily have occurred at around 18 000 years BP, for at that time (see pp. 112–21) the tropics may have undergone a period of increased aridity. Thus, in Mexico, Colombia, and possibly also Ethiopia, glacial activity may have been out of phase with global temperature trends. In Mexico the maximum of glaciation occurred between 26 000 and 39 000 years BP, while in Colombia it occurred between 35 000 and 44 000 years BP (van der Hammen *et al.* 1981).

The glacial history of the British Isles

The study of sediment cores from the floor of the North Atlantic Ocean on the Rockall Plateau (Shackleton *et al.* 1984) indicates that substantial accumulation of iceberg-rafted debris occurred around 2.4 million years ago in the Late Pliocene and this, they believe, must correlate with the first fully glacial environment in Britain. Moreover, there is evidence preserved on the floor of the North Sea for an early glacial episode, dated prior to 1.8–2.1 million years BP (Sutherland 1984).

As we have already seen, the ocean floor record would suggest that there should be evidence for multiple cycles of glaciation since that time. Unfortunately, the evidence onshore is much more imperfect, and the further back in time one goes the more imperfect the record becomes. Furthermore, reliable dating is extremely difficult in the British Isles when one goes back beyond the range of radio-carbon, and there is no site where any reasonably full sequence of sediments is preserved. The interpretation of British Quaternary stratigraphy, in spite of over 150 years of research, is still bedevilled by disagreement of a fairly fundamental sort.

The classic view (see e.g., Mitchell *et al.* 1973) is that there have been three major glacials in Britain during which ice-caps covered large areas of the country. The earliest of these (called the Anglian) is presumed to correlate with the Elster Glaciation of continental Europe. Its extent is not delimited by any well-marked end-moraine system but by dispersed till desposits, which in the case of the Plateau Drift deposits of the Oxford Region are often little more than scattered thin, accumulations of erratic pebbles, the origin of which is far from proven. A good sequence of Anglian materials occurs at the type site at Corton Cliff, near Lowestoft, Suffolk. The Anglian glaciation may have been the most extensive to affect the British Isles, reaching as far south as Oxford and south Essex.

Before the Anglian there may have been five separate glacial events, mainly represented by erratic materials (Fig. 2.8). These are attributed to the Early Pleistocene, but their precise dates are unknown (Bowen *et al.* 1986).

The Anglian was undoubtedly a very important event in the geomorphological history of Britain for it produced some of the greatest changes in the evolution of the landscape. As Bowen *et al.* (1986: 308) put it:

In particular it is the time when the drainage of East Anglia changed from an extensive series of north-eastward sloping river terraces, to a glacial landscape composed primarily of till with a radial drainage pattern. The period of the Anglian Glaciation also included the erosion of the lowland of the Severn basin, and hill ranges like the Cotswolds became prominent features. Finally it is the time when the river Thames was directed from the Vale of St. Albans and Essex into its present route through London.

The Anglian glaciation, the date of which is still uncertain, but which may have occurred at around 450 000 years BP (Eyles *et al.* 1989), was followed by the Hoxnian Interglacial. The type site is at Hoxne in Suffolk, where in a hollow in the Anglian till are lacustrine and fluvial deposits

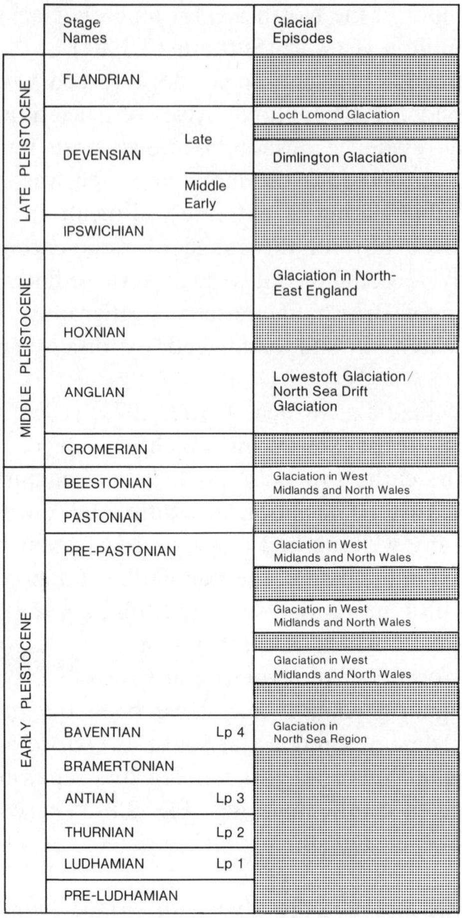

Stage Names			Glacial Episodes
LATE PLEISTOCENE FLANDRIAN			
DEVENSIAN	Late		Loch Lomond Glaciation
			Dimlington Glaciation
	Middle Early		
IPSWICHIAN			
MIDDLE PLEISTOCENE			Glaciation in North-East England
HOXNIAN			
ANGLIAN			Lowestoft Glaciation / North Sea Drift Glaciation
CROMERIAN			
EARLY PLEISTOCENE BEESTONIAN			Glaciation in West Midlands and North Wales
PASTONIAN			
PRE-PASTONIAN			Glaciation in West Midlands and North Wales
			Glaciation in West Midlands and North Wales
			Glaciation in West Midlands and North Wales
BAVENTIAN		Lp 4	Glaciation in North Sea Region
BRAMERTONIAN			
ANTIAN		Lp 3	
THURNIAN		Lp 2	
LUDHAMIAN		Lp 1	
PRE-LUDHAMIAN			

FIG. 2.8 The English Pleistocene sequence (from Bowen *et al*. 1986, Table 2). The four Lp numbers represent pollen stages in the Ludham Borehole. Note the frequency of glacial events in this interpretation.

with an interglacial flora. The Hoxnian in turn was possibly followed by the second main glacial episode, the Wolstonian cold stage, named after the type site at Wolston, in Warwickshire. The Wolstonian ice sheet, if it existed, extended as far south as the Isles of Scilly and the northern coast of the south-west pensinsula, depositing till at Fremington near Barnstaple (Devon). However, considerable controversy surrounds the Wolstonian type site and its precise stratigraphic position (Sumbler 1983, Shotton 1983).

The new interpetation of the Wolstonian has been expressed forcefully by Rose (1987: 1):

... a regional study of the Baginton/Lillington Gravels, which form the lower part of the Wolstonian at its type site, show that these sediments can be traced to East Anglia where they lie *beneath* glacial deposits of Anglian age. Thus, the sediments at the type site, and other 'Wolstonian' deposits in Midland England were found *prior to*, and *not after* the Anglian glacial. Consequently, the extent of glaciation in England between the Anglian and Devensian stages is in need of revision, and the use of the term 'Wolstonian' to describe the glaciation at this time is no longer appropriate.

There is also considerable controversy surrounding the status of the type site of the next interglacial, the Ipswichian, named after a site at Bobbitshole near Ipswich (Bowen 1978: 36). However, we reach firmer ground when we come to the last major glaciation in Britain, the Devensian.

Its maximum extent (called the Dimlington Stadial) was reached some 17–18 000 years ago. However, there was a glacial re-advance, following its wastage, which lasted for about 450 years from 11 000 years ago. This advance, the Loch Lomond Stadial, 'was the last time that England experienced the presence of glacier ice, and the last time that it experienced rapid landscape change before coming covered with the protective woodland of the Flandrian Interglacial' (Bowen *et al.* 1986: 311). This cooling event, the Younger Dryas, is described further in Chapter 4.

The classic view just outlined is the subject of continuous revision and discussion. There may well have been earlier glacial events in the Pleistocene, the evidence for which is fragmentary. The present position has been summarized by Bowen *et al.* (1986: 299):

Extensive lithological evidence for glaciation, with a sound stratigraphic basis, has been recognised in two stages of the English Pleistocene: the Anglian and the Devensian. Additionally, less extensive or indirect lithological evidence, often with a less secure stratigraphic basis, indicates that glaciation occurred in at least a further six episodes.

The changing extent of glaciers and ice-caps

The Pleistocene glaciers did not show precisely the same spread in each of the glacial periods. In Europe the Riss/Saale glaciation is regarded as representing the maximum spread of ice, and although the Illinoian (Table 2.2) probably represents the maximum spread of ice in North America, the preceding Kansan glaciation extended further in the west-central part of America. At their maximum extent the glaciers probably

Table 2.3 The former and present extent of glaciated areas

	Maximum Pleistocene Extent		Last Glaciation Extent		Present Day	(% of
	(mkm²)	(% of total)	(mkm²)	(% of total)	(km²)	Total)
Antarctica	13.20	28.0	13.20	32.8	12 650 000	84.50
Laurentia	13.79	29.3	12.74	31.6	230 250	
North American Cordillera	2.50	5.3	2.20	5.5		0.15
Siberia	3.73	7.9	1.56	3.9	—	—
Scandinavia	6.67	14.1	4.09	10.1	5000	0.03
Greenland	2.16	4.6	2.16	5.9	1 800 000	12.0
Northern Hemisphere besides above	4.07	8.6	3.45	8.6	—	—
Southern Hemisphere excluding Antarctica	1.02	2.16	0.90	2.2	26 000	0.17
Total	47.14	—	40.30	—	14 970 000	—

(Modified from data in Embleton and King, 1967)

covered 47.14 mkm², which is rather more than their extent during the Last Glaciation (40.30 mkm²) but very much more than their present extent, which is about 15 mkm². In other words during the height of the Pleistocene glaciations, ice covered about three times as great an area as it does today (Fig. 2.9) (Embleton and King 1967). However, both the Antarctic and Greenland ice-sheets, because of the restrictive effects of iceberg-calving into deep ocean water, differed little in area from their present sizes, though they were much thicker. The greatest reductions in the glacial area have taken place through the melting of the North American Laurentide and the Scandinavian ice sheets, which have now lost 99 per cent of their former maximum bulk. In terms of volume, although precise assessment is difficult, the Riss/Saale/Illinoian glaciation had a volume of 84–99 mkm³ compared to 28–35 mkm³ at the present time.

America

During the maximum glacial phases of the Pleistocene, including the late Wisconsin maximum around 18 000 years ago, ice was continuous or nearly so across North America from Atlantic to Pacific, and was composed of two main bodies, the Cordilleran glaciers associated with the Coastal Ranges and the Rockies, and the great Laurentide sheet (see Wright and Frey 1965). The former were most extensive in the mountains of British Columbia, and diminished both northward into Alaska and Yukon, and southward through the western United States. The southern limit of continuous ice was south of the Canada–USA border, down to

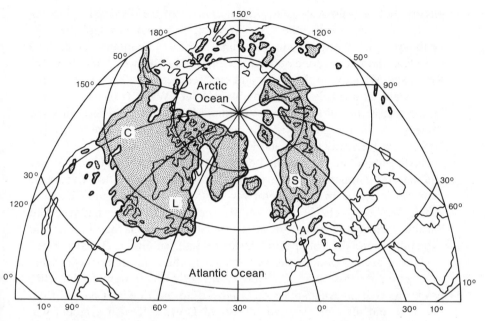

FIG. 2.9 The possible maximum extent of glaciation in the Pleistocene in the northern hemisphere. C = Cordilleran ice; L = Laurentide ice; S = Scandinavian ice; and A = Alpine ice.

the Columbia River and the Columbia Plateau. South of this there were numerous localized ice-caps and glaciers, and notably in the Sierra Nevadas the thickest ice, in the lee of the Coastal Ranges, was as much as 2300 m deep. The Laurentide ice-sheet, which, as already mentioned, was with the Scandinavian ice-sheet largely responsible for the great difference in world glacial areas between glacial phases and interglacials, reached its extreme extent in the Ohio–Mississippi basin at latitude 39° in Wisconsin time, and 36° 40′ in Illinoian times. It extended to approximately the present positions of St. Louis and Kansas City but west of this area the southern ice margin trended north-westwards, leaving western Nebraska and western South Dakota largely ice-free. On the basis of post-glacial isostatic readjustment it seems likely that the thickest ice occurred over Hudson Bay and attained a thickness of around 3300 m.

The British Isles

The British Isles ice-sheets (with an area of about 370000 km² during the Last Glaciation) merged with those of Scandinavia in some glacial phases, but also possessed large local centres of ice dispersal, including the Scottish Highlands, the Lake District, the Southern Uplands, the Pennines, the Welsh Mountains, and various mountains in Ireland including those of Connemara and Donegal in the west, and those of Kerry and Wicklow in the south and east respectively. The extent of the various glaciations (Fig. 2.10), notably in South Wales (Bowen 1973) and in Wessex, is the subject of some dispute. This became an issue of some controversy in the 1970s, largely as a result of some arguments put forward by Kellaway (1971) and Kellaway *et al.* (1975).

Having identified that ice had crossed the Severn Estuary from Wales and entered Somerset, the evidence for which is provided by erratic material exposed in new motorway sections (Hawkins and Kellaway 1971), Kellaway (1971) went on to speculate that ice had reached Salisbury Plain (providing the stones for Stonehenge in the process) and overspilled into the Hampshire Basin; though, as Green (1973) pointed out, there are no glacially derived erratics in the Hampshire Basin rivers to substantiate this view. Kellaway *et al.* (1975) later went further still, proposing that an Anglian ice-sheet moved up the English Channel from the west and affected parts of the South Downs. If such a model was correct then the bulk of the landscape of southern Britain was glaciated.

The 'evidence' put forward by Kellaway for the occurrence of very extensive glaciation includes the presence of such features as: the enclosed 'deeps' or 'fosses' of the English Channel floor (which he regarded as being due to subglacial scour); valley bulges and landslips (which he regarded as being the result of glacial loading and oversteepening); clay-with-flints (regarded as decalcified till); erratic materials in Chesil Beach, Sussex raised beaches and at miscellaneous sites (e.g. Stonehenge); and the U-shaped dry valleys of the chalk downlands. However, as Jones (1981: 50) points out: 'There is much conjecture and many circular arguments in this hypothesis, especially as most of the landforms and deposits mentioned have been explained adequately by geomorphological processes other than glacial ice.'

Another attempt to prove that the normally accepted glacial limits were too northerly was made by d'Olier (1975) as a result of investigations of enclosed depressions off the Essex coast. He suggested that the depressions were ice-scour phenomena and that an ice-sheet extended southwards across the present Thames Estuary to Thanet, possibly penetrating the Straits of Dover. However, a tidal-scour origin for such depressions is a more widely held explanation.

Uncertainty remains about the precise limits of glaciation in the London Basin, where there are some erratic-rich pebbly clays which have

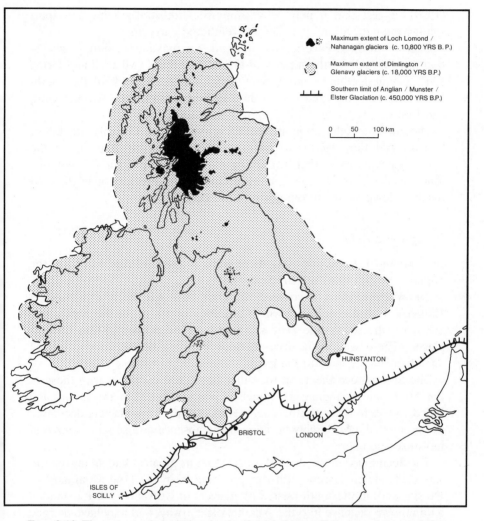

FIG. 2.10 The extent of glaciation in the British Isles, offshore regions, and the adjacent part of the Netherlands (modified after Bowen *et al*. 1986, fig. 5). Note the postulated ice-free areas of north-east Scotland during the Dimlington Stadial of the Devensian, and the lack of connection with a Scandinavian ice-sheet at that time.

been tentatively interpreted as the deeply weathered remnants of tills laid down by an early glacial advance. This material, called 'the Chiltern Drift', occurs on the Chiltern backslopes of Hertfordshire, the South Hertfordshire Plateau, and the Epping Forest Ridge. Baker and Jones

(1980) suggest that it may be a composite lithostratigraphic unit, part solifluction deposit and part deeply weathered early till.

Bearing all these uncertainties in mind, the maximum extent of glaciation ran through northern parts of the Isles of Scilly (Mitchell and Orme 1967, Coque-Delhuille and Veyret 1989), the north coast of the south-west peninsula, the area just to the south of Bristol, the Oxford region, and Essex.

The limits of glaciation in the Devensian (Last Glacial) are better known, and they appear to have been further north than during the earlier glaciations, so that in East Anglia, for example, the ice only just touched the Norfolk coast at Hunstanton. In previous glaciations it reached down to just to the north of London.

Europe and Asia

In mainland Europe and Asia there were three main centres of ice, the Alpine, the Siberian, and the Scandinavian.

The Alpine glaciers covered an area that has been estimated at $150\,000\,km^2$ and reached down to altitudes of 500 m on the north, and 100 m on the south side. The ice may have been over 1500 m thick in places. There was an ice-free corridor between this ice mass and the Scandinavian ice mass to the north.

The Siberian ice-sheet, on the other hand, was confluent with those of the Urals and Scandinavia, though it was smaller than the latter, and failed to reach as far south. The overall extent of the glaciers decreased as one moved east, largely because of the absence of a suitable source of moisture and energy.

The Scandinavian ice-sheet (Fig. 2.11), at its greatest known maximum, was probably coalescent with ice spreading from the Ural Mountains of Russia, and, in the south-west, with glaciers of British origin. It extended an unknown distance into the Atlantic off Norway, and may have merged with ice over Spitzbergen. In the south, the Elster and Saale glacial borders trend along the northern bases of the central European High-lands, and the Saale ice-sheet penetrated far down the basins of the Dnepr and Don Rivers.

Although the thickness of this ice-sheet is not known with precision, it may well have exceeded 3000 m both over the Sognefjord region of western Norway, and at the head of the Gulf of Bothnia, though the average thickness was probably about 1900 m.

The ice also extended across the present North Sea basin, which in times of maximum glaciation was largely above sea-level. The Dogger

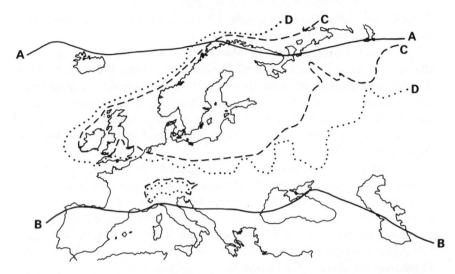

FIG. 2.11 Glacial conditions in Europe (from data in Flint 1971, and Kaiser 1969):
A. The position of the present polar timber-line in Europe.
B. The position of the timber-line during the maximum of the last (Würm) glacial.
C. The extent of north European drift deposits that can be attributed to the Last Glacial.
D. The drift borders of the Riss-Saale and Mindel-Elster in North Europe.

Bank, which rises some 20 m higher than the surrounding sea-floor, is possibly the remnant of a great moraine some 250 km long, and 100 km wide (Stride 1959).

There is a controversy as to whether large portions of northern Eurasia, and particularly the Arctic continental shelf, were covered by an integrated ice-sheet. One theory is that sea ice thickened during cold phases until it became grounded on continental shelves. Once formed these shelves might have continued to grow into domes by the addition of snow and the freezing of any meltwater draining off the land. This would be particularly the case in sheltered seas where iceberg calving was restricted, such as the Kara and Barents Seas, and in the bays and inter-island channels of the Queen Elizabeth Islands. Under this theory, sometimes called the 'maximum reconstruction' (Denton and Hughes 1981), huge ice masses would have existed over areas of the high latitudes in the northern hemisphere that are now covered by sea. By contrast the older and more firmly established 'minimum reconstruction' envisages the existence of several disconnected ice-caps and ice-sheets with continental shelves, aside from those fringing the Scandinavian ice-sheet, remaining unglaciated, as were the Arctic Seas.

The extent and sequence of glaciation in the high mountains and plateaux of central Asia (including Kashmir, the Himalayas, the Tibetan Plateau, and China) are still the subject of considerable controversy. Some workers from China and elsewhere have maintained that large ice-caps developed over wide areas (e.g. Kuhle 1987). While others suggest that the extent of glaciation was much more modest (e.g. Holmes and Street-Perrott 1989).

Much of the controversy has arisen because of problems of identification of tills, and their superficial similarity to other deposits produced by other processes (including debris flows, deep weathering, etc.). The maximalist view, as expressed by Kuhle, is that during the glacial maximum a large ice-sheet enveloped the Tibetan Plateau and surroundings extending over 2.0 to 2.4 million km^2, an area considerably larger than the present Greenland ice (1.7 million km^2).

The southern continents

In the southern continents much less work has been done on the nature of the glacial periods, and it is clear that glaciation was markedly less extensive than in the northern hemisphere. A good review of current knowledge is provided by Clapperton (1990), who suggests that there is a clear general synchroneity of events in the two hemispheres with, for example, similar dates for the last glacial maximum and evidence for the Allerød oscillation in both hemispheres. He also tentatively suggests that earlier Pleistocene glaciations in some of the southern continents (most notably in Patagonia and Tasmania) may have been more extensive than later ones.

In South America the ice-sheets developed from the great Andean cordillera were greatly expanded (Fig. 2.12), and in the far south the glaciated zone was over 200 km wide and the ice may have been over 1200 m thick. A more or less continuous zone of glaciation extended to about 30 °S, though north of 38 °S the ice tended not to expand very far from the Cordillera, either to the Pacific in the west or into the plainlands in the east. The Patagonian ice-cap was especially extensive at about 1.2 million years ago, when it covered an area of 300 000 km^2 (compared with 100 000 km^2 in the last glaciation) (Clapperton 1990). The furthest north ice body was that which capped the Sierra Nevada de Santa Marta in Colombia. Other large glaciers developed in the Sierra Nevada de Merida in Venezuela (Schubert 1984).

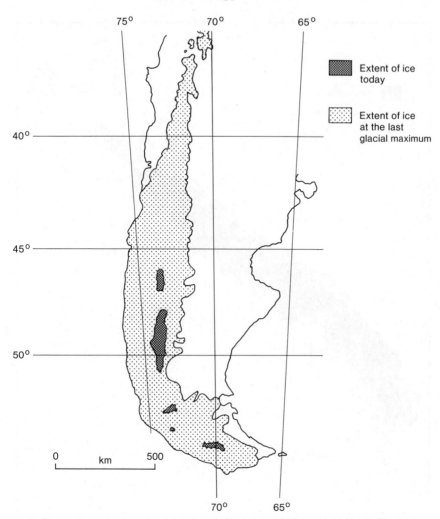

FIG. 2.12 The extent of glaciation in southern South America (modified after Broecker and Denton 1989, fig. 8).

Southern Africa, and its Drakensberg mountains, seem to have been largely unaffected by glacial activity, though signs of periglacial activity are widespread. Likewise, in Australia, because of the low relief and arid interior, there was limited glacier development. Former glaciation of the mainland was confined to a single zone in the Snowy Mountains with an area of barely $52\,km^2$. In Tasmania, however, glaciation was more

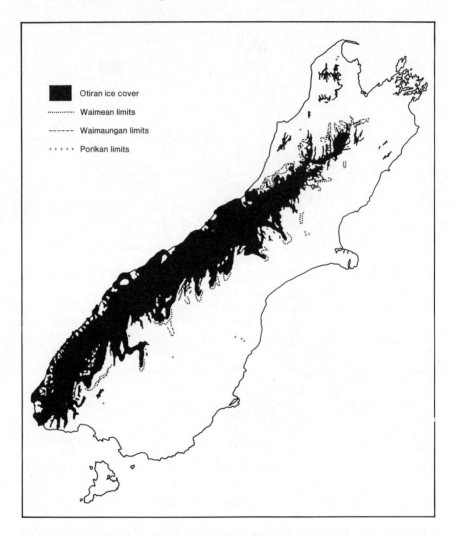

FIG. 2.13 Glacial limits in the South Island (redrawn from New Zealand Geological Survey 1973). The Otiran is the name given to the last glaciation. The other limits are shown in order of increasing age.

extensive, and a large ice-cap developed in the Central Plateau, where at least four glaciations have been identified (Colhoun 1988).

New Zealand, with its greater relief and oceanic conditions, has some major glaciers at the present day, unlike Australia, and in the Pleistocene the New Zealand Alps in the South Island were intensively and widely glaciated, though the North Island was largely unaffected. The great ice-caps and glaciers that formed in the South Island (Fig. 2.13) helped to form the extensive fjords and lakes that are such a feature of the landscape.

The glaciation of the Antarctic is as yet very imperfectly known, though it if clear that both the margins and the thickness of the ice sheet varied during the course of the Pleistocene and Late Tertiary. A large ice-sheet existed in western Antarctica as early as the Eocene, and signs of glacial action on nunataks (rocky hills rising above and through an ice-sheet) suggest that the ice may well have been 300–800 m thicker than at present, though the overall lateral extent of this great ice mass was controlled to a considerable degree by calving of icebergs into the relatively deep waters offshore. Some expansion may have resulted from lower glacial sea-levels.

Permafrost and its extent in the Pleistocene

Beyond the limits of the great Pleistocene ice-sheets there were, par-ticularly in Europe, great areas of open tundra. These areas were fre-quently underlain by permafrost. Permafrost is a frozen condition in soil, alluvium, or rock, and is currently concentrated in high northern lati-tudes, reaching thicknesses of as much as 1000 m.

The current southern boundary of continuous permafrost coincides approximately with the −5 or −6 °C mean annual isotherm. The limits of discontinuous and sporadic permafrost are rather higher, but mean annual air temperatures have to be negative. Thus in Europe continuous permafrost is restricted to Novaya Zemlya, and the northern parts of Siberia, whilst discontinuous permafrost extends into northern Lapland. However, there is very strong evidence for the former extension of such permanently frozen subsoil conditions to wide areas of Europe during the glacial cold phases. The evidence consists of the casts of ice wedges which form polygonal patterns in areas of permafrost. The casts can either be identified in sections, or detected as crop marks on air photographs. These have been encountered very widely, for instance in southern and eastern England, especially in Kent and East Anglia, and they are even fairly extensive in the valley of the Severn and Warwickshire Avon, and in parts of lowland Devon. The only part of mainland Britain to have been unaffected by permafrost is probably the extreme tip of the south-western peninsula (Williams 1975). The distribution of such features is shown in Fig. 2.14.

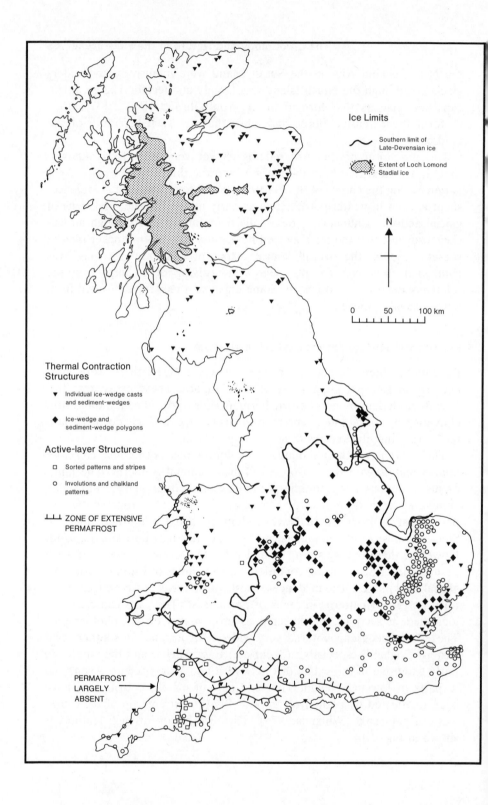

Ice Limits

Southern limit of
Late-Devensian ice

Extent of Loch Lomond
Stadial ice

N

0   50   100 km

Thermal Contraction
Structures

▼ Individual ice-wedge casts
and sediment-wedges

◆ Ice-wedge and
sediment-wedge polygons

Active-layer Structures

☐ Sorted patterns and stripes

○ Involutions and chalkland
patterns

⊥⊥⊥ ZONE OF EXTENSIVE
PERMAFROST

PERMAFROST
LARGELY   →
ABSENT

With regard to the former southern extension of permafrost in main-land Europe there is considerable dispute, though even the most northerly proposed boundary indicates that only the central and southern Balkans, peninsular Italy, the Iberian peninsula, and south-west France were largely unaffected. This indicates very forcibly the degree to which tundra and periglacial conditions were displaced southwards, and the extent to which temperature conditions were depressed in much of Europe. On the basis of a −5 °C limit for permafrost it seems likely that in eastern England, where conditions in the Pleistocene may have been made more continental by the drying out of the North Sea during glacial low stands of sea-level, mean annual temperatures were depressed by 15° (or more) during the Last Glaciation.

In North America relatively less is known about the southern dis-placement of the permafrost zone, but it seems likely that as the southern extent of the Wisconsin ice-sheets was further south than that of the Würm–Weichselian in Europe, the zone of more severe periglacial conditions was probably more restricted. None the less, Johnson (1990), on the basis of the analysis of patterned ground, suggests that in the late Wisconsin period permafrost extended as far south as 38° 30′ N.

Although the presence of permafrost suggests that mean annual temperatures were at least as low as those now experienced in tundra areas, it is probable that the periglacial climates of glacial Europe and America were different in character from any now found on Earth. Because of latitude, especially in America, days were longer in winter and shorter in summer than in any high latitude periglacial area at present. Also, the sun would have risen higher in the sky, giving both higher midday temperatures, and more marked diurnal changes. Evap-oration rates would have been higher.

The formation of loess sheets

Around the great ice-sheets, extensive spreads of loess were deposited during the course of the Quaternary. Loess is a largely non-stratified and non-consolidated silt, containing some clay, sand, and carbonate, which was deposited primarily by the wind (Smalley and Vita-Finzi 1968). It is markedly finer than aeolian sand. Over vast areas (at least $1.6 \times 10^6 \, \text{km}^2$

←

FIG. 2.14 The distribution of Late Devensian periglacial soil structures in the British Isles (after numerous authors summarized in Rose et al. 1985, fig. 18.2, and Williams 1969, fig. 2). Reproduced by permission of Routledge.

in North America and $1.8 \times 10^6 \, \text{km}^2$ in Europe), it blankets pre-existing relief, and in Tajikistan has been recorded as reaching a thickness of 180 m. In the Missouri Valley of Kansas the loess may be 30 m thick, in European Russia sustained thicknesses of 10 to 15 m are found, along the Rhine thicknesses approach 30 m, and in Argentina thicknesses, often 10 to 30 m, reach over 100 m in places, while in New Zealand, on the plains of the South Island, thicknesses reach 18 m.

The sources of loess include desert basins, but exposed outwash and areas of till recently uncovered by deglaciation are probably the most important. The winds, some of which may have blown away from the ice-sheets with great velocity, moved the finer materials, and these were then deposited as loess at some distance, especially where there was a dense vegetation cover, as along river valleys, to trap it.

The distribution of loess is now well known, and the main areas in America include central Alaska, southern Idaho, eastern Washington, north-eastern Oregon, and even more important, a great belt from the Rocky Mountains across the Great Plains and the Central Lowland into western Pennsylvania. Loess is less prominent in the eastern USA as relief conditions for deflation, and the nature of outwash materials, seem to have been less favourable than in the Missouri–Mississippi region. In Europe the loess is most extensive in the east, where, as in the case of America, there were plains and steppe conditions. The German loess shows a very close association with outwash, and in France the same situation is observed along the Rhône and Garonne Rivers. These two rivers carried outwash from glaciers in the Alps and Pyrenees respectively. The Danube was another major source of silt for loess in eastern Europe. Britain has relatively little loess, and this may result from the oceanic climate which would tend to reduce the area of exposed outwash. Indeed, in Britain windlain sediments of periglacial age are conspicuous only for their rarity—dunes are low, rather shapeless hummocks occurring only in a few localities, cover sands are thin and patchy in comparison with those in the Netherlands, and 'loess is more of a contaminant of other deposits than one in its own right' (Williams 1975). The maximum depth of loess in Britain is only about two or three metres, and sand dunes of periglacial origin are restricted to a few small areas, including the Breckland of East Anglia, and the Scunthorpe and York areas. Some sands, such as the interstadial Cheltenham Sands, may be banked up against escarpments.

In Asia, the steppes and deserts of the interior may have been the source of the great deposits in China. In South America, where the Pampas of Argentina and Uruguay has thick deposits, a combination of semi-arid and arid conditions in the Andes rain-shadow, combined with glacial outwash from those mountains, created near ideal conditions. However, in Australia and Africa, where glaciation was relatively slight, loess is much less well developed.

Of all these loess deposits, those of China are undoubtedly the most impressive for their extent and thickness, which in some cases near Lanzhou is more than 330 m. The initiation of loess formation there appears to date back to 2.4 million years BP, a fact that may reflect either the onset of severe northern hemisphere cooling or geomorphological changes associated with the uplift of the Tibetan Plateau (Pye 1987).

The degree of climatic change in glacials and pluvials

Although the presence of greatly expanded ice-sheets, and of permafrost conditions, gives a broad indication of the extent to which temperatures changed during the glacial intervals of the Pleistocene, it is possible, through a variety of techniques, to gain some more precise and quantitative measures of the degree of climatic change that has taken place.

Temperatures can be assessed through five main lines of evidence: isotopic measurements, the levels of cirques, the extent of permafrost, the limits of frost-affected sediments, and the nature of floral and faunal remains. These methods are all subject to certain difficulties and pitfalls in that temperature may be only one of the controls which influence, say, the position of the tree- and snow-lines. Similarly, the interpretation of the palaeoclimatic significance of snow-line levels, represented by cirque floor heights, depends very much on the estimation of probable local lapse rates. Lapse rates are the mean rates at which temperatures change with altitude (generally $0.6\,°C/100\,m$), but they are subject to local fluctuations.

The isotopic methods, which include an examination of O^{18}/O^{16} ratios of fossil Foraminifera, have proved fruitful, especially as they have been applied to deep-sea cores, though here too there are two main factors involved: the temperature of the ocean water, and the original isotopic composition of the ocean water. There has been much discussion as to the relative importance of these two factors (Shackleton 1967). In principle, however, there is a relation between the relative abundance of the two oxygen isotopes O^{16} and O^{18} in biogenic carbonates (mollusc shells, and tests of Foraminifera), and the water temperature at the time that the carbonate was formed. O^{18} enrichment increases 0.02 per cent per 1 °C temperature fall, and this small change in the ratios can be detected by mass-spectrometer.

The downward movement of snow-lines in glacial ages indicated lowering of temperatures, especially summer temperatures, though it has to be remembered that precipitation and cloudiness could also affect the level, as does the local lapse rate. A knowledge of local lapse rates is required to relate the altitudinal shift of the snow-line to temperature change. The position of the Pleistocene snow-line is also subject to some error in its assessment, in that it is determined by a study of the position of cirque floors. Cirque floors tend to cluster at or just below the 0° summer isotherm. In general the cirque floor measurement is only valid in areas where the former glaciers never grew beyond the corrie type. Values that have been determined by this method suggest a mean temperature depression during glacial phases of the order of 5 °C. The varying degree of snow-line depression from region to region gives a spread in temperature depression values of from only about 2.0 °C to over 10 °C. This reflects the fact that snowline depression values ranged from as little as 600–700 m in the northern Urals, the Middle Atlas, and the Caucasus, to as much as 1300 to 1500 m in the northern Pyrenees, on Kilimanjaro, in the Apennines, and in the Tell Atlas.

The former extent of permafrost has been discussed in relation to Europe on p. 65. In that the current boundary of permafrost in Siberia, Scandinavia, and North America can be related to mean annual temperatures, it is possible to infer Pleistocene temperatures from this source. The permafrost data tend to give somewhat higher values for the amount of temperature depression than do the snow-line data, with a value of 15 or 16 °C being recorded for the Midlands of England and for East Anglia, values of 10–15 °C for parts of central North America, and a value of 11 °C for Germany.

In many parts of the world, where conditions are now both too warm and too dry for frost activity to be important in rock disintegration, there are screes of angular debris, which have been widely interpreted as being the product of frost activity. These have, for example, been described from Cyrenaica and Tripolititania in Libya (Hey 1963). Periglacial deposits of this type have been used to suggest a 11 °C depression of glacial temperatures in the south-western USA (Galloway 1970), a greater than 9 °C depression in the Snowy Mountains and Canberra regions of Australia, and a depression of over 10 °C in the Cape Province of South Africa.

The data provided by organisms and plants are extremely difficult to interpret other than qualitatively, though from various sources Flint (1971) has suggested that at the height of the Last Glacial, temperatures may, on average, have been depressed by about 6 °C, a value which ties in with the snow-line evidence, though depressions of 10–15 °C have been suggested for Central Europe (Segota, 1966).

Foraminifera from the glacial segments of deep-sea cores, when compared with those of the present in the same locations indicate a temperature depression in glacials of about 5 °C for Caribbean surface-water, 4.6 °C for the Equatorial Atlantic, and 5.7 °C for equatorial waters off West Africa (Hecht 1974) and 3 to 4 °C for the south-west Indian Ocean (Van Campo *et al*. 1990). A general review of the sea-water temperature evidence (Climap Project Members, 1976) has suggested that on a world basis the average anomaly between present and glacial surface-water temperatures was of the order of 2.3 °C. However, locally, as in the North Atlantic, where the position of the Gulf Stream appears to have shifted substantially, the values may have been from 12 to 18 °C different from those of today.

On land as well, local temperature depressions may have been greater than has been suggested hitherto. Areas subjected to an ice covering, because of the temperature gradients, and highly reflective conditions associated with ice-caps, may have become as cold as Antarctica, probably cooling by as much as 60 °C to an average temperature as low as −60 °C.

The calculation of former precipitation levels is even more beset with difficulties than the calculation of temperature changes, in that most of the methods attempt in fact to measure not precipitation, but evaporation/precipitation ratios. They are thus partially dependent on temperature estimation. Decreases in temperature of the type discussed above would in many areas be sufficient by themselves to account for certain 'pluvial' or lacustral phenomena which have been interpreted in the past as being the result of increased precipitation. Phenomena that can be used to assess changed precipitation/evaporation ratios include the volumes and stratigraphy of lakes (see p. 23), the nature of cave fillings, the distribution of dune fields (see p. 97), the characteristics of fossil soils (palaeosols), and the nature of former stream regimes as deduced from sedimentological and morphological evidence. It is not easy to obtain any quantitative data from these sources though various attempts have been made.

As noted on p. 105 many lakes had higher volumes at some stage in the Pleistocene and Early Holocene. In a closed basin the level of the lake depends on the balance between rainfall inputs, evaporation, and surface area. One of the main controls of evaporation rates is temperature, so that if this can be estimated it is possible to calculate how much rainfall is needed to account for calculated volumes and areas of lakes at various levels above their current ones. On this basis, for example, it has been calculated that rainfall totals must have been about 165 per cent of totals of the present day in East Africa from 9000–6000 BP, assuming that temperatures were 2–3 °C lower during Early Holocene times than at present (Butzer *et al.* 1972).

In America, using snow-line-derived temperature data and related measures, several geologists and hydrologists have appraised the water budgets of various pluvial lakes in the Basin and Range Province during their maximum Late Pleistocene levels. Estimates of the increase in mean annual precipitation (from present average values over the drainage areas, compared with those during the lake maximum) ranged from 180–230 mm, and for the decrease in mean annual temperature, from 2.7 to 5 °C. Thus, for example, at Spring Valley in Nevada, Snyder and Langbein (1962) proposed a pluvial rainfall of 510 mm compared with about 300 mm at the present. It needs, however, to be stressed that these estimates are based on low temperature depression values. By contrast, on the basis of temperatures implied from periglacial features, Galloway (1970) has proposed that in the south-west United States temperatures were depressed by 11 °C. He calculated on this premiss that far from precipitation levels being higher during the pluvial phases, precipitation levels were only about 80 to 90 per cent of current levels. On the other hand a general survey of the evidence, notably in Australia and the United States, has led Dury (1967) to propose that shrunken lake levels and misfit streams (streams too small for their valleys) indicate an increase in mean annual precipitation during pluvials of the order of 1.5 to 2 over present totals, even when allowance is made for temperature reductions.

The vegetational conditions of the full glacials in Europe

During the various full glacial stages of the Pleistocene, the vegetation of much of unglaciated, periglacial Europe was characterized by its open nature. Trees were relatively rare, and in many respects the plant assemblages displayed many characteristics one would expect in a cool 'steppe' environment (Fig. 2.15).

In western Europe the pollen record of the Last (Würm, Weichselian) Glacial shows little arboreal (tree) pollen, and traces of *Artemisia* and *Thalictrum* are common. These plants are characteristic of open habitats. In more maritime areas, such as Cornwall and Ireland, there was also some dwarf birch and willow, but even as far south as Biarritz in south-western France the proportion of tree-pollen in full glacial sediments is

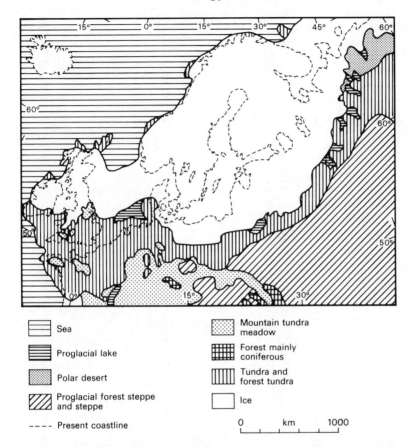

Fig. 2.15 Palaeogeographic reconstruction of northern Europe during the maximum of the Valdai (Last Glacial) (after Gerasimov 1969, fig. 3).

low, though some oak and hazel may have existed in the Gascogne Lowlands. Thus, as with some limits of permafrost, the northernmost boundaries of the major zonal vegetational types were pushed far to the south of their present-day ones (Fig. 2.16).

Further east in Europe, the areas right at the ice-fronts themselves were probably more or less completely barren, but in the belt of fine aeolian dust deposition which occurs further to the south, the loess, a more herbaceous flora seems to have been prevalent. In more favoured parts of Romania and Hungary there was even some pine present in full glacial times. Russia, on the other hand, from southern Poland across to the southern Urals, was covered by a salt-tolerant dry *Artemisia* steppe,

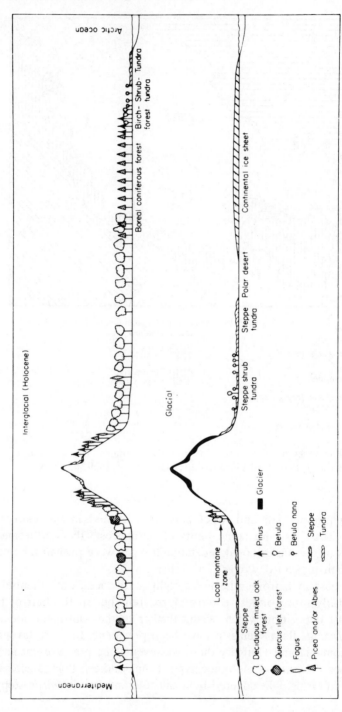

FIG. 2.16 Schematic representation of the vegetation of an interglacial and a glacial in a north to south section through Europe (after van der Hammen *et al.* 1971, fig. 6).

and south of this there was a forest tundra or forest steppe, together with small woodland areas in the Crimea, and along the shores of the expanded Caspian Sea.

The frequent occurrence of salt-demanding and salt-tolerant plants also suggests that precipitation levels were low. Such plants are known from the interstadial of the mid-Devensian (the Upton Warren) in Britain, and also from Zones I and III of the Late Glacial in the Isle of Man. 'It is tempting to see climate as in some way responsible' (Williams 1975).

Writing about the vegetation of north-west Europe during the last glacial, Birks (1986) suggests that it was of a type unknown today. It had structural and floristic affinities with steppe and tundra, and contained a mixture of ecological and geographic elements. He maintains that this 'no-analogue' vegetation type suggests a 'no-analogue' environment, with relatively warm summers, extremely cold winters, highly unstable soils (caused by extensive frost churning), low precipitation, strong winds and strong evaporation.

In the southern parts of Europe and the Levant, around the Mediterranean Sea's northern shores, the vegetation was also characteristically steppe-like and arid (Bonatti 1966), with some areas of pine. This belt seems to have extended across into the Zagros Mountains of western Iran, with *Artemisia* again characteristic or dominant at lower altitudes, and a dry alpine flora at higher altitudes. The apparent synchroneity between temperature changes in western Europe, and those in Syria and the Lebanon, is shown in Fig. 2.17. Dryness and coldness appear to coincide closely in time.

The glacial vegetation of North America

While much of the country north of the European Alps during the Last Glacial maximum supported tundra or, close to the ice, a cold rock desert, in America the available records indicate that much of the area south of the ice sheet was covered with boreal forest rather than with tundra. The reason for this difference is that the ice limit in Wisconsin times in America was much further south than it was in Europe—39° N in Illinois compared with 52° N in Germany. Further, the Alps, with their own large ice-cap, reinforced the semi-permanent area of high atmospheric pressure associated with the Scandinavian ice-sheet. This probably tended to divert warm, westerly air flows to the south of the Alps. There is no such mountain mass trending east to west in North America.

The widespread boreal forest of full glacial times in North America,

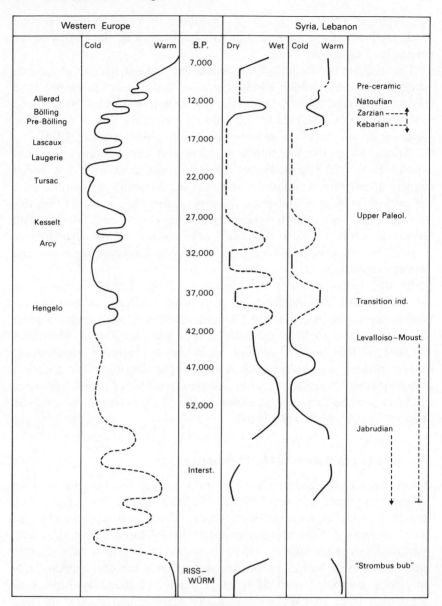

FIG. 2.17 Palaeoclimatic curves for the Last Glacial in the Near East and Western Europe, based on pollen analysis (from Leroi-Gourhan 1974, fig. 1). Some interstadial names are given for Western Europe. Arcy, Tursac, Laugerie, and Lascaux are also names of prehistoric cave sites. In the right-hand column certain names of cultural phases are given.

dominated generally by *Pinus* and *Picea*, was not found everywhere, for there were patches of tundra and of treelessness, but these were not as extensive as in Europe. The southern limit of the boreal forest is not known with any certainty, but it was probably somewhere in the south-central United States, perhaps extending westward from Georgia. Thus it may have formed a latitudinal belt as broad as it is today—1000 km from Hudson Bay to the Great Lakes. In the south-west, where pluvial lakes appear to have been synchronous with the main glacial (see p. 106), pollen evidence indicates high percentages of *Pinus* and other montane conifers during the Wisconsin, whereas now the same areas are characterized by semi-desert shrubs. In the Western Cordillera the tree-line was lowered 800–1000 m, and the extent of alpine vegetation in the mountains was very greatly expanded.

Likewise the latitudinal position of the Arctic timber-line in the Late Würm (see Fig. 2.18) was very different from that of post-glacial times (Markgraf 1974), the shift being about 24–5°.

The interstadials of the Last Glacial

One of the problems of glacial correlation is that phases of lesser glaciation and relatively greater warmth occured during the course of a major glacial phase. Such interruptions are called interstadials, but there is as yet no universally acceptable definition which differentiates an interstadial from an interglacial. Nevertheless, there are indications in many parts of Europe, and elsewhere, that the Würm-Weichsel-Wisconsin glaciation was interrupted by certain phases of less intense glacial activity (Fig. 2.19), which enabled soils and other distinctive sediments to develop. Quite a large number of these deposits have now been dated by radio-carbon means, and certain correlations seem possible (Fig. 2.20). Examination of these dates suggests that while there is a considerable spread in values, there is some clustering over a period from about 50 000 to 23 000 BP. This period was probably not a continuous phase of relatively warmer conditions, and there seems in many areas to have been a tendency for a particularly marked interstadial at the end of this time, notably around 28 000 BP (Denekamp, Plum Point, Paudorf, Kargy, Olympia, etc). There were also some relatively short-lived interstadials near the beginning of the Würm-Wisconsin-Weichsel, and these may have been sufficient to lead to deglaciation in Scandinavia (Brorup, Amersfoort, Chelford, St. Pierre). Behre (1989) suggests that in Europe there were two particularly long and warm interstadials, the Brorup and the Odderade, in which forests were widely prevalent. He attributes them

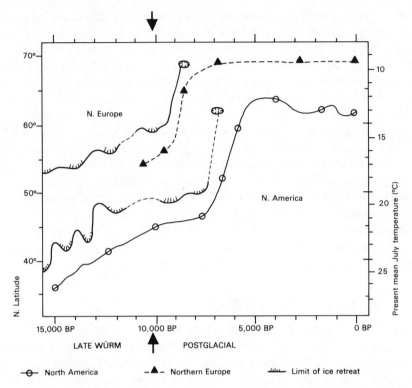

FIG. 2.18 Latitudinal changes of the Arctic timber-line since 15 000 BP (after Markgraf 1974, fig. 4).

FIG. 2.19 Climatic fluctuations over the last 100 000 years inferred from miscellaneous evidence.

A. Climatic variations expressed as changing $^{18}O/^{16}O$ ratios in an ice-core from Camp Century, Greenland (after Dansgaard *et al.* 1975).

B. Climatic variations expressed as changing $^{18}O/^{16}O$ ratios in the Byrd Station Ice-Core, Antarctica (after Epstein *et al.* 1970).

C. Climatic curve based on percentages of left-coiling *Globorotalia truncatulinoides* in a Caribbean core (A 179-4) (after Wollin *et al.* 1971).

D. Average July temperatures for the English Midlands based on the study of beetle faunas (after Coope 1975).

E. The climatic sequence in the Netherlands inferred from floral evidence (after Van der Hammen *et al.* 1967).

F. Glacial activity in the Great Lakes/St. Lawrence region according to Dreimanis and others (after Flint 1971).

G. Glacial and other fluctuations in Middle Europe (after Mörner 1969).

H. and I. Glacial fluctuations in the erstwhile USSR (after Dreimanis and Raukas 1975).

In all the diagrams colder conditions are indicated by a move of the curves to the left.

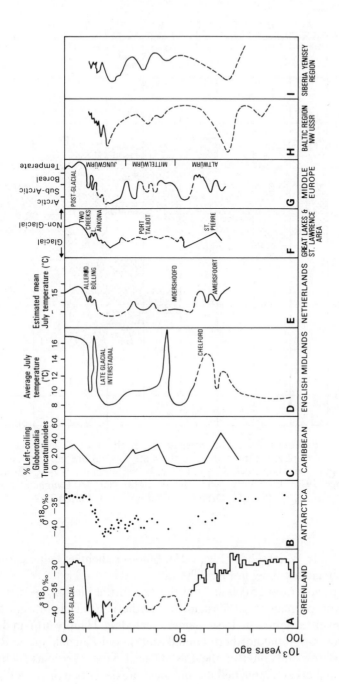

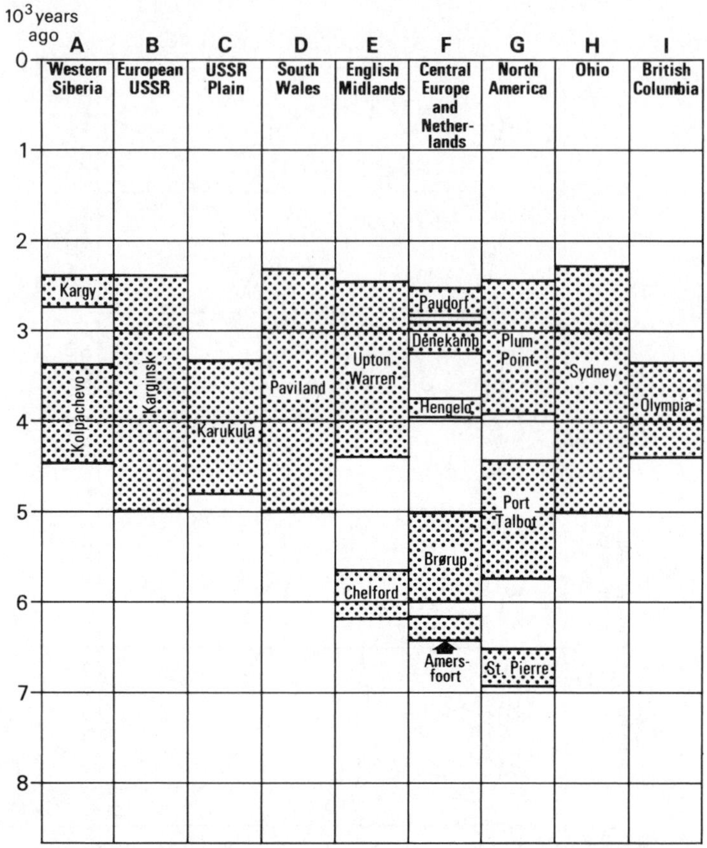

FIG. 2.20 Radio-carbon dated interstadials of the Last Glacial period in the northern hemisphere. Interstadials of the Late Glacial are not included. Drawn from data in Zubakov 1969; A. Kind 1972; B. Serebryanny 1979; C. Bowen 1970; D. Coope 1975; E. Segota 1967, Butzer 1972, van der Hammen *et al.* 1967; F. Dansgaard *et al.* 1971; G. Dreimanis *et al.* 1966; H. Fulton 1968.

to the Early Weichselian (Fig. 2.21), but sees them as being earlier than the interstadials show in Fig. 2.20.

The period from 25 000 BP until the end of the Pleistocene saw a great expansion of glaciers, at least in the northern hemisphere, and this has been given a series of local names, including Hauptwürm in Europe, Woodfordian in the northern United States, and Pinedale in the Rockies. The last few millennia of the Last Glacial were also marked by some minor stadials and interstadials, and these are described on p. 134.

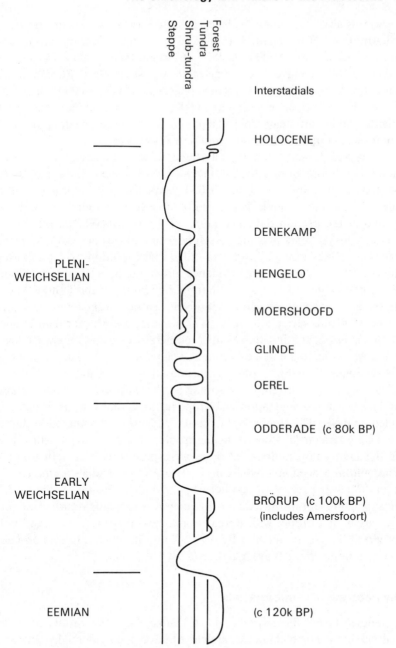

FIG. 2.21 Vegetation conditions through the Weichselian (Last Glaciation) in Europe (modified after Behre 1989, fig. 9).

There is also confirming evidence from ice-cores, both from the Arctic and Antarctic, of interstadials (Fig. 2.19). At Camp Century, Greenland, warmer periods at 19000–23000, 46000–56000, 68000–74000, and 75000–78000 BP have been suggested, whilst at Byrd Station in the Antarctic there are signs of warmer phases at 25000, 31000, and 39000 with colder phases culminating at 27000, 34000, and 46000 years BP. In Atlantic and Caribbean deep-sea cores there were maxima of warm conditions about 25000, 40000, and 65000 years ago.

Pollen and faunal evidence has been utilized for assessing environmental conditions during the interstadials. In England, for example, the Chelford interstadial of about 60000 years ago was characterized by boreal forest and a beetle fauna comparable to that found in south-east Finland at the present day. The fauna of the Upton Warren Interstadial of about 40000 years ago suggests July temperatures at least 5 °C higher than those which existed during the full glacial conditions which followed. Indeed, in a review of the evidence provided by Coleoptera, Coope (1975) has suggested that at the thermal maximum of the Upton Warren Interstadial, perhaps at about 43000 BP, average July temperatures in central England were about 18 °C, which are a little warmer than today's. He does, however, find that winter temperatures were somewhat lower than now, indicating a more continental climatic regime. The warm phase was relatively short-lived, possibly lasting only a thousand or so years.

The effects of the relative warming of the various interstadial phases can be seen in the vegetation of Europe at those times, as determined by pollen analysis. The steppe belt of southern Europe dominated in the full glacials by *Artemisia*, showed higher arboreal pollen characteristics. During the Denekamp interstadial there were pine forests in both southern Spain and in Macedonia, whilst in the Brorup interstadials a *Quercus ilex* vegetation was present in southern Spain, but a *Carpinus–Ulmus–Tilia* forest was present in Macedonia. In general woodlands returned to quite wide areas of Europe, with boreal woodlands of spruce, pine, and larch bordering the North Sea and Baltic. Oak and hornbeam forests probably occurred in northern Italy, Yugoslavia, and Albania.

The nature of the interglacials

In general terms the interglacials, to judge from the results of pollen analysis and other techniques, appear to have been essentially similar in their climate, flora, fauna, and landforms to the Holocene in which we live today. The most important of the characteristics of the interglacials was that they witnessed the retreat and decay of the great ice-sheets,

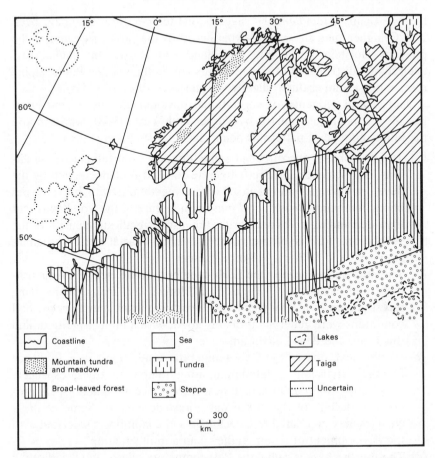

FIG. 2.22 Palaeogeographic reconstruction of northern Europe during the Last Interglacial. Compare this with the Glacial picture illustrated in Fig. 2.15 (after Gerasimov 1969, fig. 2).

and saw the replacement of tundra conditions by forest over the now temperate lands of the northern hemisphere (Fig. 2.22). Tree-lines occurred at higher altitudes and latitudes (Fig. 2.18).

The maximum temperatures attained in some or most of the inter-glacials appear to have been a little higher than those of the present, and may well have been comparable to those of the Holocene climatic optimum (see p. 156). During the Last or Sangamon Interglacial, for instance, much of North America was covered by deciduous forest as at the present, though near Toronto in Canada the presence of pollen from the sweet gum (*Liquidambar*)[1] suggests that temperatures may have been 2–3 °C higher than those now experienced in that area. Equally, in the previous Holstein Interglacial of Poland and Russia, the fauna, particularly the distribution of beech, hornbeam, holly, and the Pontic alpine rose, suggest temperatures slightly warmer than those of today.

The general sequence of vegetational development during the interglacials has been rationalized by Turner and West (1968), who propose the following type of pattern as being characteristic:

(a) The first phase, one of climatic amelioration from full glacial conditions, can be called the Pre-temperate zone. It is characterized by the development and closing in of forest vegetation, with boreal types being dominant. *Betula* and *Pinus* are a feature of the woodlands, but light-demanding herbs and shrubs are also a significant element of the vegetation. Relicts of the preceding Late Glacial periods such as *Juniperus* and *Salix* may also be present.

(b) The next phase, termed the Early-temperate zone, sees the establishment and expansion of a mixed oak forest with many shade-giving forest genera, typically *Quercus*, *Ulmus*, *Fraxinus*, and *Corylus*. Soil conditions were generally probably good, with a mull[2] condition, and this promoted dense, luxuriant cover.

(c) In the next phase, the Late-temperate zone, there is a tendency for the expansion of late-immigrating temperate trees, especially *Carpinus*, *Abies* and, sometimes, *Picea*, accompanied by a progressive decline of the mixed oak forest dominants. Some of these changes may be related to a decline in soil conditions, associated with the development of a mor[3] rather than a mull situation.

(d) The fourth phase is called the Post-temperate phase, and is indicative of climatic deterioration. There is a reduction of thermophilous genera, and an expansion of heathland. The forest becomes thinner, with temperate forest trees becoming virtually extinct, and a return to dominance of boreal trees, such as *Pinus*, *Betula*, and *Picea*.

This general sequence, whilst broadly applicable to the main interglacial phases, does vary from interglacial to interglacial. There were probably climatic differences between the various phases, different barriers to migration, differing distances to glacial refuges from which genera

---

[1] Table 2.4 is a list of common plant names, with their botanical equivalents, as used in this section.

[2] Mull is fertile, non-acidic soil humus inhabited by earthworms.

[3] Mor is an acid soil humus which accumulates at the soil surface and is too acid for earthworms.

Table 2.4 Botanical names of interglacial plants

Botanical name	Common name
Abies	Fir
Acer	Maple
Almus	Alder
Azolla	Water fern
Buxus	Box
Carpinus	Hornbeam
Corylus	Hazel
Erica	Heath
Fagus	Beech
Fraxinus	Ash
Juglans	Walnut
Juniperus	Juniper
Lemna minor	Duckweed
Liquidambar	Sweet gum
Liriodendron	Tulip tree
Osmunda claytonia	Fern
Picea	Spruce
Pinus	Pine
Pterocarya	Wingnut
Quercus	Oak
Salix	Willow
Sequoia	Sequoia
Taxus	Yew
Tilia	Lime
Trapa natans	Water chestnut
Tsuga	Hemlock
Ulmus	Elm
Vitis	Vine creeper
Xanthium	Cocklebur

expanded, changes in ecological tolerance, and variability within genera, and other changes consequent upon evolution or extinction (West 1972: 315–24).

A broadly similar scheme for an 'interglacial cycle' was proposed by Iversen (1958) and has been used by Birks (1986) (Fig. 2.23). It has largely been applied to changing conditions in north-west Europe. The *cryocratic* phase represents cold glacial conditions, with sparse assemblages of pioneer plants growing on base-rich, skeletal mineral soils under dry, continental conditions. The *protocratic* phase witnesses the onset of the interglacial, with rising temperatures. Base-loving, shade-intolerant herbs, shrubs and trees immigrate and expand quickly to form widespread species-rich grasslands, scrub and open woodlands, which grow on un-leached, fertile soils of low humus content. The *mesocratic* phase sees the development of temperate deciduous forest and fertile, brown-earth soils under warm conditions. Shade-intolerant species are rare or absent because of competition and habitat loss. The last, retrogressive phase of the 'cycle', produced by a combination of soil deterioration and cli-

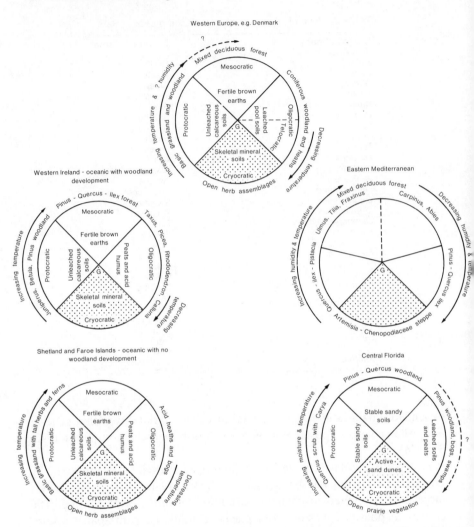

FIG. 2.23 The 'interglacial cycle' applied to various areas of north-western Europe, the eastern Mediterranean, and central Florida (USA). From Birks (1986 fig. 1.2), in B. E. Berglund (ed.). Copyright © 1986, John Wiley and Sons, Ltd. Reproduced by permission of John Wiley and Sons, Ltd.

matic decline, is called the *telocratic* phase, and is characterized by the development of open conifer-dominated woods, ericaceous heaths, and bogs growing on less fertile, humus-rich podzols and peats. This cyclic model can also be applied to areas like Central Florida and the Eastern Mediterranean.

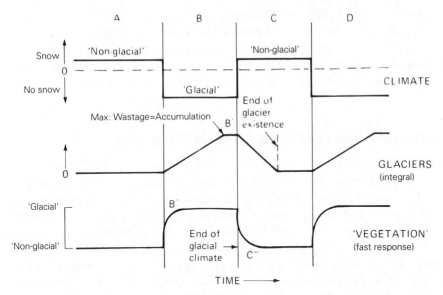

Fɪɢ. 2.24 The different response curves, with time, of climate, glaciers, and vegetation (after Bryson and Wendland, in Andrews 1975, fig. 5–6).

One very pertinent question to raise about vegetation successions in interglacial times is to ask at what speed the colonization of trees was able to progress across the country. From Sweden it would appear that in the Holocene interglacial Scots pine and pubescent birch spread at a rate of 205–60 m per year, alder at 175–230 m, elm and warty birch at around 190 m, and hazel at 130 to 190 m per year. As a whole it would seem that a rate of advance of some 200 m per year for trees with light seeds, and, rather less, say 160 m, for trees with heavier seeds, like hazel and oak, was characteristic. Thus advance of trees at the end of Late Glacial times could have taken place at about one kilometre in five years, or 1000 km in 5000 years.

Plainly, there would be very different response curves with time for climate, glaciers, and vegetation (Fig. 2.24). Ice-caps would tend to respond relatively sluggishly to a climatic amelioration because of their great mass, and because of their partial control of regional climates. Retreat rates of up to 3 km/100 years for the Greenland Ice-Cap are much lower than the rates of floral advance discussed above. The response of fauna would tend to be even quicker.

The change in vegetation at the end of an interglacial may have been very rapid. In her analysis of the Grande Pile pollen core in France,

Woillard (1978) claims that temperate forest of the Last Interglacial was replaced by a pine–spruce–birch taiga within approximately 150 ± 75 years. On the other hand, on the basis of the analysis of a core from Krumbach in southern Germany, Frenzel and Bludau (1987) believe the transition may have been rather more gradual, extending over about 3000 years.

Variations in the British and European interglacials

The interglacial temperate forests recorded in the early British inter-glacials, the Ludhamian and Antian (see Table 2.2 and Fig. 2.2 for the location of these phases in the local sequence), are mixed coniferous and deciduous, and the presence of hemlock and wingnut makes the vegetation assemblage different from that of any subsequent period in Britain. The relatively severe Baventian Glacial period led to the extinction from the British flora of hemlock (*Tsuga*), though it has remained in the vegetation associations of parts of North America until the present day. In northern Eurasia certain of the late Pliocene vegetation types do not appear to have reoccupied the area after the first severe cold of the Pleistocene, though they were present in the Tiglian (Ludhamian). These plants included *Sequoia*, *Taxodium*, *Glyptostrobus*, *Nyssa*, *Liquidambar*, *Fagus*, *Liriodendron*, and several others. Thus a marked degree of impoverishment took place in the British flora as a result of the oncoming of the first cold phases of the Pleistocene (West 1972).

With regard to the later British interglacial floras, however, the Cromer Forest Bed of the Cromerian Interglacial on the Norfolk coast resembles the present-day British flora much more closely, and *Taxus*, *Quercus*, *Fagus*, *Carpinus*, *Ulmus*, *Betula*, and *Corylus* can be identified in the bed. The Hoxnian Interglacial displays in its pollen-profiles a high frequency of *Hippophaë* at the beginning of the sequence, then a late rise of *Ulmus* and *Corylus*, and the presence of *Abies* and *Azolla filiculoides*. Irish materials of the same age have a high percentage of evergreens, indicative of a high degree of oceanicity of climate (e.g. *Picea*, *Abies*, *Taxus*, *Rhododendron*, *Ilex*, and *Buxus*). The materials also contain certain Iberian species such as *Erica scoparia*, St. Dabeoc's heath (*Daboecia cantabrica*), and Mackay's heath (*Erica mackaiana*). The last two currently have a rather limited distribution in the Cantabrian mountains of Iberia. The subsequent Ipswichian, on the other hand, seems to indicate rather more continental conditions. Its characteristic properties included an abundance of *Corylus* pollen in its early part, and then much *Acer* pollen; also present was *Carpinus*, but there was a scarcity of *Tilia* in the second part of the Interglacial. Most of the Ipswichian sites contain a number of species not now native, including *Acer monspessulanum*, *Lemna minor*, *Najas minor*, *Pyracantha coccinea*, *Trapa natans*, *Xanthium*, and *Salvinia natans*. This suggests that conditions were somewhat warmer than during the Holocene climatic optimum (West 1972: 310).

In the more continental parts of Europe the characteristic interglacial assemblages were a little modified, though the general sequence is not dissimilar. In the Likhvin (Holstein) of Russia, for example, as in Turner and West's model (see p. 84), the first stage is represented by a high incidence of *Betula* and *Pinus* pollen, with *Picea* and *Salix* pollen only making up 1 per cent or less of the total (Ananova, 1967). This birch and pine–birch forest stage was replaced by a dominantly pine and spruce–pine sequence, with *Pinus* and *Picea* predominant, with *Betula* making up about 5–10 per cent of the total pollen, and *Alnus* pollen constantly occurring with a frequency of about 15 per cent. This coniferous forest was replaced by a combination of coniferous, broad-leaved, and alder-thicket forests, with *Picea excelsa* dominant in the east, and *Pinus sylvestris* in the west. Both these types had about 40–60 per cent of the total pollen, with *Alnus* making up 25–40 per cent, and *Quercetum mixtum* (mixed oak) pollen about 10 per cent. *Abies* pollen was rare. In the next stage, *Abies* sometimes reaches 30 per cent, and *Carpinus* reached 20–30 per cent in some sections, though coniferous pollen was still dominant. Subsequently, there was a change back towards a more boreal flora, and this in turn was succeeded at the end of the Interglacial by a reduction in tree-pollen, the constant presence of various open vegetation plants (*Poaceae*, *Cyperaceae*, and *Artemisia*), and the arrival of plants of the periglacial type. Thus, although the sequence is comparable to that in western Europe, there are certain elements missing in northern and eastern Europe, including *Vitis*, *Pterocarya*, *Juglans regia*, *Abies alba*, *Carpinus orientalis*, *Buxus*, *Taxus*, *Tilia tormentosa*, and *Osmunda claytoniana*.

An overall picture of northern Europe during the Last Interglacial can be obtained from Fig. 2.22. This shows not only the vegetational characteristics of the Interglacial, including the great expansion of broad-leaved forest, but also the way in which the configuration of the Continent and of the Baltic Sea region was modified by the worldwide rise in sea-level occasioned by the melting of the great ice-caps. By contrast, Fig. 2.15 illustrates the nature of Europe during the Last Glacial.

In southern Europe the interglacials may have been associated with moister conditions, in contrast to the glacials which were essentially drier.

Palynological studies in southern Spain, for example (Florschutz *et al.* 1974) show that instead of a steppe-like vegetation such as characterized the glacial periods, the interglacials were characterized by a more humid assemblage of vegetation including *Fagus*, *Juglans*, *Quercus pubescens*, *Tsuga*, and *Cedrus*.

Faunal and floral fluctuations

The environmental changes of the Pleistocene, as already mentioned with regard to the changing nature of European vegetation in the various interglacials, led to a great impoverishment of flora, particularly of glaciated islands. It has been remarked, for example by Pennington (1969: 1), that the comparative poverty of the British flora, compared with that of continental Europe in similar latitudes, is the result of successive wiping-out of frost-sensitive species by repeated glacial episodes. After each glacial period, with its wholesale extinction of plants from Britain, migrating plants and animals followed northwards in the footsteps of the retreating ice, and combined with the descendants of the hardy species which had survived, to re-establish the British flora and fauna.

A slightly more complex situation is illustrated by the Irish fauna, where both glacial and sea-level fluctuations seem to have been important in determining the present types of animal encountered on the island. Ireland today lacks certain beasts which are encountered in England and Wales. These include the poisonous adder, the mole, the common shrew, the weasel, the dormouse, the brown hare, the yellow-necked field mouse, the English meadow mouse, and others. On the other hand, it does possess a certain proportion of the English fauna. The explanation for this seems to be that as the ice-caps retreated animals from the Continent (then connected by dry land to England because of the low eustatic level at that time), and from the unglaciated tundra area of southern England, crossed over to Ireland by a land-bridge. By the Boreal phase of post-glacial time (9500 BP onwards) however, when the climate had so ameliorated as to permit immigration of temperate species, the Irish Sea was in existence, and the dry passage to Ireland was disrupted. Thus many beasts were unable to cross.

A similar example of the role of the various Post-Glacial and Late Glacial events in creating the present pattern of fauna is provided by the distribution of bird species in the North American continent (Mengel 1970). During the late Wisconsin glaciation, which reached its peak about 18 000 or 20 000 years ago, the northern Rocky Mountains were covered by the Cordilleran ice, while to the east, the lower ground was covered by the massive Laurentide ice-sheet (see p. 56 for a further discussion of the extent of the American ice bodies). There is evidence that as these two sheets contracted in Late Glacial times a long arm of tundra and then taiga invaded the lower ground from southern Alberta to the Mackenzie River Delta. The NW to SE orientation of this corridor helps to explain what is a peculiar but recurrent feature in the distribution of North American birds and some other animals, namely the strong tendency for essentially eastern taxa, that had adapted to the taiga and its successional stages, to occur north-west to, or nearly to, Alaska, at the apparent expense of western montane kinds that had adapted to montane coniferous forest. The explanation seems to be that the western types were blocked by the persistent but dwindling Cordilleran ice-sheet, enabling the eastern taxa to get there first, and to fill the niches: a situation which they have held since.

The question as to what degree the present fauna was able to survive in areas that were glaciated is one of great interest. On the one hand some authorities maintain that the bulk of the fauna of, say, Iceland, is of post-glacial age, and has reached that island by post-glacial diffusion. On the other, there are authorities who consider that certain species were able to exist on small non-glacial peaks (nunataks), rising above the general level of the ice-caps (Gjaeveroll 1963). Other people consider that in certain favoured coastal regions there were small 'refugia' where a hardy flora might be able to live through the glacial period. The last two concepts comprise the *Overvintring* concept of certain Scandinavian botanists. That such survival is possible is attested by the flora of present-day Greenland nunataks. Moreover, various Scandinavian and Irish geomorphologists have claimed to find evidence for refugia and nunataks. One line of evidence that has been used in Arctic Norway is the presence of block fields (*felsenmeer*) and other periglacial rather than glacial features on summits. Moreover, notably in Iceland, the present distribution of the flora often shows a bicentric or poly-centric form, which ties in better with the idea of diffusion from internal refugia than with the idea that the whole flora was erased—the *tabula rasa* concept—and has been replaced by post-glacial migrations from overseas. If post-glacial migration were responsible for these plants arriving one might expect them to be more widely distributed.

The role of land-bridges in the Pleistocene should not be exaggerated, though as will be seen the fall in relative sea-level by perhaps as much as 150 m did expose large expanses of the continental shelf. Certain islands were therefore linked together, or to the mainland. Malta and Sicily, Capri and Italy, the Balearics, the Ionian Islands, and, possibly, Tunisia and Italy, are such examples. Other islands, on the other hand, remained isolated, and their fauna tends to this day to show a greater degree of endemism. This is well illustrated by an example from the Philippines. The Islands of Negros, Panay, and Masbate collectively make up Visaya. They stand together on a submerged shelf less than 50 m deep. They are, however, separated from the nearby island of Cebu by a 98 m deep strait. The main faunal result of this is that on Visaya there are 32 endemic species of non-migratory birds. They are lacking on Cebu. It thus seems likely that in this area the sea-level was low enough to permit migration among the three islands on Visaya, but not great enough to permit migration to or from Cebe (Deevey 1949).

Elsewhere in South-East Asia the effects of marine regressions and transgressions on faunal boundaries are also striking. Low stands of glacial sea-level drained most of the Sunda Platform area, consisting of Malaysia, Borneo, Java, and Sumatra (Fig. 2.25). Thus these islands and peninsulas were interconnected and more than 3 million km^2 of shallow warm seas were converted into land (Verstappen 1975). This allowed many animals to come from the Asian mainland, including beasts of Indian and Chinese type. Today the fauna of the Sunda Islands is basically a somewhat impoverished version of that on the Asian mainland, with local races of elephant, tiger, leopard, and dhole. A low stand of sea-level may also have allowed some early humans, *Homo erectus*, to penetrate Java as much as one million year ago. However, the Sunda Shelf is separated from the New Guinea–Australia Sahul Platform by deep water, albeit narrow. Pleistocene sea-levels appear not to have fallen sufficiently to allow the linking of these two realms, and so this limited belt of sea forms one of the most important of all zoo-geographic boundaries, 'Wallace's Line'.

Unlike the Sunda Platform, the fauna of the Sahul Platform is one of distinctly Australian affinity, with marsupials—kangaroos, wallabies, wombats, and koalas. In between these two platforms is the group of islands called the Celebes (Sulawesi). These appear to have been isolated from the two platforms for a considerable period, and so many types of animal are absent. Many indigenous forms such as the babirussa pig, the pygmy buffalo, and, in the fossil record, two types of pygmy elephant, have evolved.

The fall in temperature of the oceans, which was probably of the order of 3–8 °C during the cold phases of the Pleistocene, also affected the distribution of marine life. This can be illustrated from a study of coral reefs (see Fig. 2.26). The present effective limit of reef growth is approximately that of the 20 °C ocean water isotherm (Stoddart 1973). By subtracting the glacial falls in temperature for each major ocean derived from

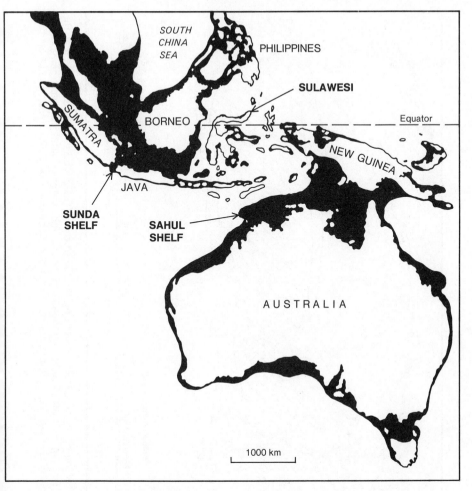

FIG. 2.25 The coastal palaeogeography of south-east Asia and Australasia when sea level was at *c.* −120 m during the last glacial maximum (modified after Van Andel 1989, fig. 4A).

miscellaneous palaeo-temperature observations a map of probable Pleistocene coral reef growth can be constructed. It shows the considerable degree of contraction which must have taken place in reef distribution. Over large areas reef corals would have died because of the relatively cool conditions. This effect would have been heightened further by the low still-stands of sea-level during glacial phases.

Similarly, the north–south migration of polar waters in the North Atlantic in response to major cycles of glaciation is shown in Fig. 2.27. In

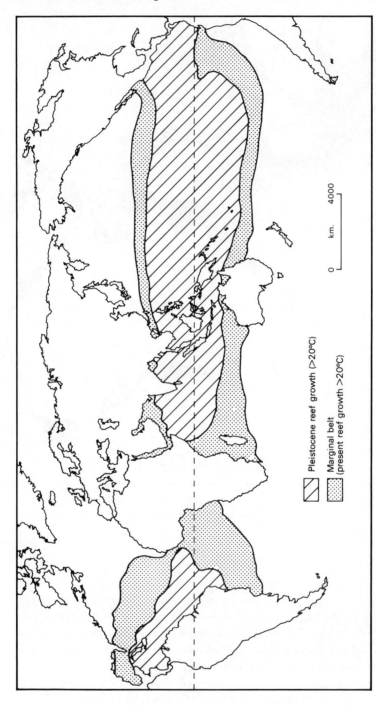

FIG. 2.26 Possible extent of the contraction of the coral-reef seas during the Pleistocene. The isotherm of 20 °C is taken as the effective limit of reef formation, and the map is constructed by subtracting the glacial falls in temperature for each major ocean derived from published palaeotemperature analysis from the present-day sea-surface temperatures of the coldest month. (From Stoddart 1973, fig. 3.)

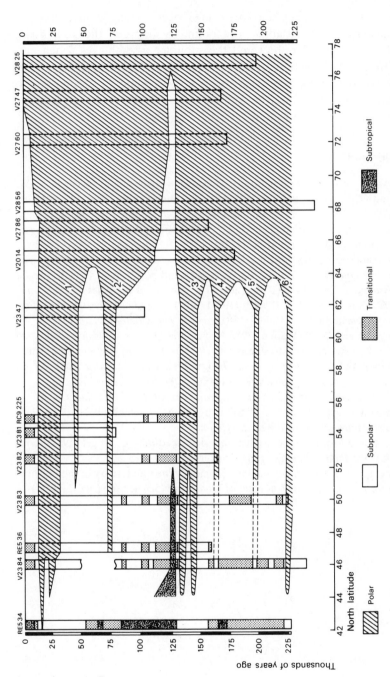

FIG. 2.27 The north–south migration of water during the past 225 000 years as revealed by evidence from deep-sea cores. The numbers 1 to 6 indicate calcium carbonate sediment minima (after US Committee for the Global Atmospheric Research Program 1975, fig. A.25).

this figure fourteen deep-sea cores have been arranged in a transect in the eastern North Atlantic. The boundary between the polar fossil assemblages (diagonal ruling) and the subpolar assemblages (open pattern) reflects the position of the oceanic polar front. At the glacial maximum about 18 000 years ago this front was some 20° farther south, while during the interglacial maximum some 125 000 years ago it had a position similar to that of today.

Selected reading

Good reviews on the general nature of Pleistocene environments occur in K. W. Butzer and G. L. Isaac (1975) (eds.), *After the Australopithecines*, and in H. H. Lamb (1977), *Climate, Present, Past and Future* (vol. ii). The ecological characteristics are well treated in R. G. West (1977), *Pleistocene Geology and Biology*, and in H. J. Birks and H. H. Birks (1980), *Quaternary Palaeoecology*.

The Pleistocene of the British Isles is well covered in F. W. Shotton (1977) (ed.), *British Quaternary Studies*, B. W. Sparks and R. G. West (1972), *The Ice Age in Britain*, C. A. Lewis (1970) (ed.), *The glaciations of Wales and adjoining regions*, C. Kidson and M. J. Tooley (1977) (eds.), *The Quaternary History of the Irish Sea*, J. Neale and J. Flenley (1981) (eds.), *The Quaternary in Britain*, and K. J. Edwards and W. P. Warren (1985) (eds.), *The Quaternary history of Ireland*.

The North American Pleistocene is discussed in three large volumes of essays: H. E. Wright and D. G. Frey (1965), *The Quaternary of the United States*; W. C. Mahaney (1976) (ed.), *Quaternary stratigraphy of North America*, and H. E. Wright (1983) (ed.), *Late Quaternary environments of the United States*.

A companion volume for the USSR is by A. A. Velichko (1984) (ed.), *Late Quaternary environments of the Soviet Union*.

Changes in the southern hemisphere are described in J. C. Vogel (1984) (ed.), *Late Cainozoic Palaeoclimates of the Southern Hemisphere*.

An attempt to provide a correlation scheme for the Quaternary in the northern hemisphere is given in a special issue of *Quaternary Science Reviews*, 5 (1985).

3

Pleistocene Events in the Tropics and Subtropics

Arid phases in the Pleistocene

The events which led to the expansions and contractions of the great ice-sheets during the Pleistocene also led to major environmental changes in lower latitudes. The positions of the major climatic belts were altered, and with them the major vegetation zones. One of the most striking and important results of such change was that the limits of the world's great tropical and subtropical sand deserts shifted. Such deserts were not the products of the Pleistocene, but many of them expanded greatly in size as the world's climate cooled (see p. 35). Sediment cores from off the western Sahara contain distinctive aeolian material that indicates that in North Africa a well-developed arid area was in existence around 20 million years ago, the Early Miocene (Diester-Haass and Schrader 1979). Siesser (1980) has demonstrated that the cool upwelling waters of the Benguela Current were promoting aridity along the Namib coast of South West Africa (Namibia) by the late Miocene, while the study of sediments from the central parts of the northern Pacific (Leinen and Heath 1981) suggests that aeolian processes became more important as the Tertiary progressed, accelerating greatly between 7 and 3 million years ago. It was, they believe, around 2.5 million years ago that there occurred the most dramatic increase in aeolian sedimentation, an increase that accompanied the onset of northern hemisphere glaciation.

One of the most satisfactory ways to assess the former extent of desert areas during the Pleistocene interpluvial or dry phases is by studying the former extent of major tropical and subtropical dune-fields as evidenced by fossil forms, often visible on air or satellite photographs.

Indications that some dunes are indeed fossil rather than active are provided by features like deep-weathering and intense iron-oxide stain-

Table 3.1 Rainfall limits of active and fossil dunes

Source	Location	Today's precipitation limit for formation of active dunes (mm)	Tody's precipitation limit of fossil dunes (mm)	Dune shift (km)
Hack (1941)	Arizona	238–254	305–80	—
Price (1958)	Texas	—	—	350
Tricart (1974)	Llanos	—	1400	—
Tricart (1974)	NE Brazil	—	600	—
Grove (1958)	West Africa	150	750–1000	600
Flint and Bond (1968)	Zimbabwe	300	c.500	—
Grove and Warren (1968)	Sudan	—	—	200–450
Goudie et al. (1973)	S Kalahari	175	650	—
Lancaster (1979)	N Kalahari	150	500–700	1200
Mabbutt (1971)	Australia	100	—	900
Glassford and Killigrew (1976)	W Australia	200	1000	800
Goudie et al. (1973)	India	200–275	850	350
Sarnthein and Diester-Haass (1977)	NW Africa	25–50	—	—
Sombroek et al. (1976)	NE Kenya	—	250–500	—

ing, clay and humus development, silica or carbonate accumulation, stabilization by vegetation, gullying by fluvial action, and degradation to angles considerably below that of the angle of repose of sand—normally 32–3° on lee slopes. Sometimes archaeological evidence can be used to show that sand deposition is no longer progressing at any appreciable rate, whilst elsewhere dunes have been found to be flooded by lakes, to have had lacustrine clays deposited in interdune depressions, and to have had lake shorelines etched on their flanks.

Sand movement will not generally take place through aeolian activity over wide areas so long as there is a good vegetation cover, though small *parabolic* (hairpin) dunes are probably more tolerant in this respect than the more massive *siefs* (linear) and *barchans* (crescentic). Indeed, dunes can develop where there is a limited vegetation cover, and vegetation may contribute to their development. It is, therefore difficult to provide very precise rainfall limits to dune development. Nevertheless, studies where dunes are currently moving and developing suggest that vegetation only becomes effective in restricting dune movement where annual precipitation totals exceed about 100 to 300 mm. These figures apply for warm non-coastal areas. Some opinions of workers from some major desert areas on the rainfall limits to major active dune formation are summarized in Table 3.1.

At the present time overgrazing and other human activities on the desert margins may induce dune reactivation at moderately high precipitation levels, and this is, for example, a particular problem in the densely populated Thar Desert of Rajasthan, India.

When one compares the extent of old dune-fields, using the types of evidence outlined above, with the extent of currently active dune-fields, one appreciates the marked changes in vegetation and rainfall conditions that have taken place in many tropical areas. This is made all the more striking when one remembers that decreased Pleistocene glacial temperatures would have led to reduced evapo-transpiration rates, and thus to increased vegetation cover. This would if anything have tended to promote some dune immobilization. Dune movement might, however, have been accentuated by apparently higher trade-wind velocities during glacials (Parkin and Shackleton 1973).

Reviewing such evidence from the different continents, Sarnthein has mapped the world distribution of ancient and modern ergs (Fig. 3.1), and summarized the situation thus (1978, p. 43): 'Today about 10 per cent of the land area between 30°N and 30°S is covered by active sand deserts. . . . Sand dunes and associated deserts were much more widespread 18 000 years ago than they are today. They characterized almost 50 per cent of the land area between 30°N and 30°S forming two vast belts. In between tropical rainforests and adjacent savannahs were reduced to a narrow corridor, in places only a few degrees of latitude wide.'

The fossil dunes of northern India

Some of the early British geologists of the Indian Geological Survey appreciated that many of the dunes in northern India were fossil forms.

As early as the 1880s W. T. Blanford remarked that in Rajasthan 'many of the sand-hills are evidently of great antiquity: despite the small rainfall of the desert region, they show signs of considerable denudation in parts, and are cut into deep ravines by the action of water'. Subsequently fossil dunes have been identified both in the Las Belas Valley area of Pakistan, in Gujarat, and in Rajasthan (Verstappen 1970). In Gujarat the dunes, showing calcification, deep gullying, and marked weathering horizons, are normally overlain by large numbers of small microlithic tools, suggesting that there has been relatively little sand movement since mesolithic man lived in the area. The fossil dunes include parabolics, transverse, longitudinal, and wind-drift types, and are now known to extend as far as Ahmedabad and Baroda in the south, and to Delhi in the east (Goudie et al. 1973) (Fig. 3.2). They occupy zones where the rainfall is now as high as 750–900 mm. In the Sambhar salt-

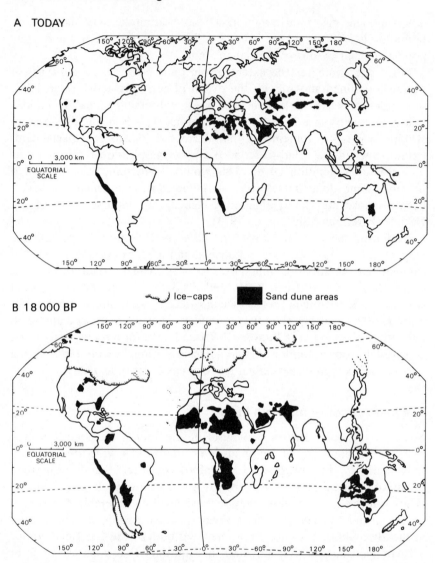

A TODAY

B 18 000 BP

FIG. 3.1 The distribution of active sand-dune areas (*ergs*).
A. Today
B. At the time of the last glacial maximum, *c*.18 000 years ago (modified after Sarnthein 1978).

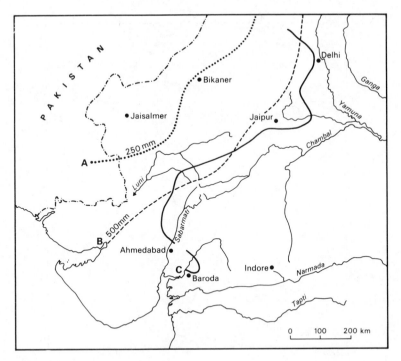

FIG. 3.2 The former extension of the Great Indian Sand Desert in the Late Pleistocene.
A. 250 mm mean annual isohyet
B. 500 mm mean annual isohyet
C. Former extension of sand desert.

lake area near Jaipur in eastern Rajasthan the dunes are overlain by lake deposits of a freshwater type. The base of this lake bed had been dated at about 10 000 years BP, suggesting that dune movement may have ceased by that time (Singh 1971). The sediments at Didwana Lake also demonstrate a period of aridity and dune building prior to 13 000 BP (Agrawal *et al.* 1990). In Saurashtra (Kathiawar), the late Pleistocene dunes, locally called *miliolite*, are heavily cemented by calcium carbonate and are used as building stone.

The fossil dunes of Africa

A basically similar picture to the Indian one comes from southern Africa, where the Kalahari sandveld is dominantly a fossil desert, now covered by a dense mixture of *acacia* and *mopane* forest, with grassland and shrubs.

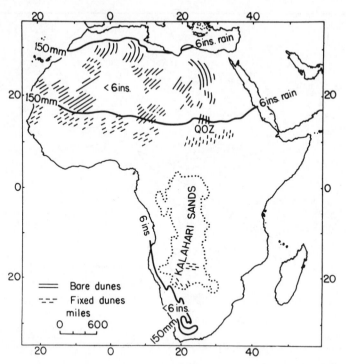

FIG. 3.3 The past and present extent of blown sand in Africa (after Grove 1967, fig. 7).

Relict dunes are widespread in Botswana, Angola, Zimbabwe and Zambia (Thomas and Shaw 1991, Grove 1969), amd may well extend as far north as the Congo rainforest zone (Fig. 3.3). Several phases of dune activity have occurred in the Late Pleistocene (Lancaster 1989).

North of the Equator, the fossil dune fields extend south into the savannah and forest zone of West Africa, and have covered lateritic and other soils and invaded palaeolake basins. The so-called 'Ancient Erg of Hausaland' (Grove 1958), extends into a zone where present rainfall is a as high as 1000 mm per annum (Nichol 1991). Many of the dunes in northern Nigeria are now cultivated, and in the vicinity of Lake Chad dunes have been flooded by rising lake waters. At one stage of the history of this area they blocked or altered the course of the River Niger. Indeed, in the middle Niger area there appear to be several ages of fossil dune, including old deeply-weathered linear dunes, and younger grey-brown and yellow dunes of lesser height. Details of age and pedogenesis are given in Völkel and Grunert (1990).

The River Niger, as we now see it, was born of two parents. In the Late Pleistocene the lower, south-east-flowing section was fed from the southern slopes of the Ahaggar Mountains, by affluents which are now practically extinct. Lower down it was augmented, as now, by the Sokoto and Benue. The Upper Niger, flowing north-eastwards from the mountains of the Guinea–Sierra Leone border, flowed during the Late Pliocene and Early Pleistocene westwards into the Gulf of Senegal. A subsequent dry period produced a barrier of sand dunes (the Erg Ouagadou), which then blocked the previous westward flow of the upper river when the last major wet phase arrived (Beadle 1974: 125). It was therefore diverted into a closed basin, Lake Araouane. The flooded basin later began to drain away (either by a breakthrough or a capture) and, with a near right-angled turn joined the lower Niger, possibly only 5000–6000 years ago.

Further east, in the Sudan, west of the White Nile, a series of fixed dunes known locally as *Qoz* covers most of the landscape up to the slopes of the Jebel Marra. The fixed dunes extend as far south as 10 °N, and merge northwards, locally, with mobile dunes at about 16 °N. They succeed in crossing the Nile, which thus probably dried up at the time of their formation. Again, as in West Africa and India, there appear to have been at least two phases of dune activity. These two phases were interrupted by a relatively wet phase when extensive weathering and degradation took place. The first phase suggests a shift in the wind and rainfall belts of about 450 km southwards, and the second phase of dune building in Holocene times represents a shift of about 200 km (Grove and Warren, 1968).

Periods of aridity of the type indicated by the fossil dunes of Africa did, in marginal areas, have a marked effect on man, as witnessed by the clear hiatuses that exist in the archaeological record. As Wendorf *et al.* (1976: 113) remark, 'There are no traces anywhere in the Nubian Desert of any occupation, spring, or lacustrine deposits that are between the Aterian sites and the Terminal Palaeolithic in age. For this period of more than 30 000 years' duration the Western Desert of Egypt was apparently devoid of surface water and of any sign of life.'

The fossil dunes of North and South America

In the USA a comparable development of fossil dune-fields has been recognized. Parts of the High Plains, for example, which are now dominated by a calcrete (caliche) caprock and numerous depressions, were formerly covered by large dune-fields displaying the characteristic anti-clockwise wheelround features of the dune systems of Australia and southern Africa (Price 1958).

This American system includes the Rio Grande Delta erg which extends about 150 km from Punta Penascal at the mouth of Baffin Bay to Oilton (Torrecillas), and about 300 km from Oilton to the southern end of the Delta. Another ancient American erg is called the Llano Estacado field, and this is outlined at least in part in the present landscape by the topographic grain of etched swales, remnant ridges, and deflated swale ponds and lakes, the latter being orientated along the swales. It is often termed as 'scabland'. It has sometimes been suggested that the lineation of the dunes and lakes indicates a former wind-pattern diverging as much as 90° from the present pattern, in addition to more arid conditions. These features probably indicate an expansion of the desert to the north and east by the order of 320 km. In Nebraska and South Dakota, the Sandhills, covering an area of 52 000 km^2 were also more active (Smith 1965). The famous Carolina Bays may have developed as deflation hollows in interdune swales.

A wide range of dates has been obtained for phases of dune activity in the High Plains. Some of the aeolian deposits are of considerable antiquity, and the aeolian Blackwater Draw Formation may date back beyond 1.4 million years (Holliday 1989). Substantial dune development also took place in the Late Pleistocene (Wells 1983), and during the drier portions of the Holocene (Gaylord 1990).

In South America Tricart (1974) has used miscellaneous remote-sensing techniques which have enabled him to identify two ancient ergs. One was in the Llanos of the Orinoco river, where fossil dunes, partly fossilized by Holocene alluvium, extend southwards as far as latitudes 6°30′ N and 5°20′ N. The other erg was in the valley of the lower-middle São Francisco River in Bahia State, Brazil. At the time of its formation the river had interior drainage and did not flow throughout its length. In addition it is likely that aeolian activity was also much more extensive in the Pampas and other parts of Argentina.

The fossil dunes of Australia

In Australia, fossil dunes associated with an enormous anti-clockwise continental wheelround, are developed over wide areas. Their great extent has only recently become apparent, and has been mapped by Wasson *et al.* (1988: Fig. 9). They are particularly well displayed on the Fitzroy plains of north-western Western Australia, where they pass under (and thus pre-date) the Holocene alluvium of King Sound (Jennings 1975). In the country to the south of the Barkly Tableland they are completely vegetated, have subdued and rounded forms, and appear broadly comparable to those of northern Nigeria. They probably represent a decrease of rainfall in the Barkly Tableland area of between 150 and 500 mm, representing an equatorward shift of the isohyets by about 8° of latitude or around 900 km (Mabbutt 1971).

Stratigraphic data from various areas indicates that a major phase of dune construction occurred in the Late Pleistocene between 25 000 and 13 000 BP (Wasson 1984), but thermoluminescence dates indicate that there were also a number of earlier phases of dune activity as well. Lunette dunes on the lee sides of closed depressions have also had a lengthy history and provide many details about both dune and lake evolution.

It is possible that some Australian dunes date back to pre-Pleistocene times. Benbow (1990) suggests that some from the Eucla Basin may have survived from the end of the Eocene, around 34 to 37 million years ago.

Pluvial phases in the Pleistocene

No less dramatic than the evidence presented for aridity by the fossil dune systems is the evidence presented for increased hydrological activity, either resulting from a temperature decrease or an absolute precipitation increase, in the Pleistocene and Early Holocene. Such phases have been called lacustral or pluvial phases.

However, in some respects the evidence is more equivocal than that provided by the dunes, for the relation between rainfall and lake levels is itself complicated by both temperature and non-climatic factors. With respect to the latter, it needs to be remembered that many lakes occur in areas of tectonic instability or volcanic activity, including the many lakes that occupy the floor of the East African Rift between the Danakil Depression in Ethiopia and Lake Malawi. Other lakes, including the Etosha Pan of Namibia, and the Makarikari and Ngami lakes of Botswana, may have been affected by the fact that the river systems in this semi-arid area are to a degree interconnected. Elsewhere the outlets of lakes may have been affected by erosion or vegetation growth at outlets, creating alternating lowering or ponding-back of lake waters.

Nevertheless, the widespread nature and similar chronologies of many large basins in many parts of the world suggest that the climate factor has perhaps been dominant in controlling lake level fluctuations.

Table 3.2 Pluvial lake dimensions

Lake	Location	Area (km²)	Depth (m) (height above present dry bed or lake level)
Bonneville	USA	51 700	335
Searles	USA	—	213
Panamint	USA	—	274
Russell (Mono)	USA	—	233
Lahontan	USA	22 442	213
Dead Sea	Israel	—	433
Tuz Golu	Turkey	—	75
Lake Van	Turkey	—	60
Izmik	Turkey	—	55
Burdur	Turkey	—	95
Kharga	Egypt	—	100
Dieri	Australia	104 000	46
Makarikari	Botswana	34 000	45
Nawait (Victoria and Bonney)	Australia	21 000	—
Aral-Caspian	Kazakhstan	1 100 000	76

(From data in Butzer, 1972; Flint, 1971; and Grove, 1969)

Both in terms of area and depth many of these lakes were extremely prominent features of the Pleistocene environment (Table 3.2) and they were in many areas favoured sites for occupation by early man.

The American pluvial lakes

The greatest concentration of pluvial lakes in the western hemisphere, and possibly also in the world, occurs in the Great Basin in the northern part of the Basin and Range Province of the United States (Fig. 3.4). Between 110 and 120 depressions, formed dominantly by late Pliocene and Pleistocene high-angle faulting, were occupied wholly or in part by Pleistocene pluvial lakes. The largest of these was Lake Bonneville, which covered $51\,640\,\text{km}^2$ at its maximum stage, had a north to south extent of around 500 km, had water to a depth of about 335 m, and was comparable in size to present-day Lake Michigan. Now it is only occupied by 2600–6500 km² of saline water. Lake Lahontan was the second largest pluvial lake in the western USA, though in contrast to Lake Bonneville it was more an interconnecting series of long, narrow lakes than open body of water. With an area of $22\,900\,\text{km}^2$ and a depth of 280 m it was slightly less than half as large as Bonneville. A third major system was the group of lakes that developed along the Owens River valley, running down from the Sierra Nevada through Owens Lake, China Lake, Searles Lake, and Panamint Lake to Lake Manly (Death Valley).

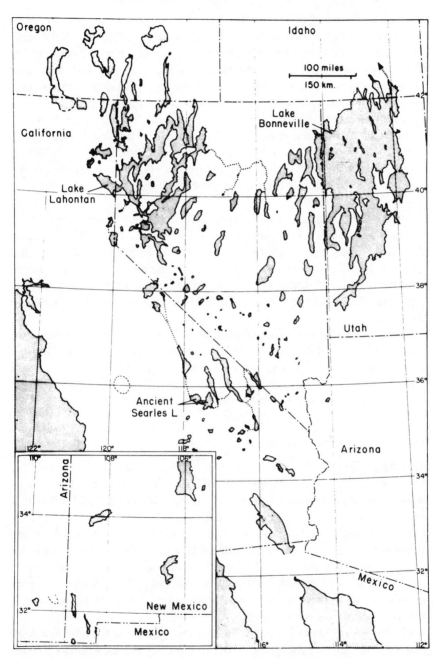

FIG. 3.4 Pleistocene pluvial lakes in the western USA. Dotted lines represent some overflow stream channels (after Flint 1971, fig. 17.3). © 1971 by John Wiley and Sons. Reproduced by permission of John Wiley and Sons, Ltd.

South and east of the Great Basin, in the Basin and Range Province, there were fewer depressions because of a lesser degree of Pleistocene deformation, while further south the amplitude of climatic change in the pluvials seems to have been reduced somewhat because of both increased distance from westerly storm tracks, and the greater aridity resulting from the higher mean annual temperatures of the southerly latitudes. Nevertheless, there were pluvial lakes in these southerly areas, especially in the Mexican Highland section from southern Arizona and New Mexico in the north to the great basin of Mexico City in the south. In Baja California there was pluvial Lake Chapala, while on the High Plains, especially the Llano Estacado, there were numerous small basins, partly of deflational origin (Reeves 1966).

Information on many of these basins, together with a discussion of their chronology, is given by Smith and Street-Perrott (1983). They suggest that most lakes were high during the period 24 000 to 14 000 BP, that between 14 000 and 10 000 years BP the basins suffered rapid, large amplitude fluctuations that may or may not have been synchronous across the region, and that between 10 000 and 5000 BP many of the basins were low and dry.

There were also numerous pluvial lakes in South America, notably in the Altiplano of Peru and Bolivia. The Altiplano lakes, in the Late Pleistocene, covered an area as much as six times that of the present, and rainfall levels have been calculated as being 75 per cent greater than today (Hastenrath and Kutzbach 1985).

The Aral–Caspian system

The Aral–Caspian–Black Sea system, formed in several broad shallow basins of Quaternary warping, received large quantities of glacial meltwater from various sources—the Caspian via the Volga and Ural rivers, and the Aral Sea via the Oxus (Amu Darya) River.

The highest shoreline was at 76 m above the present level of the Caspian Sea, and the area of the pluvial lake was then the greatest known in the world. The Aral and Caspian Seas united to inundate an area of 1 100 000 km^2, and extended 1300 km up the Volga River from its present mouth. The Caspian was also united with an expanded Black Sea through the Mantych Depression (Fig. 3.5).

Chepalyga (1984) believes that major expansions of the Caspian took place during glacial epochs, especially in the early stages, when as a result of a marked temperature depression evaporative loss was decreased. He also believes that the blocking of groundwaters by permafrost played a

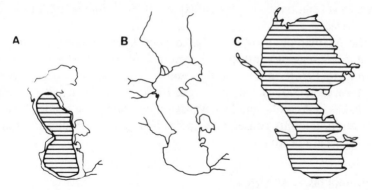

FIG. 3.5 Alternations in the extent of the Caspian Sea during the changes between warm and cold periods (after Frenzel 1973, fig. 94):
A. During an interglacial (Eemian/Mikulino).
B. At the present day.
C. During the Last Glacial.

role, and counteracted decreases in precipitation. Regressions occurred in interglacials as a result of increased evaporative losses.

The pluvial lakes of the Middle East

The present Dead Sea Rift Valley is currently a relatively dry terrain and contains three main lake basins: Lake Hula (now drained), Lake Tiberias (or the Sea of Galilee), and the Dead Sea itself. In Quaternary times a much greater area than today was covered by lakes, and in particular one great lake, the so-called Lisan Lake, extended continuously from the south shore of the present Lake Tiberias to a point some 35 km south of the south shore of today's Dead Sea. The north to south dimension of this lake was probably around 220 km, its maximum width about 17 km, and its highest shoreline was at −180 m compared to the −400 m of the Dead Sea as it exists now. In all, the water volume must have been 325 km^3 compared to the present Dead Sea with its volume of 136 km^3 (Farrand 1971).

Tectonic disturbance on a grand scale may have been partly responsible for its decline in volume, but a lake with a shoreline at −370 m, and the altitude of its floor much the same as that of the present Dead Sea, could have held only about half the water of the Lake Lisan, so that some climatic influence must also be invoked to explain the contraction to present dimensions. Begin *et al.* (in Rognon 1976) believe that the Dead

Sea was a substantial lake from 18 000 to 12 000, a time of great dryness in East Africa.

In the Arabian desert other old lakes have been recognized, and radio-carbon dates suggest two lacustral periods: 36 000–17 000 and 9000–6000 BP (McLure 1976). In Turkey the Konya basin was last occupied by a great lake at 23 000–17 000 years ago. Roberts (1983) suggests that such palaeolakes in Anatolia and Iran were predominantly the result of reduced evaporation amounts consequent upon cooling at or around the time of the last glacial maximum.

The pluvial lakes of Africa

Lake basins contained expanded bodies of water in many parts of Africa, including the driest parts of the Sahara. Large lakes occupied the salty *chotts* of North Africa, though their dates are the subject of controversy and may have been earlier than often thought (Fontes and Gasse 1989). Lake deposits have also been found in some of the hyper-arid basins of the Western Desert in Egypt, though some of these may date back to the early Holocene (Brookes 1989).

One of the largest and most spectacular of the pluvial lakes is that of Lake Chad (Fig. 3.6), but unlike the Caspian–Aral system it did not receive glacial meltwater. At more than one stage during the Pleistocene, Chad was considerably larger than it is at the present time. Chad at present stands at a height of 282 m above sea-level, but at some early stage the Chad river formed a 40 000 km^2 delta in association with a lake at 380–400 m. The lake then shrank during an arid phase of dune formation, but later again rose to 320–30 m, and formed a marked ridge or ridge complex, traceable over a distance of more than 1200 km. Between Maiduguri and Bama in north-east Nigeria this strandline is easily identifiable as a sand mound 12 m high (Grove and Warren 1968).

To the east, the rift valleys of East Africa are occupied by numerous lakes around which occur Late Pleistocene and Early Holocene high strandlines. In Ethiopia, one of the biggest of the pluvial lakes, first recognized by the Scandinavian explorer and scientist, Nilssen, is pluvial Lake Galla. This occurs to the south of Addis Ababa and is occupied by four shrunken remnants, Ziway, Langano, Abiyata, and Shala. However, when the lakes were larger, standing as much as 112 m above the present surface of Shala, the basin was occupied by one large sheet of water (Fig. 3.7) (Grove et al. 1975). In Afar, Lake Abhé attained a surface area of 6000 km^2 and a depth of more than 150 m (Gasse, in Rognon 1976).

Further south, the other lakes also display old strandlines and lake sediments. Lake Awasa shows a series of terraces cut in volcanic debris at

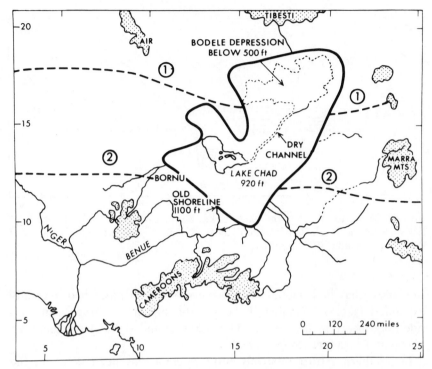

F IG . 3.6 Shifting lake shores and desert limits in the Chad basin, west-central Africa. The present extent of Lake Chad, 920 feet (280 m) above sea-level, is compared with the old shorelines of the ancient MegaChad at 1100 feet (335 m) above sea-level (thick line), which overflowed into the Benue. Dashed line (2) is the southern limit of old, vegetation-covered dunes. It lies far south of the southern limit of moving dunes of the present day, marked by dashed line (1). These extreme changes of desert limits and lake levels probably occurred in the period 20 000–50 000 years ago (after Grove and Warren 1968).

10, 22, 33, and 40 m above the present lake surface, and lakes Margherita and Chamo have a 20–30 m terrace. Oyster (*Etheria*) shells have been found at 52 m above the present swamp level. Lake Stefanie (Chew Bahir), discovered as late as 1888, is situated just to the north of the Kenya border, and increased discharge down the Sagan River appears to have led the lake, now either completely dry or seasonally flooded, to have reached a level of at least 20 m, creating fossil spits, and depositing incrustations of algal limestones on the old cliffs and islands (Grove *et al.* 1975: 183).

Even further to the south similar evidence of high lake stands has been found in Kenya and Tanzania, and the chronology seems to have been

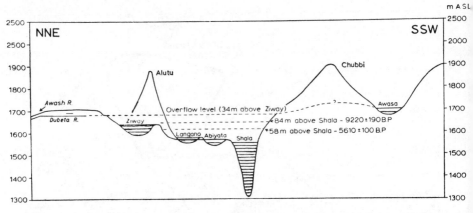

FIG. 3.7 A section through the Galla Lakes basin, Ethiopia, illustrating the high Early Holocene levels which united Ziway, Langano, Abiyata, and Shala (after Grove and Goudie 1971, fig. 2).

similar to that in Ethiopia. The Nakuru-Naivasha basin had a greatly expanded lake. In Zambia, Lake Cheshi appears to have been considerably larger in the period 8000–4000 BP and rather lower between 15 000 and 13 000 BP (Stager 1988).

In southern Africa relatively little research has been done on the pluvial lakes, and few dates are as yet available. However, aerial photographic reconnaissance does again indicate that some of the basins were very considerably enlarged in the not-too-distant past. In the northern part of the Kalahari sandveld tectonic adjustment and climatic change led to the formation of Palaeo-Makgadikgadi, encompassing the Okavango Delta, parts of the Chobe–Zambezi confluence, and the Ngami, Mababe and Makgadikgadi basins of northern Botswana (Shaw and Thomas 1988, Thomas and Shaw 1991). Palaeo-Makgadikgadi may have reached an area of 120 000 km^2, making it second in size in Africa to Lake Chad at its Quaternary maximum. However, there is still considerable uncertainty about the dating of its highest stands, but it is likely that its great volume was caused by the inflow of water from the Zambezi River. Another major lake basin in southern Africa which was greatly expanded in the Late Pleistocene is the Etosha Pan of northern Namibia (Rust 1984).

The dates of the last great lacustral phase in low latitudes

The great lakes of East Africa, as already shown, expanded and contracted greatly during the course of the Pleistocene. As seen on p. 210, they have even shown marked changes of level in the last decade. In that early man, like present man, occupied the lake basins of the Rift, their fluctuations have some importance. Many of the most important archaeological and anthropological finds in East Africa, as at Lake Turkana and the Omo valley (Ethiopia and Kenya), Olduvai Gorge (Tanzania), and Olorgesaillie (Kenya), occur in association with lake beds.

Radio-carbon dates which have become available over the last twenty years show fairly clearly that nearly all the lakes reached a maximum level around 8000–9000 BP, that is in early post-glacial time. Some of the available data are shown in Fig. 3.8. Comparative data for the other parts of tropical Africa are included. In general there appear to be few high lake deposits in tropical Africa with dates between 18 000 and 12 000–13 000 BP. Between about 12 000 or 12 500 and 7000 years ago, possibly with a peak around 9000 BP, lakes were for long periods, if not for the whole time, higher and larger than now (Fig. 3.9). It is generally believed that by 9000 BP or so, when they were at their greatest size, the temperatures were broadly similar to those of the present, and that by implication precipitation levels were between 125 and 165 per cent of those of the present day (Street 1977). This contrasts with calculated precipitation levels of 54–91 per cent of present values for the same lakes during the low lake stands of the terminal Pleistocene (14 000–13 000 BP).

The dates of early pluvial or lacustral phases are less clear, partly because there are relatively fewer dates, and partly because beyond 40 000 years or so the validity of radio-carbon dating is greatly reduced. Available dates suggest that around 40 000 to about 20 000 years ago there was a phase of relatively moist conditions both in the northern Sahara and the southern Sahara (Rognon 1976). There are very few dates for pluvial phenomena until the very Late Glacial or Early Holocene. This does imply that the glacial maxima of the Last Glaciation, dated about 23 000 to 11 000 BP, were times of relative drought in much of Africa north of the Equator. Equally, the wet phase from 40 000 to about 20 000 may correspond very roughly with the interstadial period encountered in many parts of the northern hemisphere. In the Rudolf Basin of Kenya and Ethiopia, however, the last dry phase seems to have been of greater duration, 35 000–10 000 BP, and such a dry phase is also recognized in the sequence from Lake Nakuru in Kenya.

The Australian evidence (Bowler 1976) seems to confirm this general picture. From about 40 000 to 25 000 BP, lake levels were high, and dunes relatively stable. After 25 000 the last major arid phase began, causing the lake levels to fall. Increasing alkalinities assisted in the early construction

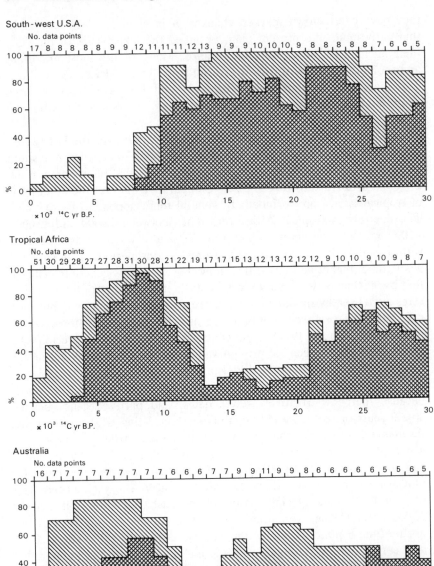

FIG. 3.8 Histograms showing lake-level status for thousand year time periods from 30 000 BP to the present day for three areas: south-western USA, intertropical Africa, and Australia (after Street and Grove 1979).

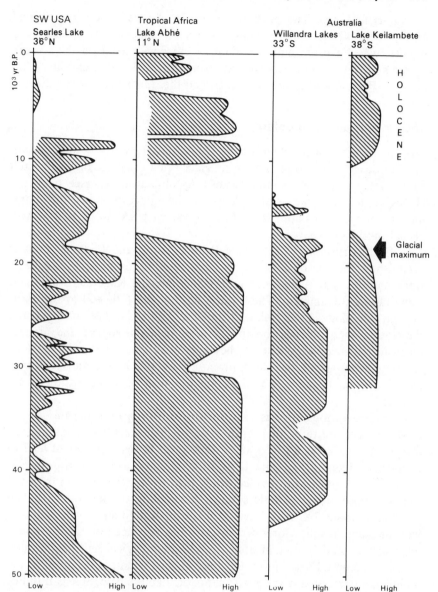

FIG. 3.9 Lake-level curves for four basins in the late Quaternary. Note the high level of Searles Lake at the times when those from Africa and Australia were low. The curves are modified from Smith (1968), Gasse (1977), and Bowler *et al.* (1976).

of clay-rich dunes. The peak of aridity in Australia occurred around 18 000 to 16 000 BP (approximately the same time as the glacial maximum), when gypsum and clay dunes were constructed on the eastern margins of lakes simultaneously with expansion of desert linear dunes. By about 13 000 BP, aridity was declining, and the dunes became stabilized.

The glacial–pluvial problem

One of the greatest problems of the Quaternary is to assess the question of what desert areas were like during glacial phases. Were they subjected to pluvials in glacials or interglacials? As already noted (see p. 8) the classic model sees glacials and pluvials as being synchronous, the shorelines of pluvial lakes appearing, in the western USA, to be tied in with glacial moraines.

In East Africa the work of Wayland, Nilsson, and L. S. B. Leakey established the existence of a variable number of pluvials. Leakey's work showed that there were four main pluvials, Kageran, Kamasian, Kanjeran, and Gamblian. These were, he thought, followed by two post-pluvial wet phases, the Nakuran and the Makalian. For many people these four main pluvials, on the basis of palaeontological and archaeological evidence, were seen as being broadly correlated with the four classic glaciations of the Alps proposed by the Penck and Brückner model, and the sequence became applied almost throughout the African continent.

There are various reasons for doubting and even rejecting this simple model. There are three theoretical points which suggest that full glacials may in fact have been drier: an eustatic drop in sea-level would lead to greater continentality of climate and thus greater aridity; a drop in sea-level and an extension of sea-ice would lead to less evaporation from the ocean surface, leading to less rain; the cooling of the oceans by an average of about 5 °C would lead to less evaporation and fewer cyclones, with the same result. Moreover, reservations produced on these grounds are substantiated in certain areas by sedimentological and geomorphological evidence. First, during the Last Glacial, siltation occurred in the middle courses of many great tropical rivers like the Nile, Senegal, Indus, Ganga, and Narbada, and the rivers appeared to have had insufficient discharge to move their load. Second, deep-sea cores taken from the South Atlantic off Brazil contain much more feldspar (25–60 per cent) in the Late Pleistocene than in the Holocene (17–20 per cent) suggesting that chemical weathering processes were less intense in the Late Pleistocene, possibly because of a reduction in rainfall.

Bonatti and Gartner (1973) studied a core from the central Caribbean, and employed the ratio of kaolin to quartz to ascertain past precipitation records, arguing that relatively high proportions of kaolin would occur during wetter phases when chemical weathering was more intense, and that high quartz levels would be characteristic of dry phases. On this basis they stated (p. 564) 'During Pleistocene cold stages (ice ages) conditions of relatively high aridity prevailed in the Caribbean basin, while more humid conditions prevailed during the interglacials.'

Likewise, Rossignol-Strick and Duzer (1980) have analysed cores from off Senegal in West Africa for their pollen content and report: 'From 22 500 to 19 000 BP, more general aridity than today is indicated by very small amount of tropical pollen and moderate amount of Mediterranean pollen. From 19 000 to 12 500 BP, a very arid phase brought more pollen from north of the Sahara, and eliminated the tropical pollen input.'

It needs to be stated that such aridity was not characteristic of this time further north off Morocco (Diester-Haass, 1980). In an analysis of the sedimentology of the Atlantic cores off West Africa Sarnthein and Koopman (1980: 247) find that at 18 000 BP river-borne sediments are completely absent along the whole African continental margin from some 10–12° to over 27° N, but that the input of aeolian silt was greatly expanded in both area and quantity compared to the present. This is confirmed by Kolla *et al.* (1979) who report that in Holocene sediments of the eastern Equatorial Atlantic a band of high percentage quartz in ocean sediments exists directly off the present Sahara desert and Sahel region and reflects the trade-wind transport of dusts from these semi-arid regions. During the last glacial maximum they believe that this high quartz band expanded southward by about 8° of latitude. An attempt to extend this sort of evidence back further into the Pleistocene has been made by Parmenter and Folger (1974) who examined the content of biogenic detritus in cores from the equatorial Atlantic and compared it with the oxygen isotope temperature record. They concluded not only that aridity and low temperatures went together but that this relation goes back at least 1.8 million years.

In their study of quartz in cores from the northern Indian Ocean, Kolla and Biscaye (1977) found that around 18 000 years BP high amounts of quartz were transported in the form of atmospheric dusts from the Arabian and Australian deserts into the adjacent ocean. During the Last Glacial Prell *et al.* (1980) believe that up-welling in the Arabian Sea was decreased and salinities in the Bay of Bengal increased, suggestive of a weakened south-west monsoon. A similar picture emerges for the south-west Pacific off Australia. Thiede (1979) has identified a greatly expanded zone of aeolian quartz together with relatively higher quartz concentrations at 18 000 BP.

Another form of deep-sea core evidence that has been brought to bear on this problem is the isotopic composition of planktonic Foraminifera from the Red Sea and the Gulf of Aden. Their study (Deuser *et al.* 1976) indicates that during periods of maximum continental and polar glaciation in the Late Pleistocene, the Red Sea was subject to strong evaporation. Between glacial maxima, the salinity of the Red Sea was equal to or below that of the open ocean. This suggests that high-latitude interglacial periods coincided with pluvial stages in the area.

In recent decades, the isotopic dating of groundwater reserves has provided new information on those periods when groundwater was being recharged, and those when it was not. In the Sahara the studies of Sonntag and collaborators (1980) have indicated that very little recharge occurred between 20 000 and 14 000 BP, indicating once again the existence of a period of very greatly reduced hydrological activity at the time of maximum glaciation (Fig. 3.10). Confirmatory dates come from the Sokoto basin of northern Nigeria (Geyh and Wirth 1980), where there was a low rate of recharge from 20 000 to 10 000 BP.

Another major advance in recent years in our understanding of intertropical environments has been provided by the pollen analysis of peat and lake cores. In Australia, Kershaw's (1976) analysis of material from Lynch's Crater in Queensland provides a sequence back to about 60 000 years BP. From that time to about 38 000 BP rain forest of a slightly drier type than today is implied, dominated by *Araucaria*, *Podocarpus*, and *Dacrydium*. Then follows a period, covering the last glacial maximum, dominated by sclerophyllous plants with a high representation of *Eucalyptus* and *Casuarina*. This persisted until rain-forest angiosperms became dominant between 9500 and 6000 BP. The picture which Kershaw puts forward for Queensland is confirmed by Dodson's work from the Wyrie Swamp of south-eastern South Australia. He believes on the basis of his pollen analysis (Dodson 1977: 97) that 'It probably was drier during the period from 50 000 to 10 500 BP than in the Holocene. The driest period was from 26 000 to 11 000 BP.'

Information of relevance to this issue can also be obtained from pollen encountered in ocean core sediments. For example, Hooghiemstra and Agwu (1988) found that in the equatorial Atlantic off Africa the Last Interglacial (around 124 000 years BP) was a time of greatly expanded humid tropical rainforest, with low-velocity trade winds. Once again there is a general correspondence in time between moist conditions and warmth in low-latitude situations.

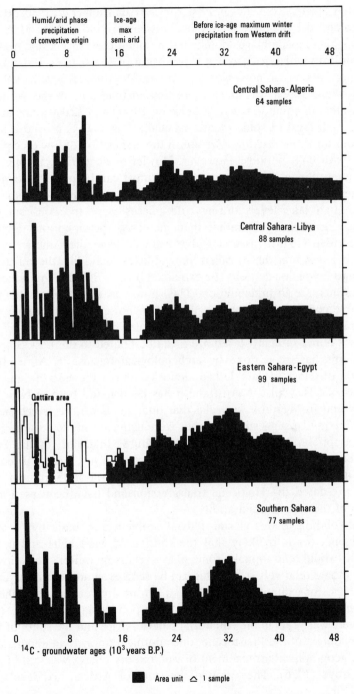

FIG. 3.10 Groundwater ages for the Sahara (after Sonntag *et al.* 1980).

Pollen spectra from sites in East Africa (Adamson *et al*. 1980: 50) point to a colder, drier climate than today between around 26 000 and 12 500 BP with highland assemblages dominated by small tree, shrub, and grass pollen, indicative of a widespread suppression of forest trees and a vegetation more open than now. Cores from Lake Victoria (Kendall 1969) show that lowland forest was much more limited than now in the White Nile headwaters in equatorial Africa between 15 000 and 12 000 years ago.

The high level of some Quaternary lakes has always proved a ticklish problem for those seeking to establish the association of low precipitation levels with cold periods. However, Bowler *et al*. (1976), talking about Australian pluvial lakes, proposed that 'most can be probably explained by increased rainfall efficiency resulting from depressed evaporation rates. High lake levels during cold phases were the consequence of reduced or unchanged, rather than increased, precipitation.' Similarly, Brakenridge (1978), discussing the evidence from the American southwest, believes that the so-called 'pluvial lakes' may, like the relict snowlines and cryogenic deposits, be explained by a 7–8 °C cooling rather than by any increase in precipitation. This, however, is still a matter of considerable controversy. These arguments do not apply to the same extent to the early Holocene (interglacial) lakes of Africa.

Fossil dunes in many desert areas can be related to the Last Glacial. In India this has been done by archaeological means, in Africa by the relation of dunes to datable lake deposits, and in the case of Senegal, to low sea-levels, whilst in Australia it has been noted both in New South Wales and in Western Australia that dune-fields can be traced beneath estuarine muds, suggesting that they were active during low stands of sea-level, supposedly of a glacio-eustatic nature. Likewise, in the Persian Gulf and the Gulf of Oman, submarine research has indicated the presence of seif dune remnants present on the sea-floor (Sarnthein 1972). They pre-dated the Holocene transgression and have been used as evidence of the Late Glacial aridity.

The solution to this glacial–pluvial problem may basically be a locational one, for it is likely that the shifting of wind belts and pressure systems would tend to make some places relatively wetter, whilst it would make others relatively drier. This can be seen even at the present day. In the 1970s, for example (see p. 198), a run of dry years in the Sahel and certain districts of north-west India corresponded in time with a run of relatively wet years in the equatorial belt of Africa and the south of India. In the context of the Pleistocene, one approach towards an understanding of the geographical arrangement of wet and dry phases was that of Street and Grove (1976). They plotted the state of African lake-levels for a series of different times for which radio-carbon dates were available, and found some striking patterns (Fig. 3.11). Thus at the glacial maximum around 18 000 years ago the northern shore of the Mediterranean was dry and dominated by an *Artemisia* steppe over wide areas (see p. 75). In that area the glacial maximum coincided with a high degree of aridity. However, some radio-carbon dates for high lake levels on the north side of the Sahara (the south side of the Mediterranean basin) would suggest that in that area the reverse may have been true. However, the tropical zones of Africa (and probably in other continents as well) witnessed low lake-levels and aridity around 18 000 years ago. Pluvial conditions came to that area together with other parts of Africa around 8000 to 9000 years ago, and would appear to have been remarkably widespread.

In various areas it seems that just as there were stadials and inter-stadials in glacial areas so there were short pluvial or non-pluvial stages in non-glacial areas. These further complicate the simple attempts to relate glacials to dryness or wetness. Indeed, there is some support for the view that in Africa at least, pluvials were of relatively short duration (often only 2000–5000 years). There is little or no support for the view of a pluvial (or inter-pluvial) spanning the whole of the Last Glacial. Thus the classic East African pluvial sequence developed during the inter-war period to correspond with the European glacial sequence should be totally abandoned.

Attention needs to be drawn to the very marked contrast in the lake-level histories of tropical Africa (and similar environments) and of the south-western USA. The American lakes were dominantly expanded at the time of the last glacial maximum, possibly because of the effects of reduced evaporation consequent upon reduced temperatures. Also of significance in explaining the contrast may have been the position of jet streams and associated storm tracks. The COHMAP members (1988) have suggested that at the glacial maximum the great Laurentide ice-sheet split the westerly jet stream into northern and southern branches over North America, and that the southern branch brought storms into the south-west USA.

Faunal and flora changes in the tropics

The massive environmental changes described so far in this chapter have led to changes in floral and fauna distribution in the tropics, and to curious or anomalous patterns. The classic example of this is the distri-bution of the crocodile in Africa. It was formerly ubiquitous in the rivers of that continent from Natal to the Nile. Today it is found in pools in the

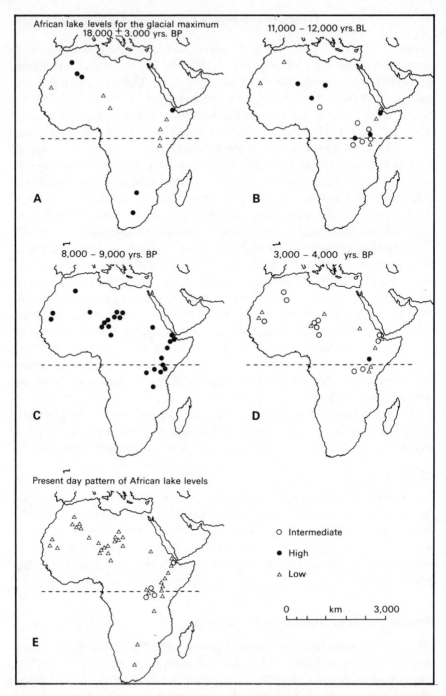

FIG. 3.11 Radio-carbon dated lake level fluctuations in Africa from *c.*18 000 BP to the present (modified after Street and Grove 1976). A. At the glacial maximum, 18 000 ± 3000 BP; B. At 11 000–12 000 BP; C. At 8000–9000 BP; D. At 3000–4000 BP; E. At the present day.

Tibesti Massif in the heart of the Sahara, 1300 km from either Niger or Nile, and clearly isolated. There is no likelihood of natural migration there across the arid Saharan wastes, given present hydrological conditions, so that pluvial conditions presumably played a role.

Another example from Africa illustrates the way in which the flora of the East African mountains has become isolated. The distinctive tree heath (*Erica arborea*) occurs in disjunct areas including the mountains of Ruwenzori, the Ethiopian mountains, the Cameroon Mountains in West Africa, and the peaks of the Canary Islands. In addition to this highly fragmented distribution in Africa, the plant has an extensive range in Europe from Iberia to the Black Sea. Once again it seems likely that post-glacial changes in temperature and rainfall have led to this position. In general, because of its height characteristics, the African continent would have been particularly severely affected by temperature depression in the glacials. The effect of a temperature depression of 5 °C would have been to bring down the main montane biomes from around 1500 m to 700 or 500 m (Moreau, 1963). Instead of occupying a large number of islands as it does now, and as it must have done in interglacials, the montane type of biome would have occupied a continuous block from Ethiopia to the Cape, with an extension to the Cameroons. The strictly lowland biomes, comprised of species today that do not enter areas above 1500 m, would, outside West Africa, have been confined to a coastal rim and to two isolated areas inland (the Sudan and the middle of the Congo Basin).

The substantial degree of change in the altitudinal zonation of vegetation on tropical mountains during cold phases can also be demonstrated from the highlands of New Guinea and the Colombian Andes of South America (Fig. 3.12). Detailed pollen analysis from numerous lakes and swamps (Flenley, 1979) shows that over the past 30 000 years the major vegetation zones have moved through as much as 1700 m. The boundary between the forest and the alpine zone above it was low before 30 000, shows a slight peak (of uncertain height and imprecise date) between 30 000 and 25 000, reaches an especially low point at 18 000 to 15 000 (more or less equivalent to the Glacial Maximum in higher latitudes), shows a steep climb as climate ameliorated between 14 000 and 9000 BP, and reaches modern altitudes or slightly above them by about 7000 BP.

Further massive changes in African biomes would have been occasioned by changes in humidity as well as of temperature. In West Africa, where the great sand ergs of the Sahara encroached as much as 500 km on the more moist coastal regions, the southward movement of the vegetation belts cannot have failed to have had powerful effects on flora and fauna. Indeed, since at present the West African rain forests nowhere

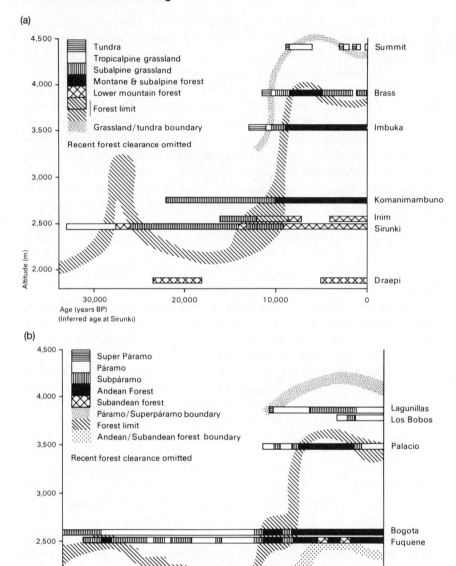

FIG. 3.12 Summary diagram of the Late Quaternary low-latitude vegetational changes (after Flenley 1979, figs. 7 and 3).
A. New Guinea Highlands
B. Colombian Andes

reach inland by so much as 500 km, if the entire system of vegetation belts had shifted south as much as did the Saharan dunes, then the whole of the West African forests would have been eliminated against the coast-line. The present richness, however, of the West African forests, and the existence of so many endemic species there, makes it virtually certain that this did not happen, but the effect of the southward advance must have been formidable. Moreau believes that it is probable that during the arid period in which the dunes were established to the south of their present limits, the savannah on the coast, now restricted to a few small areas, stretched so far as to eliminate the forests of western Nigeria, and, joining the Dahomey Gap, produced a gap of over 1100 km between the forests of Upper Guinea and the neighbourhood of the Cameroons.

An indication of the timing and degree of rain-forest disruption in West Africa is given by a consideration of pollen analyses undertaken in Lake Bosumtwi in southern Ghana (Maley 1989: Fig. 3.13). These show that at the time of the last glacial maximum, and especially between 20 000 and 15 000 BP the lake had a very low level. Moreover, the principal pollen results show that before about 9000 BP forest was largely absent in the vicinity. Indeed, between the present and c.8500 BP arboreal pollen percentages oscillated from 75 to 85 per cent, while before 9000 BP they were generally below or close to 25 per cent. During the period from 19 000 to 15 000 BP arboreal pollen percentages reached minimum values of about 4 and 5 per cent. Trees had in effect been replaced at that time by herbaceous plants, *Gramineae* and *Cyperaceae*.

Such environmental changes, by producing new environments, and by isolating species in a restricted area, lead to the development of some species which are 'endemic', that is, which are entirely confined to an area. It also follows that the longer the period of isolation, and the more effective the barriers to dispersal, the more the local species would diverge from the original population. In extreme cases endemic genera, or even families, might evolve and be restricted to a relatively small region. Speciation is complete when the divergence of two or more portions of an original population involves the reproductive system, and thus prohibits interbreeding, so that they remain distinct and may continue to diverge, even if circumstances should bring them together yet again.

Examples of such basic evolutionary processes related to Pleistocene environmental changes in the tropics can be provided by a study of the faunas of the Amazonian rain forest and the lakes of East Africa.

In the rain-forest zone of South America there are some curious speciation patterns at the present day involving birds, trees, butterflies, and

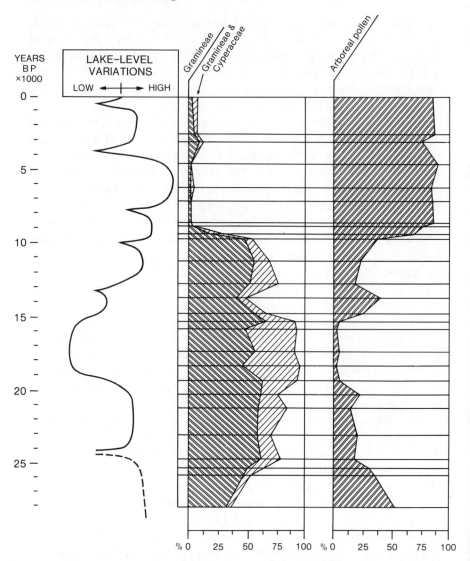

FIG. 3.13 Changes in lake level and pollen at Lake Bosumtwi in Ghana during and since the last glacial maximum. Note the low arboreal pollen content in pre-Holocene times, and the very marked expansion of arboreal pollen after *c*.10 000 years BP (after Maley 1989, figs. 2 and 3). Reprinted by permission of Kluwer Academic Publishers.

lizards (Haffer 1969, Prance 1973, Brown *et al.* 1974). These might result from changes in the nature and extent of the rain forest in the Quaternary. Areas of what is called secondary contact have been established between distinguishable forms of birds and lizards. These are recognized as stepped clines, hybridization belts, places with character displacement, or areas with narrow sympatry of closely-related fauna. In the case of Amazonia there is a striking coincidence in the location of the areas of such secondary contact between unrelated groups of birds: cracidae, tucanets, parrots, contingids, and manakins. In most sexually reproducing, outcrossing organisms, such as birds, differentiation can only occur if populations are isolated from one another (reduced gene flow). Consequently we can assume that an area in which we find secondary contact and overlapping of differentiated forms, indicates where the two forms had previously been separated. If the modern zone coincides with a discernible physiographic feature, such as a mountain range, a large lake, a river, and so on, it is probable that the feature is, and was, a barrier to gene flow. However, in the Amazon rain forest the present area of overlap between different species does not seem to coincide with any visible physical or ecological feature. Thus one can postulate that a barrier existed in the past which is no longer operative. Recent geomorphological work indicates that during parts of the Pleistocene the great Amazonian rain forest, which at present shows a considerable uniformity over areas, was fragmented into small isolated patches by greatly increased aridity (Fig. 3.14). The area of savannah was greatly extended (van der Hammen 1974). It was these small isolated patches of rain forest, concentrated in areas of favourable hydrological conditions, which enabled the differentiation to take place in the various species of the area (Vuilleumier, 1971). The return of pluvial conditions has enabled the rain forest to spread once again, and for the formerly isolated species to merge together again in zones of secondary contact.

The effectiveness of such disruption of the rain forest in leading to distinct speciation of the Amazonian forest birds would depend to a large extent on the rate at which the evolutionary process might operate. Research by Haffer (1969) suggests that under favourable circumstances the speciation process in birds may be completed in 20 000–30 000 years or less. This estimate refers mainly, though not exclusively, to passerine birds with a high reproductive rate and evolutionary potential. If this order of magnitude is approximately correct it means that within one long interpluvial the necessary degree of speciation could take place, and that during the whole length of the Quaternary the Amazonian birds may have speciated repeatedly.

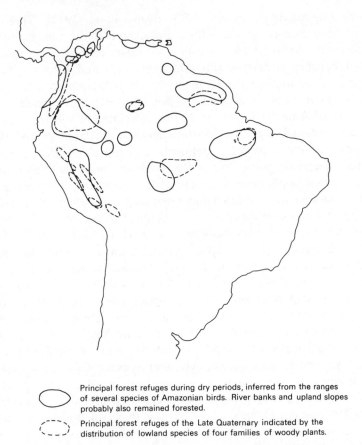

Principal forest refuges during dry periods, inferred from the ranges of several species of Amazonian birds. River banks and upland slopes probably also remained forested.

Principal forest refuges of the Late Quaternary indicated by the distribution of lowland species of four families of woody plants.

FIG. 3.14 Late Pleistocene refugias during dry periods, inferred from the range of species of Amazonian birds and woody plants (after Meggers 1975, figs. 3 and 4).

However, it needs to be pointed out that the postulated explanation of species diversity in Amazonia by means of the 'forest refugia hypothesis' outlined above, is not without controversy. In particular, Colinvaux (1987) has argued that there is little completely convincing evidence for very marked precipitation diminution in Amazonia in the Pleistocene, and he believes that a better explanation is that the species diversity results from ongoing geomorphological disturbance and fire in an area which is far from homogeneous. He concluded (p. 112):

That the Amazon basin supports the highest species diversity of any ecosystem on Earth is in part due to its great size which allows subtly varied climatic regimes in its different parts. Also this rain forest, more than the rain forests of Asia and Africa, may experience an added diversity-inducing system of disturbance because of the extraordinary erosion pressures of its great rivers driven by Andean precipitation.

African fish and changing African waters

For about sixty years work has been undertaken on the environmental changes of East Africa as they might have influenced the distribution characteristics of the water fauna of the great lakes. During pluvial or lacustral phases the rivers and lake basins of East Africa would have been linked up to a greater extent than they are now. During dry interpluvial phases on the other hand the lakes would have become partly or wholly desiccated, and the linkage of water bodies would have been reduced. Such fluctuations would lead to alternations of contact and isolation for fauna, and complete lake desiccation would have led to the extermination of many species in any one basin. By a study of present and fossil fish types, and also of crocodiles, the consequences of these changes can be identified, and this helps to unfathom the anomalies of zoo-geography (Beadle 1974).

Lake Turkana, between Kenya and Ethiopia, is now not connected with the Nile, having no outlet. On the other hand, it has a fauna very similar to that in the Nile. The explanation seems to be that Turkana once stood at a much higher level. There are shorelines to prove this. When this was so the lake could have overflowed via a gorge, now dry, to the Sobat River, and hence to the Nile. This former connection would explain the faunal similarities. There are, nevertheless, twelve endemic species of fish in Turkana, and the Nile perch has divided into two subspecies. The Nile perch also occurs in Lakes Stefanie, Abaya, and Chamo. These too were formerly connected to Turkana (Grove *et al.* 1975: 183).

A more complex situation is illustrated by Lake Kivu. Tectonic changes have led to its loss to the Nile but adoption by the Congo. Formerly, Kivu was connected with Lake Edward and the Nile via the River Ruchuru. However, in comparatively recent times this outlet for Kivu was dammed by volcanic outpourings from the Birunga Range, and so the Lake increased in size and overflowed southwards towards Lake Tanganyika. It is for this reason that Lake Kivu contains some characteristic Nile fish like the barbel (*Barbus altianalis*), even though it is no longer connected with that river.

Desiccation of the lakes explains the differences between fossil and current faunas and also helps to explain the absence of some species from some of the lake basins. For example, Lake Edward in Uganda possesses no crocodiles, even though they are found in Lake Victoria, and in the Semliki River. It is difficult to explain the fact that the crocodiles have not managed to pass beyond the Semliki Gorge to Lake Edward, but it is probable that the falls of the Gorge, and the dense forest on either side, have acted as barriers. However, in fossil beds which border the shores of the Kazinga Channel, numerous teeth, scales, and bones of crocodiles have been found. Thus these reptiles once lived in large numbers in the lake, and so it appears likely that their present absence can be explained by the desiccation of the lake, and by the natural barriers which have prevented recolonization. Violent volcanic eruptions may also have led to their destruction.

The work of Kendall (1969) and others has shown that Lake Victoria too largely dried up in the Late Pleistocene, and cores in other lake beds also indicate that many of the lakes became highly alkaline, if not actually dry, at some point. The only fish that could survive the very dry conditions of the interpluvials would be the lung-fish and the mud-fish, for they can burrow into mud and live there for long periods. It is therefore interesting that it is these two species which are most widely distributed in the Nile and the lakes at the present time, for they still live both above and below both the Murchison and the Semliki Falls. Many other fish were killed off by the aridity, unlike the lung- and mud-fish, and only partial recolonization has been possible because of the barriers presented by these two falls.

One of the most striking demonstrations of the speed with which the process of speciation occurs is provided by Lake Mabugabo, a small shallow basin of the western shore of Lake Victoria. It was disconnected from Lake Victoria about 4000 years ago by the growth of a sandbar. Three new species of *Haplochromis* have evolved during this relatively short span of time.

Elsewhere in Africa there are further zoo-geographical anomalies of interest. For instance, there are some species of fish which are common to all the major basins of the Sudan belt, Senegal, Gambia, Volta, Niger, Chad, and Nile. The fish fauna is remarkably uniform over this enormous area, which is all the more surprising when one realizes that there are now more than 1600 km of desert separating the Nile from Lake Chad. This similarity may be explained by the presence of more rivers and lakes during humid phases. These would provide the necessary linkages (Beadle 1974: 147). Likewise, it was a surprise to zoologists when French expeditions during the early years of this century discovered that the

Sahara itself supports a considerable, though a widely scattered, fresh-water vertebrate fauna. In permanent but isolated water holes from Biskra to Tibesti there are the remains of a former tropical African fish fauna such as *Tilapia zillii*, *Astotilapia* (*Haplochromis*), *Desfontainessi*, and *Clarias lazera*—species that are now very widespread in tropical Africa. They have become isolated as a result of a decrease in humidity (Beadle 1974: 157).

The general circulation of the atmosphere

Having described in the last two chapters the nature of the environmental changes of the Pleistocene (both in high and in low latitudes) it is worth considering the general atmospheric circulation pattern with which they were associated. Thermal factors play a dominant role in determining its form, so that the thermal variations provoked by the growth and decay of the great ice-sheets decisively influenced the patterns of the atmospheric circulation. For example, in theory an increased temperature gradient resulting from the presence of a greatly expanded northern hemisphere ice-sheet would result in stronger westerlies, an equatorial desplacement of major circulation features, and an intensification and shrinking of the Hadley Cell and associated Subtropical High Pressure zone. Temperature gradients would also influence the location of the transition between the Hadley Cell zone and the zone of extra-tropical Rossby Wave Circulation. It might also influence the number and position of the Rossby Waves—the dominant control of mid-latitude weather.

At the glacial maximum there would also have been a decreased thermal contrast between the two hemispheres. At present the southern hemisphere, in comparison with the northern, is much cooler and its temperature gradient greater. This is because of the varying amounts and distribution of land and ocean in the two hemispheres. In the southern hemisphere the stronger temperature gradient produces a more intense circulation, and is probably largely responsible for the asymmetry that exists whereby the meteorological equator (or Intertropical Convergence Zone—ITCZ) lies in the northern hemisphere. During a glacial phase intense continental glaciation in the northern hemisphere should have led to a displacement of this meteorological equator to a position more coincident with the geographical equator (i.e. southwards).

Furthermore surface temperature changes would effect precipitation levels through their influence on evaporation rates and on the stability of the atmospheric column.

What then is the likely situation round about 18 000 years ago?

Nicholson and Flohn (1981) propose that the large changes in the northern hemisphere and the smaller changes in the southern hemisphere would have reduced the thermal contrast between the hemispheres. Moreover, the great continental ice-sheets in the north would have increased the hemispheric temperature gradients causing circulation features to be displaced equatorwards, and intensifying the subtropical high-pressure zone. The result would have been a greater degree of westerly flow and mid-latitude cyclones over an area like North Africa. A corresponding displacement of the subtropical high pressure zone would have displaced the aridity maximum into West Africa, while increased oceanic upwelling resulting from the stronger subtropical high pressure (together with lower sea-surface temperatures) would both have reduced evaporation levels and the energy available for convection and storm formation. Furthermore, the decreased thermal contrast between the hemispheres would have displaced the ITCZ southwards, disrupting the annual march of the monsoon.

Selected reading

There are very few books that deal specifically with the lower latitudes in the Pleistocene. However, J. R. Flenley (1979), *Equatorial Rain Forest: a Geological History*, is immensely informative on the oscillations that affected that particular environment. Also of great value, especially for its information on drier areas, is M. A. J. Williams and H. Faure (1980) (eds.), *The Sahara and the Nile*.

The Quaternary in arid India is discussed in B. Allchin (1978), *Prehistory and Palaeogeography of the Great Indian Desert*. A comparable volume on Australia is D. J. Mulvaney and J. Golson (1971) (eds.), *Aboriginal man and environment in Australia*. A fascinating discussion of the effects of African environmental changes can be found in L. C. Beadle (1974), *The inland waters of tropical Africa: an introduction to tropical limnology*.

A summary of available information on Quaternary changes in southern Africa is provided by J. Deacon and N. Lancaster (1988), *Late Quaternary Palaeoenvironments of Southern Africa*.

4

Environmental Change in Post-glacial Times

A stable Holocene?

The ending of the Last Glacial period was not the end of substantial environmental change, though from time to time the existence of any major changes of climate has been doubted. A hydrologist, Raikes (1967), has put forward the hypothesis that 'From at latest 7000 BC, and possibly earlier, the world-wide climate has been essentially the same as that of today.' He believed that with the single exception of localized changes induced by large-scale, eustatic–isostatic marine transgression, all changes since about 7000 BC have been 'local, random, and of short duration'.

Raikes was especially sceptical about pollen, zoological, and historical evidence for Holocene climatic change, and pointed out correctly that man had influenced vegetation, that animals are often poor ecological indicators, and that the evidence for a dense population in the Indus Valley in prehistoric times (the Harappan civilization) could be explained by non-climatic factors. Nevertheless, Raikes ignores much of the evidence that has been put forward in many countries for Holocene climatic changes.

There are a whole series of faunal and floral remains which indicate, for instance, the relatively higher temperature conditions of the Hypsithermal interval (p. 156), there are a large number of radio-carbon dates which illustrate the fluctuations of glaciers (p. 162), there are a large number of dates of lacustrine sediments indicating Holocene lacustral phases in the tropics and subtropics (p. 112), and there are the more recent meteorological and hydrological records which indicate considerable changes and fluctuations in the last two centuries (Chapter 5). Such evidence taken as a whole shows quite clearly that the concept of a stable Holocene environment is quite untenable. This chapter is concerned with both the evidence for Holocene changes, and with the nature and influence of the changes themselves. It starts off by considering the nature and effects of the transition from a glacial to a non-glacial environment, and then it considers some of the major events of the Holocene itself.

The transition from Late Glacial times

The maximum of the Last Glacial, as we have already seen, occurred at around 18 000 years BP. Studies of the oxygen isotope composition of deep-sea cores suggest that deglaciation started at around 15 000 to 14 500 years BP in the North Atlantic and at 16 500 to 13 000 years BP in the Southern Ocean (Bard *et al.* 1990). The years between the glacial maximum and the beginning of the Holocene are usually termed the Late Glacial, and they were marked by various minor stadials and interstadials. In Europe, for example, a number of distinct cool phases interrupted the retreat of the Scandinavian ice-margins, and there were a number of short interstadials during which retreat was accelerated.

Some attempts to record the fluctuations of the Late Glacial are shown in Fig. 4.1. This shows the oscillations of glaciers in the Russian Plain, and the trend of temperatures in the English Midlands. An event called the Allerød figures large in all three curves, and precedes the great increase in temperature evident at about 10 000 BP.

The character, identification and correlation of the Late Glacial interstadials is, however, a matter which is still in need of clarification. The classic threefold division (Table 4.1) into two cold zones (I and III) separated by a milder interstadial (II) emanates from a type section at Allerød, north of Copenhagen, where an organic lake mud was exposed between an upper and lower clay, both of which contained pollen of *Dryas octopetula*, a plant tolerant of severely cold climates. The lake muds contained a cool temperate flora including some tree birches, and the milder stage which they represented was called the Allerød interstadial. The interstadial itself, and the following Younger Dryas temperature reversal, are sometimes called the Allerød oscillation. A somewhat earlier minor interstadial, the Bölling, has also been recognized in parts of Europe. On the basis of flora studies attempts have been made to reconstruct the nature of the European landscape as it probably appeared in the Allerød interstadial. It is worth comparing this figure (Fig. 4.2) with that for the maximum of the Last Glaciation (Fig. 2.9). The ice-sheets have been greatly reduced in extent in comparison with their maxima, but sea-levels are still low, Britain is connected with the con-

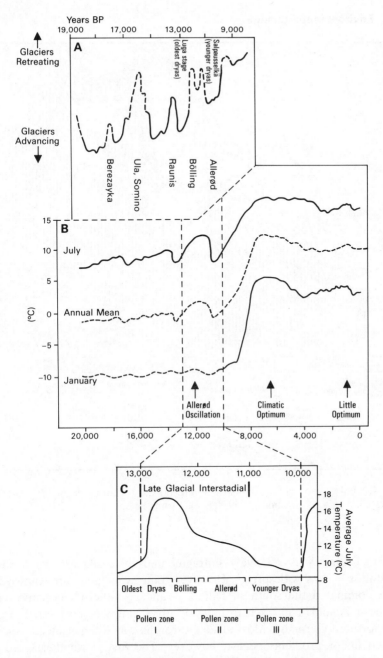

FIG. 4.1 Late Glacial climatic fluctuations revealed from various sources:
A. Glacial oscillations in the Russian Plain.
B. Estimated trend of the summer and winter mean temperatures during the past 20 000 years in the English Midlands (after Manley 1964).
C. Average July temperatures for Central England based on Coleoptera remains (after Coope 1975).

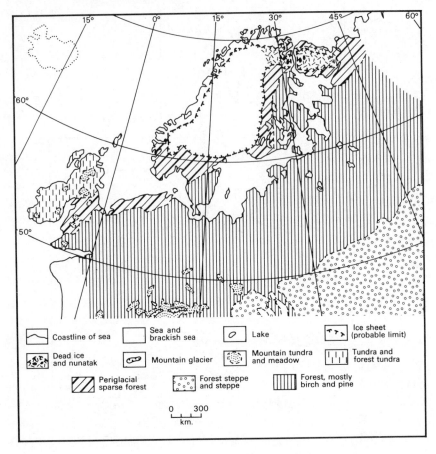

FIG. 4.2 Palaeogeographic reconstruction of northern Europe during the Allerød Interstadial (after Gerasimov 1969, fig. 4). Compare figs. 2.9 and 2.15.

tinent, Denmark is relatively unfragmented into islands, and tundra vegetation is less widespread. However, pine woodlands are known to have dominated the southern half of France, southern Germany, and northern Poland, and birch to have occupied much of northern France and northern Germany. Most of Fennoscandia was still glaciated.

Two major problems persist, however. The first is whether there is evidence for the Allerød oscillation outside Europe. For example, while there appeared to have been various oscillations in ice-margins in North America in Late Glacial times, Mercer (1969) found it difficult to prove their contemporaneity with the Allerød oscillation, though pollen-analysis in East Africa and South America (Coetzee 1964, van der Hammen 1974) does suggest a direct comparability with the European sequence. High resolution ocean core evidence from as far away as the western Pacific tends to suggest that the Younger Dryas cooling event was a world-wide phenomenon (Kudrass *et al.* 1991).

A second problem is posed by a comparison of the palynological and the Coleopteran evidence in Britain (Coope 1975). The Coleoptera do support the general interpretation of a Late Glacial climatic oscillation, but they also indicate that perhaps its thermal maximum differed both in timing and intensity from that inferred from the floral evidence. The main phase of warmth suggested by Coope is between 13 000 and 11 000 BP (Fig. 4.1C), with the peak occurring during Zone I time. He believes that this 'Late Glacial interstadial' cannot be correlated with the Bölling or Allerød oscillations of continental authors, since only one oscillation is indicated by the Coleoptera and this does not correspond with either. The interstadial began well before, and also reached its thermal maximum before, the generally accepted date of the Bölling oscillation (Zone Ib). Furthermore, the climate indicated by the Coleoptera during the thermal maximum was apparently warm enough to support a mixed deciduous forest, but the pollen spectra at this time indicate a much more open country which has been interpreted as having a tundra-like climate. This apparent difference between the climates indicated by the insects and the plants may perhaps be explicable by the greater facility and speed with which the insects would respond to rapid climatic fluctuations in comparison to, say, birch trees. The arrival time of birch trees in a particular area is likely to be more closely related to the distance of the glacial refuge from which they spread when the climate ameliorated, rather than to the time of the onset of climatic conditions acceptable to them.

Lowe and Gray (1980) have attempted to bring some order to the confusion surrounding the events of Late Glacial times, and have produced a table of events for north-west Europe (Table 4.1). Their Younger Dryas stadial was a time when glaciers readvanced in Scotland—the so-called Loch Lomond Readvance (Sissons 1976)—and it may have been caused by a rapid cooling of the ocean caused by the breaking-up of large ice-shelves from the Arctic (Ruddiman and McIntyre 1981), or by a sudden influx of meltwater from the Laurentide ice-sheet into the North Atlantic via the St. Lawrence system causing salinity changes in the ocean. These in turn affected the sea-water density, currents, and thus climate (Broecker and Denton 1989).

The Loch Lomond stadial saw cold conditions return to Britain, and glaciers reformed in many upland areas. Over Rannoch Moor in the

Table 4.1 Lowe and Gray's climatostratigraphic subdivision and terminology for the Late Glacial of north-west Europe

Radiocarbon years BP	Climatostratigraphic Units	
	FLANDRIAN INTERGLACIAL	
10 000	---	
	transition	} Time of thermal improvement
10 500	---	
	YOUNGER DRYAS STADIAL	
11 000	---	
	transition	} Time of thermal decline
12 000	---	
	LATE GLACIAL INTERSTADIAL	
13 000	---	
	transition	} Time of thermal improvement
14 000	---	
	LATE WEICHSELIAN/LATE DEVENSIAN/LATE MIDLANDIAN MAIN GLACIAL	

(Modified after Lowe and Gray, 1980, Table 2)

north-west Highlands of Scotland the ice thickness may locally have exceeded 400 m. Smaller ice-caps developed in the Grampians, on Mull and Skye, in the Southern Uplands of Scotland, in the English Lake District, in the mountains of North Wales, and in the Brecon Beacons. The end moraines from this short sharp cold period are still very clear in the present landscape, as are a whole range of periglacial landforms that developed down-valley from them, including blockfields, solifluction lobes, screes, rock glaciers, and protalus ramparts (Sissons 1980).

The Younger Dryas climatic deterioration may have come to an abrupt end about 10 700 years ago. From a study of ice-core heavy isotope and dust concentrations in Greenland, Dansgaard *et al.* (1989) suggest that in southern Greenland a 7 °C warming may have been completed in just 50 years.

After the Allerød oscillation, whatever its exact status, the traditional division is made between the Late Glacial (Pleistocene) and the post-glacial (Holocene, Recent, or Flandrian). The classic terminology of the Holocene was established by two Scandinavians, Blytt and Sernander, who, in the late nineteenth and early twentieth centuries, introduced the terms Boreal, Atlantic, Sub-boreal, and Sub-Atlantic for the various environmental fluctuations that took place. These terms are still widely used for subdivisions of the Holocene (Table 4.2). The fluctuations which Blytt and Sernander and subsequent workers have established, though they have sometimes been contested by workers who believe the sequence of events has been less complex, with only a simple climatic

Table 4.2 The classic European Holocene sequence

Period	Zone number	Blytt–Sernander Zone name	Radio-carbon years BP
Post-Glacial	IX	Sub-Atlantic	post 2450
	VIII	Sub-Boreal	2450–4450
	VII	Atlantic	4450–7450
	VI	Late Boreal	5450–8450
	V	Early Boreal	8450–9450
	IV	Pre-Boreal	9450–10 250
Late Glacial	III	Younger Dryas	10 250–11 350
	II	Allerød	11 350–12 150
	Ic	Older Dryas	12 150–12 350
	Ib	Bölling	12 350–12 750
	Ia	Oldest Dryas	

(After Embleton and King, 1967, Table 3, p. 14, and other sources)

optimum and then deterioration, have been remarkably durable. Nevertheless, it has to be remembered that it is essentially a scheme of vegetation change, and not a scheme of climate change. The bulk of the evidence used by Blytt and Sernander was provided by plant remains, and especially macro-fossils. Thus is terms of climatic reconstruction certain inaccuracies may creep in because of non-climatic factors affecting vegetational associations. Such factors include the intervention of man (see p. 150), the progressive evolution of soils through time (see p. 151), and the passage from pioneer to climax species during the course of succession. As already noted with regard to the Allerød oscillation, plants may not be able to respond with great alacrity to climatic change: migration and colonization take time. Consequently, although the terminology may still be used, there have been substantial changes in interpretation of the classic Blytt–Sernander model in recent years.

For man the time of waning of the ice-sheets brought a very great and apparently rapid alteration of the environment throughout the world. For the most part these changes did, according to Sauer (1948) and others, offer increased opportunities, though in some areas a reduction in rainfall may have reduced the possibilities of life in marginal desert areas, while increasing tree-cover in northern Europe may have adversely affected the Upper Palaeolithic hunting communities.

The recession of the ice-sheets uncovered millions of square kilometres of land in higher latitudes, which thus became available for human occupation and colonization. The world population of migrant waterfowl possibly increased hugely with the addition of the great breeding and feeding grounds of the northern hemisphere. Further, the transgression of the sea, resulting from the melting of the ice-caps, though it may have led to a considerable inundation of the continental shelves (see p. 235) did in some ways improve the sea-shores for man. A more diversified and sinuous coastline would give a wider choice of environments, while the drowning of valleys to give rias (sinuous inlets of sea) would tend to lead to an increase in tidal ranges which would be most valuable for food-collecting peoples. The formation of quiet landlocked bodies of water might also provide a favourable setting for early trials in navigation. Many alluvial valleys grew in length and breadth and offered optimal sites for plant growth. As Sauer (1948: 258) has put it:

A new world took form, developing the physical geography that we know. The period was one of maximum opportunity for progressive and adventurous man. The higher latitudes were open to his colonization. In mild lands rich valleys invited his ingenuity. It was above all a rarely favourable time for man to test out the possibilities of waterside life, and especially of living along fresh water.

Environmental change and cultural transition

The transition from the Pleistocene to the Holocene also saw the transition from an Upper Palaeolithic technology to a Mesolithic and Microlithic technology. J. G. D. Clark believes (1970: 90) that in terms of this technological change, 'Beyond a doubt the most important factors involved were the complex changes in the physical environment that marked the onset of Neothermal conditions at the close of the Ice Age and adjustments to these made by the hunter–fishers themselves.'

The main environmental change involved was the dramatic change in temperate Europe whereby forest trees were able to expand from their refuge areas and to colonize the relatively open spaces of the Late Glacial landscape. Such a change, resulting from the increasing temperatures of the Holocene, was by no means advantageous to the European hunting peoples. Indeed, the reverse may well have been the case, for the late Magdalenians and their counterparts on the North European Plain were adapted to hunting animals in a relatively open and unforested environment. This environment was one which was highly favourable to herds of reindeer, bison, and horse, and the outstanding development of certain forms—for example, the Giant Irish Deer (*Megaceros giganteus*) attained an antler span of up to 3.4 m—emphasize how suitable grazing conditions must have been. The onset of forested conditions in post-glacial times must have been little less than catastrophic. The spread of forest reduced the density of grazing animals and meant that, instead of being hunted in herds, they had to be killed individually by the hunters in the forest. This reduction in the supplies of easily killed game probably led to the intensification of methods which characterized the change from an Upper

Palaeolithic method of hunting to that of the Mesolithic. The bow came into much more general use, and the microlith used to barb and tip arrows became a veritable symbol of the Mesolithic phase.

Clark (1970: 96) has written that

The coincidence between the passing of Late-Glacial and Late-Pluvial climate and the emergence of Mesolithic societies is more than merely temporal: it must surely have been causal, even if the precise links are not always apparent. Traditions formed under ecological conditions that had passed away had either to disappear or to undergo the modifications needed to accommodate them to new ones.

It was only in the far north of Europe, where conditions of temperature became markedly more favourable to human activities, that post-glacial warmth brought very marked advantages, so that there was in mesolithic times a dramatic expansion of people into Scotland, northern Ireland, and as far as the White Sea coasts in Norway and Finnmark.

For some considerable time it has been suggested that supposed climatic desiccation in the Near East at the end of the Last Glaciation, which has already been referred to, might have played a role in man's adoption of a food-producing economy. The archaeologist Gordon Childe (1954) expressed the opinion that 'Enforced concentration by the banks of streams and shrinkage of springs would entail a more intensive search for means of nourishment. Animals and man would be herded together in oases that were becoming increasingly isolated by desert tracts.' Likewise, East (1938), a historical geographer, supposed that the response of man to the desiccation of the Afrasian grasslands as storm tracks moved north could have been in four main ways: he could emigrate to more congenial areas; he could stay where he was and if he survived the hard conditions he would have to modify his life; he could think up an entirely new means of livelihood, through cultivation and animal husbandry; and he could explore the possibilities of the formerly neglected riverine lands.

Certain researches in the mountains north of Mesopotamia throw light on this relation between early food production and environment. Wright (1968) and his co-workers have shown that the high parts of the Zagros Mountains were glaciated in Pleistocene times and that the snow-line must have been 1200–1800 m lower than today. Beneath the snow-line, conditions would have been cold, and the main vegetation association would have been a bleak steppe. The environment was too cold for man to live in the mountains between 28 000 BP and about 13 000 BP. The environmental change from a cool steppe to a warm oak-pistachio savannah about 11 000 years ago, as determined by pollen and lake sediment studies, occurred at the same time as the first manifestations of domestications of plants and animals. Emmer and barley probably arrived in the area at this time, following the climatic amelioration, and man was able to live elsewhere than in caves. Wright (1968) has written:

Although I have always felt that cultural evolution—gradual refinement of tools and techniques for controlling the environment—is a stronger force than climatic determinism in the development of early cultures, the chronological coincidence of important environmental and cultural change in this area during initial phases of domestication is now well enough documented that it cannot be ignored. A much greater problem, of course, will be to prove that the environmental change was the cause of the cultural revolution.

Solecki (1963), on the other hand, has been more bold: 'A kind of trigger was needed to make him [man] depart from being a perpetual "lotus-eater" forever dependent upon hunting and gathering for his existence. In the area under discussion the rise in temperature could have served as just such an indirect stimulus.' This theme has been investigated in greater depth by Butzer (1972).

In other parts of Asia there are also records of an end to the hiatus which separated the Mousterian and the Upper Palaeolithic from the Mesolithic. In Rajasthan (India), for example, the sudden appearance of large numbers of microliths seems to mark the end of a marked phase of pleistocene aridity, and judging from the evidence of freshwater lakes the humid phase started around 9000 to 10000 years BP (see p. 99). Also Solecki has related how after 11000–12000 BP 'there was a rash of Mesolithic settlements'. As in the Near East, 'they blossomed over what is now Soviet Asia like desert flowers after a rain, taking advantage of an apparent cultural vacuum . . .'.

The great extinction problem of Late Glacial and Early Holocene times

Another major event associated with the transition from Late Glacial to post-glacial times was the demise of many of the world's mammals. As Alfred Wallace put it in 1876, 'We live in a zoologically impoverished world, from which all the hugest, and fiercest and strangest forms have recently disappeared.' The great geologists and zoologists, men like Darwin, Lyell, Owen, and Cuvier, were fascinated by this enduring problem of the great Pleistocene extinctions or 'overkill' of the Earth's mammalian population (Martin 1966). For recent workers, however, the problem goes beyond explaining the massive reduction in species, par-

Table 4.3 The dates of the major Pleistocene and Holocene mammalian extinctions

Location	Date (BP)
North America	11 000
South America	10 000
Northern Eurasia	13 000–11 000
Australia	13 000
West Indies	Mid-post-glacial
Madagascar	800
New Zealand	900
Africa and South-East Asia	40 000–50 000

(After Martin, 1967)

ticularly of big-game mammalian and avian herbivores *per se*: the concern is why the biggest wave of extinctions occurred only once, and at a time within the last 15 000 years, except, that is, in Africa and South-East Asia, where the extinction probably occurred before 40 000 to 50 000 years BP.

Table 4.3 gives the dates for the start of the major extinctions as proposed by Martin (1967).

The real cause of the controversy is whether the extinctions resulted dominantly from the actions of man the hunter or whether they were caused by the sudden environmental changes, which it is known, occurred around 11 000 BP.

There is a good deal of weighty evidence to support the anthropogenic hypothesis (Krantz 1970). First, outside continental Africa and South-East Asia, massive extinction is unknown before the earliest arrival of prehistoric man. In America, for instance, there is as yet little conclusive evidence that man had arrived from Asia via the Bering Land Bridge before 12 000 to 13 000 BP, and certainly, if he had, his numbers were small or relatively localized compared to the time of the so-called Clovis hunters (11 000–12 000 BP). The extinctions in North America thus seem to coincide in time with the arrival of man in sufficient quantity and with sufficient technological skill in making suitable artefacts (the bifacial Clovis blades) to be able to kill large numbers of animals. Equally, early man and his dog, the dingo, only arrived in Australia at a time of low sea-level in the latter part of the Würm glacial period. In Africa the massive extinction of animals coincides with the final development of Acheulean hunting cultures which were widespread throughout that continent. In Europe, the efficiency of Upper Palaeolithic hunters is attested by such sites as Solutré in France, where a late-Perigordian level is estimated to contain the remains of over 100 000 horses. The restricted orientation of the subsistence economy of these people is evinced by the concentration on animal representation in their art form to the exclusion of virtually all other naturalistic motifs, with the exception of women.

The hunting activities of man would be made more efficient as he developed more sophisticated tools and learnt to use fire for driving game. Moreover, as Darwin noted on his voyage of the *Beagle*, many beasts which have not known man are remarkably tame and stupid in his presence. It would probably have taken many animal species a considerable time to acquire the knowledge to flee or seek concealment at the sight or scent of man. In addition to the direct effects of hunting activities in reducing their numbers, man may have competed with mammals for a particular food or water supply.

Certain objections have been levelled against the climatic change hypothesis and these tend to support the anthropogenic model. First, it has been maintained that changes in climatic zones are generally sufficiently gradual for beasts to be able to shift with the shifting vegetation and climatic zones of their choice. Second, the climatic changes associated with the multiple glaciations, interglaciations, pluvials, and interpluvials do not seem to have caused the same abrupt degree of elimination.

Nevertheless, the climatic change model of Pleistocene extinctions still has its adherents, and certain major arguments can be advanced in its favour. Guilday (1967), for instance, has written that '. . . in the absence of man much the same pattern of extinction would have occurred' and he proposed that in the Western Hemisphere man 'may have delivered no more than the final *coup de grâce* to isolated remnants already doomed by rapid post-glacial environmental changes'. In other words the view is being put forward that some of the environmental changes were rapid, and that man's arrival coincided very largely with them.

A second argument that is raised against any simple anthropogenic model is that in some localities boundaries of a natural type, such as high mountain ranges, did not enable beasts to migrate with the gradually changing conditions (or more rapidly changing conditions) resulting from climatic change. The relatively unaltered state of the African fauna, which still has a fairly large number of large mammals, is due, according to this point of view, to the fact that the African biota is not and was not greatly restricted by any insuperable geographic barrier.

Darwin, in his *Origin of Species* (1936 edn.: 290), suggested in a similar manner that 'As the cold came on, and as each southern zone became fitted for the inhabitants of the north, these would take the places of the former inhabitants of the temperate regions. The latter, at the same time, would travel further and further southward, unless they were stopped by barriers, in which case they would perish.' In Europe the great mountain ranges from the Pyrenees to the Carpathians could have constituted such a barrier, as would the Mediterranean Sea.

Another possible way in which the climatic changes could lead to extinctions is through their effects on mammalian mating habits. Animals with inflexible mating habits are often restricted in range by the season in which their young are born. Slaughter (1967) has maintained that animals with gestation periods of several months would be adversely affected by lengthened winter seasons such as were characteristic of the period 11 000–9500 BP. They would tend to mate in the autumn and the offspring would then be likely to arrive when there was no grass to sustain them. They would then perish. Animals with relatively short gestation periods (usually, but not always, smaller beasts) tend to await clear signals for optimal weather before mating. It is therefore perhaps significant that it was the larger mammals that were most reduced in numbers during the period of Pleistocene overkill.

Another mechanism of extinction worthy of consideration is disease. This would be especially effective in the case of large mammals because of their lower reproduction rates which would make recovery in population numbers much slower. This again ties in with the relatively higher proportion of large mammals which became extinct. It has been suggested that during glacials animals would be split into discrete groups cut off by ice-sheets, but that as the ice melted (before 11 000 BP in many areas), contacts between groups would once again be opened up and diseases to which immunity might have been lost because of isolation would spread rapidly. An analogous situation is that by which early European explorers introduced virulent and unaccustomed diseases into the Americas with devastating effects on the native population. Thus any time of rapid population movement resulting from a marked environmental change such as that which marked the end of the Pleistocene and the start of Holocene could lead to an increased disease risk.

The detailed dating of the European megafauna's extinction lends further support to the climatic hypothesis (Reed 1970). The Eurasiatic boreal mammals, mammoth, woolly rhinoceros, musk-ox, and steppe bison, were associated with and adapted to the cold steppe which was the dominant environment in northern Europe during the glacial phases of the Würm–Weichselian glaciation. Each of these forms, especially the mammoth and the steppe bison, had been hunted by man for several tens of thousands of years, yet managed to survive through the Last Glacial. They appear to have disappeared within a space of a few hundred years—

woolly mammoth, woolly rhinoceros, steppe bison, together with horse, saiga, and reindeer, were still present in parts of south-western France during the Bölling (13000–12500 BP), but woolly mammoth, woolly rhinoceros, musk-ox, steppe bison, giant deer, and the various cave predators were missing from the Western European fauna of the Younger Dryas (around 10800–10150 BP) when the climate and general environment were otherwise broadly similar to those of the Bölling. Thus the disappearance of this group from western Europe can be pinpointed fairly accurately to the warm period of the Allerød, with its restriction and near disappearance of their habitat. Moreover, as the radio-carbon dates for early man in countries like Australia are pushed back it becomes increasingly clear that humans and several species of megafauna were living together for quite long periods, thereby undermining the notion of rapid overkill (Gillespie *et al*. 1978). Likewise, Grayson (1977) has added to the doubts expressed about the anthropogenic overkill hypothesis. He suggests that when the fossil record in North America is examined in detail for the terminal Pleistocene it is found that the rate of generic extinction of the great mammals is much the same as for other classes of vertebrates. In other words the extinction of the types of beast that man would have slaughtered for food is comparable to the extinction of types that he would not have hunted for food.

Thus the role of man in the great Late Pleistocene extinction is still a matter of debate. The problem is complicated because certain major cultural changes in man may have occurred in response to climatic change, including the adoption of mesolithic technology. The cultural changes in turn may have assisted in the extinction process, and increasing numbers of technologically competent humans may have delivered the final *coup de grâce* to isolated remnants already doomed by post-glacial environmental changes.

The vegetation and climatic conditions of Holocene Britain

In spite of the role of man in altering vegetational characteristics in the European Holocene, the role of climatic changes in causing some of the observed patterns is considerable.

After the sub-Arctic conditions of the Pre-Boreal (Table 4.1) the Boreal period saw a marked increase in warmth, and conditions were probably relatively dry and continental compared to those of today. The immigrants of the early Boreal included hazel (*Corylus avellana*), and this probably created a scrub beneath a canopy of pine and birch, or else, in some localities, a pure hazel woodland. In the Late Boreal more warmth-loving plants such as elm (*Ulmus*) and oak (*Quercus*) appeared in greater numbers, so that the Late Boreal appears to have been the last time at which pines grew generally in England on soils of all types. Later in England, Wales, and Ireland pine seems only to have grown locally, probably on the poorer soils on which it is familiar today. At the transition between the Boreal and the Atlantic elm and oak spread further, and warmth-loving plants such as the lime (*Tilia*) appear. Rather dry conditions, which had led to the reworking of deposits marginal to lakes and to the drying-out of some mire surfaces, were replaced at the very end of the Boreal by wetter conditions conducive to the growth of wet, peat-forming communities of *Eriophorum* and *Sphagnum*. In the Atlantic itself, when warm, wet conditions prevailed in Britain, the extreme oceanity of climate brought about replacement of the deciduous forest on flatter surfaces over about 360 m in altitude. In their place blanket-peat-forming communities developed, though on steeper slopes with well-drained soils deciduous forests were able to extend to at least 760 m, and in Ireland the forest extended over great lowland areas now covered by blanket-peat and raised bogs. At this period grassland was rare except above around 900 m, and open habitats must have been greatly restricted to such specialized niches as unstable scree, limestone pavement, and coastal shingle, sand, and silt. In addition, over much of England pine practically disappeared, and the forest consisted of oak, elm, alder, and lime, with birch in the north and west but little in the south and east. This was the period of the widest spread of *Tilia*, the most exacting in its climatic requirements of the British forest trees. Most forests were of *Quercetum mixtum* type, in places pure oakland, but in places a more complex deciduous mosaic with elm and lime. Pine was restricted to the Scottish highlands.

The warming of climate in post-glacial times thus seems to have set off the successive return of species of tree with different tolerances of cold and different powers of colonization. At first there were a relatively few arctic tree species in Britain; later larger numbers of species of more tolerant type arrived and the woodlands became more complex in their composition (Rackham 1980). The various trees did not, in general, stream in from the south in successive waves of massive invasion. Rather, each crept up in small numbers and became widespread (though rare) long before it increased to its full abundance. Both the date of a tree's first known presence and the date at which it received full abundance are correlated with its 'arcticity' as measured by its present northern limit in Europe. In general the lag between the two dates was longer for the later

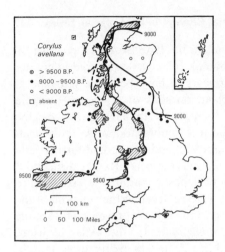

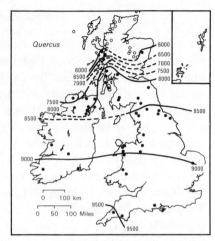

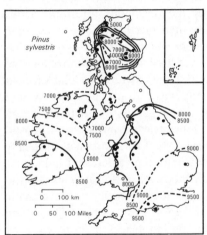

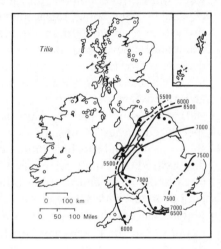

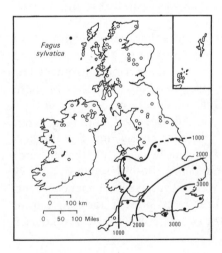

arrivals such as ash and beech, for these had to displace existing trees and not merely to occupy vacant ground.

Birks (1990) has synthesized the broad-scale changes in vegetation that occurred across Europe in response to Holocene warming. His work is based on consideration of dated pollen diagrams from many sites. At 12 000 BP *Pinus* was mainly in southern and eastern Europe, but by 6000 BP was abundant in northern, central, and Mediterranean Europe but absent from much of the Western European lowlands. *Quercus* spread progressively northwards from southern Europe and reached its maximum range limits by 6000 BP. *Ulmus, Corylus* and *Tilia* also reached their present-day range limits by the same time. However, not all forest trees had reached these limits by 6000 BP. *Picea* has spread westwards through Finland, across Sweden, and into central and eastern Norway in the last 6000 years. *Fagus* had a rather small range in southern and central Europe by 6000 BP, which is in contrast to its present extensive range in western Europe today. Similarly, *Carpinus*, which is today widespread in lowland Europe was, at 6000 BP, very largely confined to Bulgaria, Italy, Yugoslavia, Romania, and Poland.

The post-glacial colonization of the British Isles also shows considerable complexity in terms of vegetation response (Fig. 4.3). The northward spread of trees varied as to direction, rate, and timings, and this may have been controlled by the place and time where colonization first occurred, which may in part have been a matter of chance occasioned by the vagaries of dispersal and establishment (Birks 1990: 142).

In Britain there are some good examples of how the warming of post-glacial times and the associated spread of forest has led to the fragmenta-

---

FIG. 4.3 Some examples of the spread of various tree types through the British Isles (after Birks 1990):

A. Spreading of *Corylus avellana* (hazel). The contours, in years before present (BP), show its likely range limit at 500-year intervals. The shaded area is where hazel was present prior to 9500 BP.

B. Spreading of *Quercus* (oak). The contours, in years BP, show its likely range limits at 500-year intervals. Note the slowing in its spread after 8000 BP.

C. Spreading of *Pinus sylvestris* (pine). The contours, in years BP, show its likely range limits at 500-year intervals in England and Wales and in Ireland and at 1000-year intervals in Scotland.

D. Spreading of *Tilia* (lime). The contours, in years BP, show its likely range limits at 500-year intervals. Note the slowing in its spread after 7000 BP.

E. Spreading of *Fagus sylvatica* (beech). The contours, in years BP, show its likely range limits at 500-year intervals. Note the absence of any slowing in its spread.

tion of the distribution of certain cold-loving flora types which covered a much wider area in the Pleistocene and Early Holocene. Two of the most remarkable 'disjunct areas' are the Burren of County Clare and the Teesdale area of the northern Pennines (Seddon 1971). Although ecologically very different they both possess certain types of flora which are found scarcely anywhere else in the British Isles. One such plant is the Shrubby Cinquefoil (*Potentilla fruticosa*) which is intolerant of a forest cover and is found currently in its most continuous form in central and eastern Siberia. On the unique limestone pavements of the Burren, and on the river banks and gravels of Upper Teesdale, the plant has been able to withstand the post-glacial warming trends and afforestation.

The effects of post-glacial warmth in creating relict clumps of certain plants in isolated areas is well displayed by the dwarf birch (*Betula nana*). It has been recorded at sites in many parts of Britain in late-glacial and post-glacial deposits, but is now found only in Upper Teesdale and on Scottish mountains. Equally, in north-western Europe there are similar relict areas in the French Jura, in the Harz mountains and on peat from Luneberg Heath. It seems clear that this plant, which is from the Arctic–alpine group and is currently found at high altitudes in the Alps, was once widespread over the lowlands of Late Glacial, north-western Europe, but that it has been removed from all localities except on mountains and on some very special habitats with distinctive ecological and microclimatic conditions. This removal has been caused by the spread of forest trees into environments which were formerly suitable.

Man and the classic sequence of Holocene climatic change

Although we have so far outlined some of the ways in which the marked environmental changes at the transition from the Late Glacial to post-glacial times greatly affected man, plants, and animals, it becomes increasingly clear in the Holocene that man himself was an increasingly potent agent of environmental change (Pennington 1969). For a considerable time it was believed that palaeolithic and mesolithic man was relatively ineffectual either because of his small numbers at this stage in his development, or because he did not possess the technological wherewithal. The characteristic palaeolithic hand-axe, for example, was apparently either a weapon or grubbing tool, and not until the development of the polished stone axe was man equipped with a tool to attack the forest cover of Europe and elsewhere (Smith 1970).

However, pre-neolithic man did possess the so-called tranchet axe which could have been effective in forest clearance, but more importantly mesolithic man may have utilized fire deliberately for driving game and clearing woodland. Sparks and West (1972) believe that fire may have been important as an agent of ecological change even earlier: '. . . the regularity with which hearths are found associated with the middle and upper palaeolithic sites leaves little doubt that Neanderthal Man and his successors were capable of fire production.' More recently, Gowlett *et al.* (1981), working near Lake Baringo in Kenya, at a site called Chesowanje, have proposed that fire may have been used deliberately by man at least 1.42 million years ago.

In the British Isles a marked abundance of hazel appears in mesolithic time and there is no doubt that the European *Corylus avellana* is fire resistant. Of particular interest is the decline of the linden *Tilia*, the pollen of which tends to disappear in many British sites at the same time as charcoal and other evidence of human activity appears (Turner 1962). This means that the classic VII/VIII zone boundary between the Atlantic and the Sub-boreal as originally defined by changes in tree-pollen frequencies may have little or no climatic significance. In Switzerland, the first marked fall of the beech curve in the pollen sequence is synchronous with the oldest agriculture in that country (Older Cortaillød culture), and the decline of elm in Denmark coincides with the arrival of the Younger Ertebolle culture (Troels-Smith 1956). In many parts of Europe the decline of elm around 5000 BP may be explained by the use of elm leaves for feeding stock. The leaves were collected for stalled animals. This phase was followed by more intensive clearances for shifting agriculture—the 'Landnam' clearances.

Certainly, the elm decline was the most rapid vegetational change in post-glacial history with elm pollen falling in a matter of just a few hundred years to levels that were typically reduced by half. As Rackham (1980: 265) has pointed out, a deterioration of climate is inadequate to explain so sudden, widespread, and specific a change. Neither does climatic change explain the decline of elm rather than other species. However, it is still not proven that the spread of agriculture was totally responsible, and it seems unlikely that neolithic man could have pollarded elm so effectively that its pollen yield was halved over an area of such extent so quickly. It is therefore possible that elm disease may have contributed to the decline.

Changes in post-Atlantic vegetation owe relatively less to climate than the changes of the Boreal and the Atlantic, and man and soil deterioration assume still greater importance. The role of soil deterioration is less easy to assess, though intense leaching of glacial sediments under the warm, wet conditions of the Atlantic may have led to the development of podzols and other soils relatively inimical to the deciduous forest (Pearsall 1964). Hardpans of the podzolic type may also have led to some waterlogging of soils by impeding their drainage, and they would also have been relatively acid in type.

It is conceivable that new agricultural techniques could, from neolithic times onward, have accelerated this podzolic condition (Mitchell 1972). Podzolic soils would in turn encourage blanket bog formation. Burning and ploughing would help to release minerals which would accumulate as hardpan. This hardpan, by impeding drainage, would give ideal conditions for blanket-peats to accumulate, and in Ireland and western Wales mesolithic fields, occupation sites, and megalithic tombs are sometimes found buried by peat. However, peat bogs or blanket mires, the development of which coincides with the elm decline (5300–5100 BP), do not always occur above well-developed podzol horizons (Moore 1975). Indeed, sub-peat profiles are often immature, suggesting that the soil is not always the predominant cause of mire development. Another possible factor in the timing of peat-bog development was the removal of the natural tree canopy by mesolithic and neolithic man. This would reduce the transpiration demand of the vegetation and would also reduce the degree of interception of precipitation. Hence more water would be available to raise groundwater and soil-water levels, thereby favouring peat development. In the southern Pennines the basal layers of all marginal peats (Tallis 1975) contain widespread evidence of vegetation clearance by burning, either in the form of microscopic carbon particles (similar to present-day 'soot'), small charred plant fragments, or larger lumps of charcoal.

Thus peat bogs, in that their development could be promoted by climatic change, soil maturation, and by human interference in the highland ecosystem, illustrate the complexity of factors that may be involved in any environmental change.

The North American Holocene sequence

It is interesting to compare the North American sequence with that of Europe and Britain. Although a simple cold–hypisthermal–cool sequence has found some favour, it now seems that the sequence was at least as complex as that in Europe.

In Canada (Table 4.4) it has been suggested that, as human interference has been less in the Canadian Holocene, the Canadian sequence gives a more realistic impression of the role of climatic change in the development of post-glacial vegetation associations, and that by using the

Table 4.4 Holocene environmental changes in central Canada and north-west Europe

Central Canada	Yrs. BP	North-western Europe
Forest retreat, expansion of tundra, peat growth ceases at Ennadal Lake		Recurrence surfaces, Greenland colonists perish, Little Ice Age
Small northward extension of forest	700	Retardation layers in peat. Exploration of north Atlantic
Retreat of forest to south of Ennadal	1500	Recurrence surfaces, Alpine glaciers advance
Alternations of cool and warm climate	2500	Alternations of cool and warm climate, recurrence surfaces and retardation layers in peat
Small retreat of forest	3500	Ulmus decline
Forest extended far north	5000	Continuation of climatic optimum
Rapid deglaciation, swift immigration of forest	6500	Beginning of climatic optimum. Warmest period of post-glacial
	8000	

(After Nichols, 1967)

Canadian sequence as a standard one can assess the importance of man as opposed to climate in certain major vegetational changes such as the *Ulmus* decline. In central Canada there appear to have been changes that are broadly comparable to those in Europe: the extension of the forest between 6500 and 5000 BP for example correlates with part of the classic Atlantic in Europe, while the retreat of forest after 2500 BP appears to correlate with the cold, wet, and oceanic conditions of the European Sub-Atlantic.

Bernabo and Webb (1977) have produced an interesting summary of vegetational changes in north-eastern parts of North America by summing the observed changes in the pollen record from sixty-two sites. As Fig. 4.4A shows there have been major changes in the relative importance of spruce, pine, oak, and herb pollen (non-arboreal types characteristic of temperate grasslands) between each 1000-year level from 11 000 BP to the present. The largest changes occurred in the early Holocene between 11 000 and 7000 BP with another peak in the last 1000-year interval when European settlement took place. One of the most important events was the decline of the spruce from 11 000 to 8000 BP as it gradually moved northwards (Fig. 4.4B). Another important feature of the Holocene vegetational history has been the fluctuating position of the boundary between prairie and forest (Fig. 4.4C). The signs of prairie development in the western Midwest are visible in the pollen record over

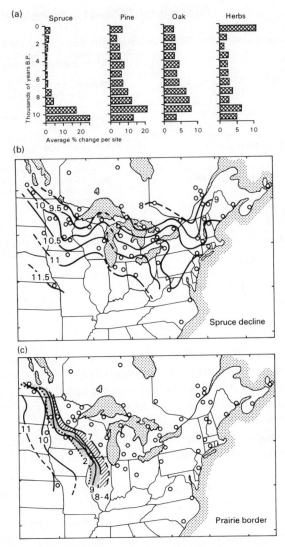

Fig. 4.4 Holocene vegetation change in eastern North America.

A. Graph depicting the average percentage change per site, between each 1000-year level from 11 000 BP to present, for spruce, pine oak, and herb pollen. The figure shows important shifts in the amount of change these major pollen groups underwent during the Holocene. Values were obtained by summing the total changes (regardless of signs) seen from all mapped sites and then dividing by the number of sites.

B. Isochrones plotting the time, in thousands of years BP, when spruce pollen declined to below 15 per cent.

C. Isochrones in thousands of years BP illustrating the movements of the prairie/ forest ecotone. The position of the prairie border is based on isopoll maps for herb pollen. Shaded areas show the region over which the prairie retreated after reaching its maximum post-glacial extent at 7000 BP (after Bernabo and Webb 1977, figs. 2, 21, and 26).

11 000 years ago as the vast region formerly occupied by the Late-Glacial boreal forest began to shrink. The largest eastward shift of the prairie took place between 10 000 and 9000 BP. It reached its maximum eastward extent in about 8000, but receded somewhat from 7000 to 2000 BP.

Bartlein *et al.* (1984) have attempted to estimate past climatic values by reference to the vegetational changes indicated by the pollen data. In particular they suggest that there was a major precipitation decline in the Midwest from 9000 to 6000 BP, amounting to 10 to 25 per cent, which was combined with a mean July temperature increase of 0.5 °C to 2.0 °C.

Other evidence of Holocene vegetational and climatic fluctuations has been derived from a study of aeolian stratigraphy in some of the drier areas of the USA. In Wyoming, for example, Gaylord (1990) identified pronounced aridity and aeolian activity from 7545 to 7035 BP, and from 5940 to 4540 BP.

Post-glacial times in the Sahara and adjacent regions

In the Sahara desert it has for long been suspected that climatic conditions were wetter at some stage or stages in the Holocene than they are at present. This was deduced from facts such as the widespread distribution of rock paintings, and of human stone and other tools, in areas which are currently far removed from waterholes. Certain of the species represented in rock painting, notably elephant, rhino, hippo, and giraffe were regarded as being representative of a moderately to strongly luxuriant savannah flora. Pollen analysis, though so far on a limited scale and subject of many doubts, has confirmed this essentially subjective archaeological evidence, and pollens of Aleppo pine and other trees have been found in sediments of Holocene age in the Hoggar and other massifs. There are also now a large number of radio-carbon dates for lacustrine sediments in various parts of the desert which enables one to establish the sequence of events with a little more certainty than hitherto. It seems likely on the basis of dates from Chad, Ténéré, the Nile valley, the Saoura valley, and the Hoggar that there were three lacustral phases in the early Holocene (before about 8500 BP from 7050–4150 BP and from 3550–2450 BP), and that during them vegetation was denser than at present.

In the desert to the west of the Nile there are numerous tree stumps of acacia, tamarisk, and also of sycamore fig (*Ficus sycomorus*). It is significant that these tree stumps, with diameters of 34–40 cm and a density of 5–11 per ha indicate that an open savannah existed in this sub-pluvial some 200 km farther north than this vegetation can survive today (Butzer 1961).

At the Kharga Oasis there are immense deposits of lime-rich spring tufas around or in which neolithic tools have been found in great numbers. This indicates higher groundwater levels and a considerable population. The Neolithic was a time particularly favourable for human activities in the Sahara (Faure 1966).

A good example of mid-Holocene humidity in the dry heart of the eastern Sahara is provided by a study undertaken at Oyo by Ritchie and Haynes (1987). Their pollen spectra (Fig. 4.5) dating from 8500 years BP until around 6000 BP show that there were strong Sudanian elements in the vegetation, and they identified pollen of tropical taxa such as *Hibiscus*. During this phase Oyo must have been a stratified lake surrounded by savannah vegetation similar to that now found 500 km to the south. After 6000 BP the lake became shallower and acacia-thorn and then scrub grassland replaced the sub-humid savannah vegetation. At around 4500 years BP the lake appears to have dried out, aeolian activity returned and vegetation disappeared except in wadis and oases. Roberts (1989) suggests that in effect the Sahara did not exist during most of the early Holocene. This is a point of view that is supported by the work of Petit-Maire (1989) in the western Sahara (p. 652): 'Biogeographical factors implicate total disappearance of the hyperarid belt at least for one or two milleniums before 7000 BP. . . . The Sahel northern limit shifted about 1000 km to the north between 18 000 and 8000 BP and about 600 km to the south between 6000 BP and the present' (see Fig. 4.5,C).

In addition to these humid phases there may also have been some phases which were somewhat drier than today to the extent that dune reactivation took place in semi-arid areas. Thus, for example, studies of the erg in the Accra plains of Ghana (Talbot 1981) have demonstrated a phase of marked aridity combined with strong south-westerly winds at about 4500–3800 BP.

The post-glacial climatic optimum and neoglaciation

Whatever may have been the sequence of environmental changes which took place in different areas during the course of the Holocene, one of the most contentious but interesting problems is that of the so-called climatic optimum (Manley 1966).

That climatic conditions were appreciably warmer during a section of the Holocene than they are now, was discovered by Praeger, who detected it in connection with his investigation of the fauna of the estu-

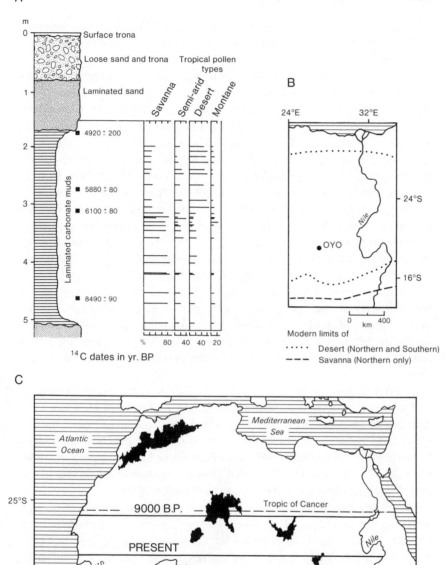

A

m

Surface trona

Loose sand and trona   Tropical pollen types

Laminated sand

Savanna Semi-arid Desert Montane

4920 ± 200

5880 ± 80

6100 ± 80

Laminated carbonate muds

8490 ± 90

%   80 40 40 20

^{14}C dates in yr. BP

B

24°E   32°E

24°S

16°S

Nile

•OYO

0   400
       km

Modern limits of

· · · · · Desert (Northern and Southern)

— — — Savanna (Northern only)

C

Mediterranean Sea

Atlantic Ocean

25°S   — — — 9000 B.P. — — —   Tropic of Cancer

PRESENT

Nile

Senegal   Niger   L. Chad

15°S   18000 B.P.

FIG. 4.5 A. Pollen diagram from Oyo in the eastern Sahara (after Ritchie *et al.* 1985) showing main features of stratigraphy and pollen sequence in the Holocene. Reprinted by permission from *Nature*, 314: 352–5. Copyright © 1985, Macmillan Magazines Ltd.
B. The location of Oyo.
C. The changing position of the Sahara–Sahel limit (after Petit-Maire 1989, fig. 8). Reprinted by permission of Kluwer Academic Publishers.

arine clays of the north of Ireland. This was in 1892. From a comparison of the fauna of the Belfast estuarine clays with that of the present shores of Ireland, he gave the first proof of a definitely higher temperature during their deposition. Scandinavian workers subsequently confirmed this finding by an examination of the fauna of the Tapes Submergence (see p. 247) in the Oslofjord. They also discovered that in Lapland the pine forest had at some stage in the Post-Glacial moved into zones which are now dominated by birch or alpine associations. This period of extended distribution has been called the post-glacial climatic optimum. The tree-line extended 500 m higher than it does today in northern Europe, and the treeless tundra almost disappeared from northern Siberia. In Norway, where the maritime influence was stronger, the tree-line displacement was less—only 300 m. One of the most important markers of this optimum was the European land tortoise (*Emys orbicularis*) which spread into Denmark at this time, but disappeared in the Sub-Atlantic. Cool and damp summers are highly unsuitable for the animal, especially to the development of its eggs (Godwin 1956). Another important indicator of the post-glacial warm period is the hazelnut (*Corylus avellana*). Its present distribution in Scandinavia, shown in Fig. 4.6, is markedly different from its distribution 5000–6000 years ago, for at that time it extended further north, and higher in altitude than at the present time.

In Greenland too there is evidence that at some point during the Holocene conditions were more favourable to life than they are now. The edible mussel, *Mytilus edulis*, which now has a northern limit in Greenland waters at about 66°N, is found in raised beaches, which have been dated at about 5000–7000 BP, which occur at a considerably higher latitude—73°N. Another 'southern' bivalve, *Chlamys islandica*, is also found in areas outside its present range (Funder 1972).

However, some dissatisfaction has been expressed from time to time with the use of the word 'optimum', especially as in rather drier areas the increased temperature would be far from advantageous for plant growth. Various other terms have therefore been introduced including 'altithermal' and 'hypsithermal'. The latter was proposed in 1957 by Deevey and Flint as a term to cover four of the traditional pollen zones (V through VIII in the Blytt–Sernander system) embracing the Boreal through to the Sub-boreal (8950–2550 BP). However, the dates which other workers give for the 'optimum' or 'hypsithermal' do not always tally with this. Lamb (1969) for instance gives dates of 6950–4350 BP.

Another problem is that in different parts of Europe the maximum temperature was recorded at different times. In Sweden, for instance, the

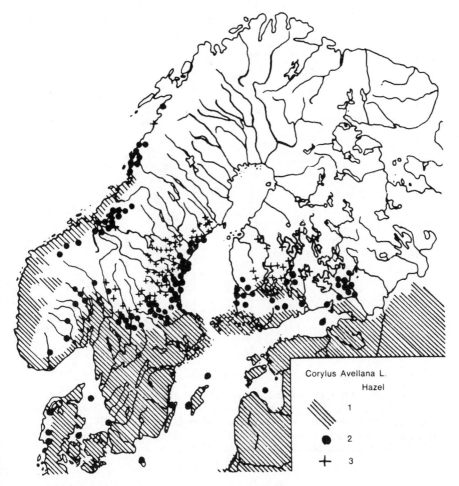

Fɪɢ. 4.6 Indications of the post-glacial warm period illustrated by the present and post-glacial distribution of the hazelnut (*Corylus avellana*) in Scandinavia.
1 = Present general distribution
2 = Current records of individual occurrences
3 = Hazelnut fossils in sediments of the post-glacial warm period.
(From Frenzel 1973, fig. 2.)

'optimum' appears to be associated with the Atlantic period (about 6000 ʙᴘ), when the Littorina Sea (see p. 247) would have reinforced the mildness of the winters. In Denmark, on the other hand, the optimum appears to have been about 4000–3000 ʙᴘ (the Sub-boreal). In Scotland too, trees reached their highest limit on the Cairngorms in the Sub-boreal (Manley 1966).

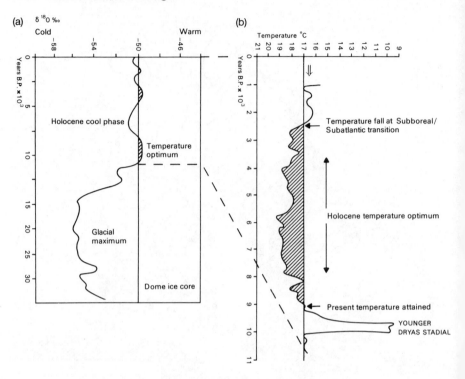

FIG. 4.7 Two different temperature records for the Holocene, giving two different dates for the Holocene temperature 'optimum', (A) the oxygen isotopic curve for the Dome Ice-Core, Antarctica (after Lorius *et al.* 1979), and (B) the oxygen isotopic curve (converted directly to summer water temperature) for a lake in Sweden (after Mörner and Wallin 1977).

The hypsithermal has also been recorded through oxygen-isotope measurements of an ice-core at Camp Century on the Greenland Ice-Cap. This had indicated that a warm phase lasted from about 8000 to 4100 BP with extremes close to 5000 and 6000 BP (Dansgaard *et al.* 1970). It was replaced by colder conditions between 2500 and 2100 BP: dates which correspond fairly closely to a neoglacial glacial advance, and the early part of the classic, Sub-Atlantic deterioration. However, when a comparable study was made of the Dome ice-cap in Antarctica (Lorius *et al.* 1979) it was found that the warmest phase of the Holocene was between 11 000 and 8000 BP, and that it was relatively cool between 8000 and 4000 years ago (Fig. 4.7).

However, although the nature and timing of the optimum may be the subject of dispute, it does appear from many sources that temperatures have been appreciably higher during certain parts of the Holocene than they are at present, though the degree of temperature amelioration is rather variable. A positive deviation over the present mean of 1–3 °C is suggested for many areas in the temperate zone.

The spread of early man through Europe may have been partially influenced by climatic factors at this time. A major phase of agricultural colonization in Europe went forward under mild Atlantic conditions from the sixth to the middle of the second millennium BC and largely involved the cultivation of wheat. The Campaignien peoples entered from North Africa, and the Danubian peoples from the steppes of eastern Europe during this phase (Demougeot 1965). Subsequent climatic deterioration led to the decline of wheat cultivation, which in many areas was supplanted by oats and rye. Indeed one oat, *Avena fatua*, was particularly favoured by the increasing length of cold winters for its seed is not viable if winters are not cold (Isaac 1970).

A related factor which greatly influenced the spread of the neolithic agricultural revolution through Europe was the great belt of wind-blown loess (see p. 67) which had been deposited extensively over the plainlands. It was at one time thought that these soils, because of their relatively dry soil climate when compared to, for instance, the damp, heavy, clay soils developed on the boulder clays of the North German Plain, might have supported a much more open heath-like steppe vegetation which would have favoured early man. The clays on the other hand would tend to support a denser forest which it would be relatively difficult to clear for agricultural purposes. However, more recent pollen analysis of deposits in the loess terrains do not fully substantiate this point of view, for they indicate that the loess too was covered by a forest.

For the same reasons it has become necessary to lay aside another early theory, which tried to relate a dry or 'xerothermic' phase during the classic Sub-boreal of the Blytt–Sernander model with a decline in forest cover. Under this theory, Gradmann's 'Steppenheidetheorie', it was held that drought conditions were such that forest could not be maintained on the lighter loessic soils of central Europe, so that there existed a corridor of open steppe along which the fauna, flora, and prehistoric peoples of south-eastern Europe migrated in to the oceanic west.

Nevertheless, the correspondence between settlements of Danubian type and the loess soils is remarkable. The explanation for this is probably that whilst these soils were not related to open vegetation conditions except locally, they were well-drained, fertile, and easy to till, whereas other types of soil would in general have been heavy, cold, and ill-drained. Moreover, recent investigations suggest that neolithic man was much more competent technologically to clear woodland than had previously been believed. Thus woodland may not have been quite the barrier envisaged at first sight.

In the Americas there has been a considerable amount of discussion about the effects of the Holocene climatic fluctuations, for although man came late to the Americas, once he became established, he built up quite large populations. The south-west dry zone of the United States, for example, was extensively populated between 14 000 and 10 000 BP. However, relatively few prehistoric sites are encountered in that area between 10 000 and 4500 BP (Griffin 1967), and some archaeologists take the probably extreme view that the prairies and western plains were almost abandoned between 6000 and 4500 BP, only to be recolonized after that time. The period of greatest population sparseness appears to coincide with the allegedly dry Altithermal (Stephenson 1965, Irwin-Williams and Haynes 1970), while periods of demographic advance (such as the Folsom occupation of 10 800–10 300 BP and the post-4500 resurgence of population) coincide with improved moisture conditions. Although there is some doubt as to how dry the Altithermal was (see, for example, Martin 1963), it is widely regarded as being a major factor in the prehistory of the drier parts of America. As Malde (1964: 127) has written, 'As the Altithermal drew to a close, relatively wetter and colder conditions returned, and the tempo of human life accelerated, along with evident growth in population.'

Subsequent studies have suggested that a concept of one hypsithermal interval needs some modification. There is now abundant evidence that during the hypsithermal as originally defined there were some renewals of glaciation and some cold conditions. Denton and Porter (1970), for example, have written that 'It is now known that rather complex low-order changes of climate characterized the hypsithermal interval, resulting in several early neoglacial episodes of glacier expansion. Therefore, in some regions at least, neoglaciation and the hypsithermal interval, as they are currently understood, partly overlap in time.'

Attempts to detect order in the complexity of Holocene neoglacial fluctuations have been the subject of debate, and it is still difficult to say with certainty whether the Holocene glacial advances and retreats were contemporaneous in different parts of the world.

In 1973, Denton and Karlén put forward a model in which they postulated that the Holocene had been punctuated by at least three phases of Alpine glacial expansion: the Little Ice Age of the last few centuries, the period 3300–2400 BP, and 4900–5800 BP. They believed that in general the periods of glacier expansion had a duration of 600 to 900 years, and that they were separated by periods of contraction with a duration of up to 1750 years.

Grove (1988), however, believes that information is still too 'imprecise to affirm that glaciers of similar size expanded synchronously or, indeed that they did not' (pp. 358–9), but she none the less felt able to make certain generalizations, including some about differences in different hemispheres with respect to the timing of the extent of maximum Holocene glacial expansion (p. 359):

There seems to be a contrast between the Holocene behaviours of mid- to high-latitude glaciers in the northern and southern hemispheres. In both Patagonia and New Zealand, Little Ice Age advances were on a smaller scale than those of the Middle Holocene, whereas in Iceland, Greenland, and Spitzbergen, data at present available point to Little Ice Age advances, and notably those of the nineteenth century, as having been the greatest to have taken place since the end of the last glacial period.

A very detailed study of glacial chronologies in many parts of the world by Röthlisberger, based on personal study of similar sizes of glacier using uniform sampling techniques, gives us the best available information on this contentious matter. He did detect a considerable degree of synchroneity of glacier extension and shrinkage in both hemispheres, and found especially good accordance where the information was best, as in the European Alps and New Zealand. Fig. 4.8 shows his regional curves. Other detailed information on Holocene glacial fluctuations is provided by Davis and Osborn (1988).

The Little Optimum, AD 750–1300

After the climatic optimum of the Middle Holocene, conditions once again became cooler in many regions, but in early medieval times there may have been a return to more favourable conditions, particularly in North America and Europe—the so-called 'Little Optimum' or 'medieval warm epoch'.

Proponents of this idea, including Lamb (1966a), would argue that from about AD 750 to 1200–1300 there was a period of marked glacial retreat which on the whole appears to have been slightly more marked than has been that of the twentieth century. The trees of this phase, which were eventually destroyed by the cold and glacial advances from about AD 1200 onwards, grew on sites where, in our own time, trees have not had time, or the necessary conditions, to grow again. In terms of a more precise

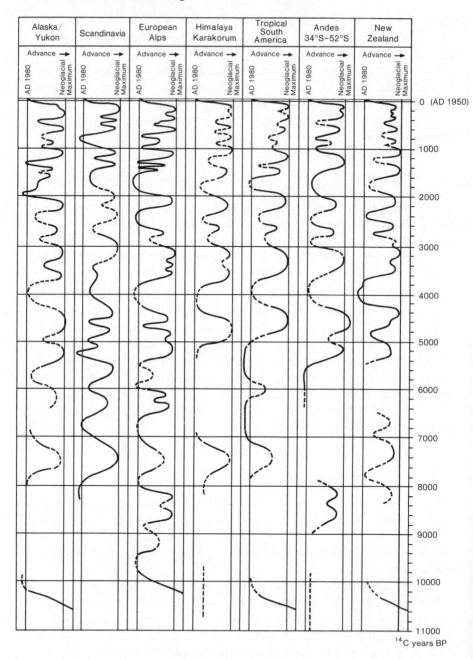

FIG. 4.8 Fluctuations of glaciers in the northern and southern hemispheres during the Holocene. (From Röthlisberger 1986, in Grove 1988.)

date the medieval documents that are available place the most clement period of this optimum, with its mild winters and dry summers, at AD 1080–1180. At this time the coast of Iceland was relatively unaffected by ice, compared to conditions in later centuries, and settlement, as will be seen, was achieved in now inhospitable parts of Greenland. It is also believed that the relative heat and dryness of the summers, which led to the drying-up of some peat bogs, was responsible for the plagues of locusts which in this period spread at times over vast areas, occasionally reaching far to the north. For instance, during the autumn of 1195 they reached as far as Hungary and Austria. In northern Canada, west of Hudson Bay, fossil forest has been discovered up to 100 km north of the present forest limit, and four radio-carbon dates from different sites show that this forest was living about AD 870–1140. It is also interesting that the Camp Century ice-core from Greenland has revealed to American and Danish workers that a cold wave is evident after about 1130–60, but that for five centuries preceding this there was a phase of appreciable warmth. This has also been confirmed by a more recent core at Crete (Central Greenland) (Dansgaard *et al.* 1975).

One additional line of evidence that has been utilized to gain an appreciation of the nature of this phase is the presence of vineyards in various parts of Britain. Domesday Book (1085) records thirty-eight vineyards in England besides those of the king. The wine was considered almost equal to the French wine in quality and quantity as far north as Gloucestershire and the Ledbury area of Herefordshire, the London Basin, the Medway Valley, and the Isle of Ely. Some vineyards even occurred as far north as York, and Lamb (1966a) regards this as being indicative of summer temperatures 1–2 °C higher than today, a general freedom from May frosts, and mostly good Septembers. In China, at about the same time, lychees, sensitive trees which succumb at temperatures below −4 °C, were an economic crop in the Szechuan Basin in western China, but today they are limited to the south of Nanling. Miscellaneous evidence of this type was used to construct Fig. 4.9, the pattern of which suggests a striking correspondence with the fluctuations derived from the Greenland ice-core (Hsieh 1976).

It has also been maintained that the favourable conditions of the proposed Little Optimum coincided with radical changes in the fortunes of agricultural peoples in various parts of the USA. Considerable growth and development occurred from AD 700 to 1200 (Malde 1964), whereas shortly before the thirteenth century a very rapid withdrawal of agricultural people took place as a result of cold, dry conditions (Woodbury 1961).

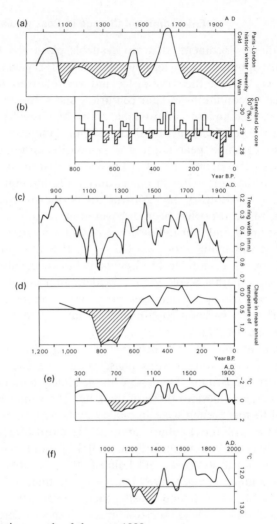

FIG. 4.9 Climatic records of the past 1000 years.

A. The 50-year moving average of a relative index of winter severity compiled for each decade from documentary records in the region of Paris and London (Lamb 1969).

B. A record of the oxygen-18 values preserved in the ice-core taken from Camp Century, Greenland (Dansgaard *et al.* 1971).

C. Records of 20-year mean tree growth at the treeline of bristlecone pines, White Mountains, California (La Marche 1974). At these sites tree growth is limited by temperature with low growth reflecting low temperature.

D. The 50-year means of observed and estimated annual temperatures over central England (Lamb 1966*a*).

E. Chinese temperature patterns based on miscellaneous phenomena (appearance of frost, freezing of rivers, blossoming of trees and flowers, migration of birds, etc), gazetteers, and instrumental observations (after Hsieh 1976).

F. Cave temperatures derived from oxygen isotope studies in a New Zealand cave (after Wilson, *et al.* 1979).

However, some recent tree-ring analysis for Fennoscandia by Briffa *et al.* (1990) has failed to find any unambiguous evidence for this warm phase, or for very extended runs of warm years. They conclude (p. 438) that their reconstruction 'dispels any notion that summers in Fennoscandia were consistently warm throughout that period. Although the second half of the twelfth century was very warm the first half was very cold. For most of the eleventh and thirteenth centuries, summers were near normal (relative to the mean for 1951–70).' They suggest that the significance of the Little Optimum has been overstated and that much of the historical evidence upon which the concept was based was essentially sketchy.

The last Little Ice Age (neoglaciation)

One of the most significant Holocene environmental changes, not least because of its effects on the economies of highland and marginal areas in Europe, was the renewed phase of glacial advance since the medieval warm epoch. This phase has often been called the Little Ice Age, but recently the term 'neoglaciation' has been proposed 'to encompass the interval of rebirth or renewed growth, and all subsequent fluctuation, of glaciers after the time of maximum hypsithermal glacier shrinkage' (Denton and Porter 1970: 102).

The term 'Little Ice Age' was first introduced by Matthes (1939) to describe an 'epoch of renewed but moderate glaciation which followed the warmest part of the Holocene', but it is now more widely used 'to describe the period of a few centuries between the Middle Ages and the warm period of the first half of the twentieth century, during which glaciers in many parts of the world expanded and fluctuated about more advanced positions than those they occupied in the centuries before or after this generally cooler interval' (Grove 1988: 3).

Its effects have been noted all over the world. In China, for example, it was at its peak from 1650–1700 (Chu Ko-Chan 1973), but the date at which the late-medieval Little Ice Age began is variable from area to area. The Great Aletsch Glacier in the Swiss Alps advanced over part of an aqueduct used to transport meltwater to a local village as early as the thirteenth century. Similarly, the Chickamin Glacier in the Cascade Range of Washington State (USA) reached its maximum in the thirteenth century. In the South Tyrol there was a major advance in AD 1150–1250 (Mayr 1964). In most areas, however, the maxima were reached at various times from the middle of the fourteenth century to the middle of the nineteenth century, though conditions were far from stable and there were complex patterns of warming and cooling throughout the period. The climate was not continuously cold and year to year fluctuations were evident then as now. Thus the seventeenth century, a time of frequent severe winters in many parts of the world also experienced a number of hot summers such as that before the Great Fire of London in 1666. Grove (1988: 107) finds no convincing evidence of expansion of Scandinavian glaciers before the seventeenth century, notes that the enlarged state of glaciers and ice-sheets in the eighteenth century was common to both north and south, and draws attention to the great recession in the twentieth century.

In Norway, where the glacial advances are relatively well chronicled by tax records and other sources, the advances appear to have begun between 1660 and 1700. The first half of the eighteenth century was marked by a general advance which amounted to several kilometres for some glaciers and culminated between 1740 and 1750. After this there was some recession, interrupted by re-advances, notably in 1807–12, 1835–55, 1904–5, and 1921–5. These have tended to leave small moraines. These later advances did not usually manage to reach the position gained in 1750.

An examination of land rent assessments from the Jostedalsbre region of western Norway, and of documents concerned with applications for their reduction, has provided J. M. Grove (1972) with detailed information about the incidence of landslides, rockfalls, floods, and avalanches during the Little Ice Age in Norway (Fig. 4.10). The evidence makes it clear that there was a much increased incidence of major mass rock movements and floods in the late seventeenth century and on until the nineteenth century. Moreover, this environmental change began abruptly with a marked clustering of disastrous incidents between 1650 and 1760, and in certain years during that period, such as 1687, 1693, and 1702. Conditions for farming were thus very much less favourable than they had been in previous centuries.

The Icelandic glaciers and ice-caps show a broadly similar history during this period. From the time of the first colonization of the island about AD 900 until at least the fourteenth century, the glaciers were considerably less extensive than they were after about 1700. There was a general advance in the early eighteenth century which reached its maximum around 1750. From 1750 to 1790 the ice tended to be relatively stagnant or to be in a state of retreat, but it advanced again in the early nineteenth century and in some cases had, by 1840–60, reached a more forward position than the previous maxima of 1750. A general recession towards the present position took place after about 1890.

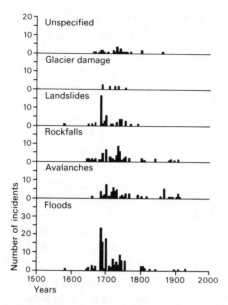

FIG. 4.10 The incidence of Little Ice Age mass movements and other natural hazards in various Norwegian parishes as revealed by Landskyld (land rent) records (after Grove 1972, fig. 2).

In the Alps the situation is extremely well documented compared to other areas. From about 1580 onwards tracts of cultivated land and forests were covered by advancing ice, and the local people were also subjected to greater flood risk. The local economies suffered and a series of supplications for tax relief were made. The Rhône Glacier advanced strongly from 1600 to 1680. From the mid-seventeenth century to the mid-eighteenth the glaciers were relatively quiescent, though still at a more forward position than in 1600. From the mid-eighteenth century there was a major phase of advance, divided into three main stages: 1770–80, 1818–20, and 1835–55. Retreat was then fairly general between 1850 and 1880, only to be replaced by some advance from 1880–95. From 1895 to 1915 recession continued, but was temporarily reversed from 1915 to 1925. Thus the picture in the Alps broadly corresponds to that in Norway and Iceland. The various stages are often visible as moraines, whilst the positions of hotels and other tourist facilities tell the story of retreat since the nineteenth-century maxima. As Jean Grove (1966) has said, 'All over the Alps mountaineering huts stand high above the ice and must be approached by steep moraines, fixed ladders, or even ropes.

Table 4.5 Glacial fluctuations in the Dome Peak area, Washington State, USA

	South Cascade	Glacier Le Conte	Dana	Chickamin
Century of maximum advance	16th/17th	16th	16th	13th
Area of glacier (km^2) at maximum	4.15	2.98	3.99	7.80
Area of glacier (km^2) in 1963	2.72	1.58	2.46	4.87
Ratio (min./max.)	1:1.5	1:1.9	1:1.6	1:1.6
Altitude of terminus at neo-glacial maximum (m)	1490	1340	1270	1100
Altitude of terminus in 1964 (m)	1615	1829	1768	1525

(After Miller, 1969)

These were not built to ensure an awkward scramble at the end of the day; their isolation is due to the wasting of ice during the last century'.

The American Little Ice Age pattern shows that the glacial advances were broadly contemporaneous across the northern hemisphere, although an advance has been suggested for the Sierra Nevada about 1000 BP (Curry 1969). The maximum advance occurred in the middle 1600s and lasted until about 1700, from which there was some recession, then an advance, and then recession, which lasted until about 1785. Advances were characteristic of the 1880s, and in some areas the maxima of the 1700s were exceeded. In Alaska maxima were recorded between 1700 and 1835.

Some data from the North Cascade Range, Washington, allow one to compare the state of glaciers during their Little Ice Age maxima with their state at the present. One can see that in general their temini are now on average about 300 to 400 m higher in terms of altitude, and that their areas have been reduced by 50 to 60 per cent (Table 4.5).

The glaciers of Greenland were also affected by the Little Ice Age, and advanced strongly between AD 1700 and 1850. The maximum extensions were, however, reacher rather later than in many parts of the world. In south-west Greenland the maximum was around 1850. At the inland ice margin the maximum was reached about 1890–1900, whilst in the north-west the maximum was not reached until 1915–25. As elsewhere, however, there was a general retreat along the whole Greenland west coast between 1920–5 and 1940–5. Since the latter date there has been a decelerating rate of retreat.

The Antarctica area seems to have escaped the cold epoch until rather late, a factor which doubtless helped the explorations of Captain Cook and others in the Southern Ocean. Between 1770 and 1830 the edge of the Antarctic ice appears to have been perhaps a degree of latitude south of its position in the years 1900–50 (Lamb 1967). On the other hand, the climatic deterioration seems to have persisted rather longer in Antarctica, and to have lasted until 1900 or later (Lamb 1969*a*). The data for the southern hemisphere in general are still meagre, but Salinger (1976) has been able to show that New Zealand experienced a climatic deterioration starting at about AD 1300, which was most severe between 1600 and 1800. This ties in well with the situation in Europe and North America.

After her analysis of the evidence in many parts of the world, Grove (1988: 355) suggests that at a global scale

there is a striking consistency in the timing of the main advances. In Europe they have been dated to around 1600 to 1610, 1690–1700, in the 1770s, around 1820 and 1850, in the 1880s, 1920s and 1960s. Where information is available it appears that this timetable was also followed in many parts of the world outside Europe . . .

The French economic historian and historical geographer Braudel (1972) has proposed that the marked climatic decline of the late sixteenth century had various significant effects on life in the Mediterranean lands, and refers to a high incidence of floods on rivers like the Rhône. The Guadalquivir iced over at Seville, and at Marseilles the sea froze in 1595 and 1638. Of more especial economic significance was the series of frosts which killed the olive trees of Languedoc in 1565, 1569, 1571, 1573, 1587, 1595, 1615, and 1642. Likewise, Ladurie (1972) has shown how the fluctuations of the Little Optimum and of the Little Ice Age affected traditional societies through their influence on both wine quality and on harvest dates. 'In these societies,' he writes (p. 23), 'mainly agricultural, and dominated by the frequently difficult problem of subsistence, the relation between the history of climate and the history of man had, in the short term, an urgency it has now lost.'

The Greenland settlements

Against the background of climatic changes involved in the deteriorating environment of the Little Ice Age or neoglaciation, it is perhaps not surprising that settlements in highly marginal areas like northern Norway, Greenland, Iceland, and highland Britain, suffered reverses to their economies, which led to grave social effects, and population decrease (see Wigley *et al.* 1981).

There are, of course, dangers in any simple deterministic interpretation, but the climatic effects noted above should not be ignored. As Utterström (1955), a proponent of the role of climatic change as a determinant of economic advance and decline, has written:

Ever since Malthus and Ricardo, all discussions of the pressure of food supplies have started from the assumption that population is the active factor and Nature the fixed. This interpretation, however, can hardly be reconciled with modern scientific thought, especially if the problem is viewed in the long term.

Of particular interest is the question of the settlements in Greenland, which were established by the Vikings, reached a reasonable size, and then declined almost catastrophically in the late-medieval period. It has been argued that both sea conditions and the actual environment in Greenland itself must have been favourable for the initial settlement to take place. Drift-ice rarely appeared near Iceland and Greenland south of 70°N in the 900s and was apparently unknown between 1020 and 1194 (Lamb 1969a). Other lines of evidence in addition to this one have led Lamb to propose an optimum. Viking burial grounds of the period in question in south-west Greenland, in which graves were dug deep and trees rooted, are now permanently frozen, and this has suggested to Lamb (1966a) that since that time, mean annual temperatures may have fallen by 2 to 4°C. However, after 1410 there was no regular communication between Europe and any part of Greenland, and no one seems to have reached anywhere on the east coast of Greenland from the sea between 1476 and 1822. Gradually the settlements failed, and it seems highly likely that deteriorating climatic conditions would have led to a highly unfavourable environment, and to a growth of sea-ice which would both limit communication with the outside world and restrict fishing activity.

However, Sauer (1968: 157) disputed the role of climatic change in these events and made his position abundantly clear. 'The arguments that the settlements failed because the climate changed seem incompetent or irrelevant. They failed as far and small outposts that slowly lost the ability to live in the European manner.' He proposed that various other factors could be responsible for the decline, including a deterioration in drainage; a degeneration of the people (the more vigorous and venturesome being lost by emigration); a state of cultural rigidity imposed by the population being too small, too remote, too little diversified in skills, and too unenterprising; a series of raids by European 'pirates'; and a lack of wood for building their own ships. The last of these would lead to increased isolation so that life became very unattractive. Nevertheless, certain of these factors, including emigration, isolation, the absence of wood, and lack of enterprise, could be caused by environmental changes. Moreover, the evidence of changes in sea-ice limits cannot be disputed.

Medieval highland cultivation

In much of northern Europe, including Britain, the little climatic optimum and some of the preceding centuries were times of extension of settlement to highland areas. About the 1230s medieval villages and their strip cultivation systems were spreading so far on to the higher ground as to cause anxiety for the preservation of enough pasture. In Central Norway, the limits of settlement, forest clearance, and farming were pushed 100 to 200 m farther up the hillsides and valleys in Viking times (AD 800–1000), whilst in many parts of Britain there is evidence of medieval tillage on high ground far above anything that would be reasonable now, even in wartime (Lamb 1966a). For instance, the thirteenth-century limit of tillage in Northumberland appears to have been around 300–350 m above sea-level, 120–150 m above the limit of any worthwhile possibility at the present day. This was also a period when medieval vineyards spread into a number of localities in southern and eastern England. After a while, however, these extended settlements underwent a fairly considerable decline, and much of this decline started before the Black Death. Of nearly fifty deserted villages in Oxfordshire and thirty-four in Northamptonshire only about 10 per cent were attributable to the Black Death of 1348. All appear to have suffered serious decline in the years of disastrous summers and famines between 1314 and 1325 (Lamb 1967).

In Scotland, Parry (1975) has proposed that secular deterioration of climate since the early Middle Ages caused much of the high-lying cultivation is south-east Scotland to become profoundly sub-marginal in the seventeenth century. The consecutive harvest failures of the 1690s and 1780s may have been the immediate stimulus to abandonment, but the response to these stimuli, he maintains, would neither have been so widespread nor so permanent if the potential for cropping in the upland areas had not been so severely reduced over the preceding three centuries. On the basis of the data presented by Lamb and others for the degree of climatic change, and by a study of the present-day climatic limitations on oat-ripening, Parry (1975) suggests that in the early medieval warm period the chances of crop failure were only about one year in twenty in the Lammermuir area. By the mid-fifteenth century this had grown to one year in three, and in all about 4950 ha of land seem to have been abandoned in the Lammermuir–Stow Uplands.

Fig. 4.11 plots the progress of the climatic deterioration in that area using two parameters. One measure of the intensity of summer warmth is the accumulated temperature calculated over a base of 4.4 °C. The

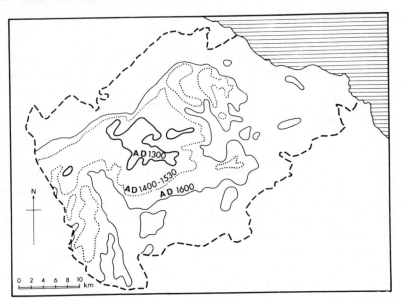

FIG. 4.11 Climatic deterioration AD 1300–1600 in the Lammermuir–Stow uplands of Scotland represented by the fall of continued isopleths of 1050 day-degrees C and 60 mm PWS. (From Parry 1975, fig. 5.)

isopleth of 1150 day-degrees C showed a marked correspondence with the 1860 cultivation limit. PWS (Potential Water Surplus) is a measure of summer wetness, expressed as the excess of a middle and late summer surplus (up to 31 August) over an early summer deficit.

In Denmark, the villages most often abandoned were those terminating in '-thorp'. These were just those villages that had been established relatively late—from the tenth to the twelfth century (Steensberg 1951). In Iceland, corn-growing decreased greatly not long after 1300, ceased altogether in the sixteenth century, and was not to be re-established until after the Little Ice Age. It appears to have reached its maximum in the tenth century. In the first half of the fourteenth century the centre of gravity of economic affairs moved away from the interior of the island to the coast, where fishing was to become the main economic activity in place of cereal-growing and vadmal (homespun) production (Utterström 1955).

In Sweden, the so-called 'Golden Age' of Gustav I ended in the mid-sixteenth century, and this prosperous era was followed by an era from which there is a plethora of reports of natural catastrophes, crop failures, and famine, persisting for a hundred or so years from the 1590s.

Another economic consequence of this phase of deteriorating climate was that the silver mines of Central Europe were subjected to increased water problems after 1300, leading to the closure or run-down of many mines.

Selected reading

The transition from full glacial conditions to the warmer conditions of the Holocene is recorded in J. J. Lowe *et al.* (1980) (eds.), *Studies in the Late Glacial of North-west Europe*. The vegetation changes in Britain are described by W. Pennington (1969), *The History of British Vegetation*. Some of the historical implications of Holocene changes are discussed in T. M. L. Wigley *et al.* (1981) (eds.), *Climate and History*, a theme explored in great detail by E. Le Roy Ladurie (1971), *Times of feast, times of famine: a history of climate since the year 1000*, and by M. L. Parry (1978), *Climatic Change, Agriculture and Settlement*.

The massive Late-Glacial and Holocene extinctions of fauna are dealt with in P. S. Martin and H. E. Wright (1967) (eds.), *Pleistocene extinctions* and the growing influence of man on the British Holocene environment can be followed in I. G. Simmons and M. J. Tooley (1981), *The environment in British prehistory*.

A useful and wide ranging textbook that deals with many aspects of the Holocene is H. H. Lamb (1982), *Climate, history and the modern world*.

The Holocene in the USA is discussed in all its facets in H. E. Wright (1984) (ed.), *Late-Quaternary environments of the United States*, ii, *The Holocene*.

Undoubtedly the most useful general reference to the Holocene is N. Roberts (1989), *The Holocene: an environmental history*.

5

Environmental Changes During the Period of Meteorological Records

Introduction

Although various scholars, especially Brooks, Ladurie, Manley, and Lamb, have adroitly interpreted past climatic conditions from documentary records (see Bradley 1985, ch. 11 for a review of techniques), thereby greatly adding to our knowledge of climatic conditions in pre-industrial Britain and Europe, it is not until the nineteenth century that there was an organized growth of instrumental observations from stations all over the world. It is on the basis of these relatively reliable instrumental records that most of our knowledge of the latest chapter of environmental evolution is founded.

Such records, while infinitely more reliable than the historical methods utilized for previous centuries, are not without their limitations: instruments need to be replaced and recalibrated from time to time, and sites and locations are liable to be affected by such factors as urbanization or vegetational change. However, by careful selection of the more reliable stations available, and by taking averages for several stations within an area, a valuable picture of changes can be obtained.

The extent of changes in climate over the last 100 years or so is greater than was formerly believed: both temperature and rainfall have shown trends which have led periodically to great fluctuations in glaciers, lakes, and river discharges. A comparison of these climatic changes with others in Britain since the 1690s is made in Fig. 5.1. It is, however, dangerous to generalize too much about the nature of the changes on a world basis, as even over quite short distances trends may have been in opposite directions, or may have shown a time-lag.

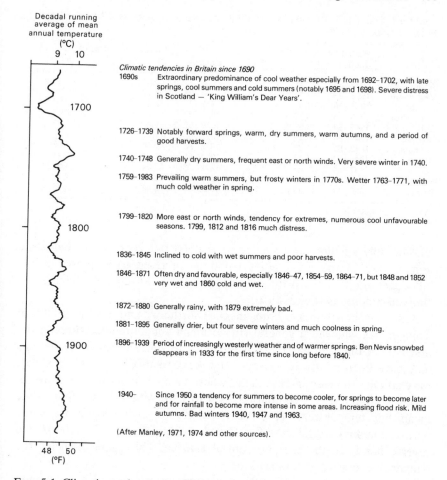

Decadal running
average of mean
annual temperature
(°C)

Climatic tendencies in Britain since 1690

1690s   Extraordinary predominance of cool weather especially from 1692–1702, with late springs, cool summers and cold summers (notably 1695 and 1698). Severe distress in Scotland — 'King William's Dear Years'.

1726–1739 Notably forward springs, warm, dry summers, warm autumns, and a period of good harvests.

1740–1748 Generally dry summers, frequent east or north winds. Very severe winter in 1740.

1759–1983 Prevailing warm summers, but frosty winters in 1770s. Wetter 1763–1771, with much cold weather in spring.

1799–1820 More east or north winds, tendency for extremes, numerous cool unfavourable seasons. 1799, 1812 and 1816 much distress.

1836–1845 Inclined to cold with wet summers and poor harvests.

1846–1871 Often dry and favourable, especially 1846–47, 1854–59, 1864–71, but 1848 and 1852 very wet and 1860 cold and wet.

1872–1880 Generally rainy, with 1879 extremely bad.

1881–1895 Generally drier, but four severe winters and much coolness in spring.

1896–1939 Period of increasingly westerly weather and of warmer springs. Ben Nevis snowbed disappears in 1933 for the first time since long before 1840.

1940–   Since 1950 a tendency for summers to become cooler, for springs to become later and for rainfall to become more intense in some areas. Increasing flood risk. Mild autumns. Bad winters 1940, 1947 and 1963.

(After Manley, 1971, 1974 and other sources).

FIG. 5.1 Climatic tendencies in Britain since 1690.

Warming in the early twentieth century

In many parts of the world, a warm trend occurred in the late nineteenth century and the first decades of the twentieth century which effectively brought to an end the Little Ice Age (see p. 167).

Urbanization almost certainly accounts for some local variations. The Japanese evidence, for example, suggests that between 1910 and 1950 the most rapid rises of temperature occurred in large cities such as Tokyo, Osaka, and Kyoto with amounts of 0.9, 0.6, and 0.9 °C respectively. Japanese scientists found that rural stations showed a rise but that it was

Table 5.1 Changes in temperature conditions (April to June) at European stations between 1860 and 1960

Station	Warmest decade	Mean temperature (°C)	Coldest decade	Mean temperature (°C)	Difference (°C)
Angmagssalik	1926–35	2.14	1899–1908	0.09	2.05
Vestmanno	1889–98	7.69	1948–58	5.63	2.06
Spitzbergen	1951–60	−2.99	1912–21	−5.99	3.00
Haparanda	1945–54	6.22	1873–82	4.08	2.14
Bodø	1945–54	6.53	1873–82	5.29	1.29
Helsinki	1945–54	9.22	1873–82	7.13	2.09
C. England	1943–52	11.82	1879–88	10.46	1.36
De Bilt	1940–9	14.12	1951–60	11.71	2.41
Zurich	1942–51	13.85	1879–88	12.34	1.51
Milan	1943–52	18.98	1879–88	16.80	2.18
Barnaul	1938–47	11.36	1882–91	8.61	2.75

(After Harris, 1964)

considerably smaller, and suggested that 60 per cent of the increased temperature in the great cities could be ascribed to increased urban influences on the microclimate rather than to any general change in climatic conditions (Fukui 1970).

Although most areas, both in the northern, and in the southern hemispheres, showed a general rise in mean annual temperatures in the first half of this century, there is evidence to suggest that some seasons may have been relatively warmer, and some relatively cooler. This was, for example, the case in East Asia, where mean January temperatures in Hong Kong fell 0.8 °C comparing 1884 to 1910 with 1911 to 1940, while mean July temperatures rose 0.2 °C. In Kyoto, over roughly the same period, the January fall was 0.2 °C and the July rise 0.9 °C. Almost the reverse has been the case in Central Europe. This again indicates the danger of excessive generalization.

The date when the amelioration in temperature reached its maximum has varied from area to area. In the British Isles 1931–40 was the warmest decade in the extreme north-west (Stornoway), whereas the warmest decade did not occur until 1943–52 in the south-east (Kew). The maximum was reached in the 1930s in the Middle East, but in the 1920s in Alexandria (Roseman 1963). In Japan the warming went on to 1961, but declined thereafter. Table 5.1 indicates the dates of the warmest and coldest decades, together with their temperature characteristics for Europe.

One consequence of the warming trend can be seen when one looks at the dates of the first and last snowfalls in London from 1811 to 1960. As Table 5.2(c) shows, whereas in the early years of the nineteenth century the mean dates of the first and last falls were separated by over 150 days,

Table 5.2 Snow, ice, and frost frequencies in the nineteenth and early twentieth centuries

(a) Frost days, ice days, and cold days in Sweden

No. of	1861–70	1871–80	1881–90	1891–1900	1901–10	1911–20	1921–30	1931–40
Frost days[1]	121.8	122.3	123.4	124.7	125.0	115.4	117.2	103.4
Ice days[2]	55.9	58.2	57.2	57.9	56.3	55.9	57.2	47.1
Cold days[3]	43	48	24	33	8	19	21	19

1 = days with a minimum temperature $< 0°C$
2 = days with a maximum temperature $< 0°C$
3 = days with a maximum temperature $< 10°C$

(After Liljequist, 1943)

(b) Days of ice cover in Norway and Sweden

Lake	1900–	1910–	1920–	1930–	1940–
Femund (Norway)	176.3	158.8	168.2	156.4	161.8
Mjosa (Norway)	—	71.6	65.2	22.8	49.8
Rossvatn (Norway)	—	164.4	159.0	138.9	144.6
Bolmen (Sweden)	—	105.2	89.4	86.0	93.9
Siljan (Sweden)	131.1	120.2	110.0	108.5	102.0
Storsjon (Sweden)	168.4	155.8	149.9	145.5	144.6
Mean value	—	129.33	123.62	109.68	116.1

(From data in World Weather Records processed by author)

(c) Dates of the first and last snowfall in London from 1811 to 1960

	Autumn	Spring
1811–40	18 Nov.	22 Apr.
1841–70	21 Nov.	17 Apr.
1871–1900	23 Nov.	12 Apr.
1901–30	25 Nov.	15 Apr.
1931–60	8 Dec.	1 Apr.

(After Manley, 1964)

by the period 1931–60 this figure had declined to only 113 days. Even when one compares 1931–60 to 1901–30, the period during which one might expect snow was reduced by around four weeks. This may partly result from the effects of urbanization.

Another consequence of the greater warmth was that the length of ice-cover on rivers and lakes in high latitudes declined appreciably until the 1930s or later. Some data for Norway and Sweden are shown in Table 5.2(b), and of the lakes considered, Mjosa in Norway is the one which shows the greatest decline in ice-cover, with an average of 71.6 days of ice per year for the decade after 1910 falling to only 22.8 days in the 1930s.

Similar data are available for Oxford where, on a temperature basis associated with the first occurrence of five days with temperatures high enough for plant growth, an operational definition of spring has been made (Fig. 5.2,E). For the first fifty years of the period 1869–1970 the ten-year moving mean date of the first day of spring (as defined) fell

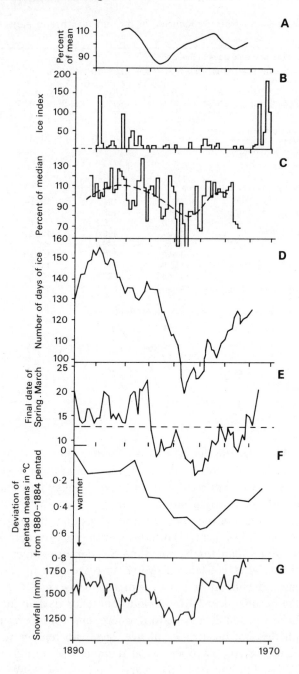

between 13 and 22 March, but as a result of the warming trend of the 1930s and 1940s the date around 1940 was as early as 3 March.

The warming led to a general diminution in the ice-cover in the Arctic Sea, which had major implications for navigation. Off Iceland, both in the 1860s and 1880s, there had been on average between twelve and thirteen weeks of ice in a year. By the 1920s the incidence was down to only 1.5 weeks per year, though by the decade 1947–56, because of the temperature fall already described, this had increased slightly to 3.7 in the year. Equally, the area of drift-ice in the Russian sector of the Arctic was reduced by no less than 1 million km^2 between 1924 and 1944 (Diamond 1958). The ice also tended to become less thick, so that whereas Nansen found that the average thickness of ice in the Polar Sea was 365 cm in the period 1893–6, the Sedov expedition of 1937–40 found that it was down to 218 cm (Ahlmann 1948). Iceberg frequencies off Newfoundland also declined. The annual average for 1900–30 was 432 whereas for 1931–61 it was 351, a decrease of 19 per cent (Schell 1962). The coast of Greenland also became less subject to ice as illustrated by the frequency of years in which the Polar Ice, which comes round Cape Farewell, reached as far north as Godthaab. From 1870 to 1879 it was over 70 per cent; since 1910 it has always been less than 25 per cent (Beverton and Lee 1965).

As a result of the improved ice conditions, the shipping season for the coalfields of West Spitzbergen lengthened from three months at the beginning of the century to about seven months in the 1940s.

The changes of sea-temperature associated with these changes in ice-cover were of a high order. The changes were generally positive, though some areas, notably those affected by the Irminger Current off Iceland,

---

FIG. 5.2 Changes in climatic parameters since 1900.
A. 20-year running means of the mean winter–spring rainfall at 14 stations in North Africa and the Middle East. (From Winstanley 1973.)
B. Extent of ice off Iceland (duration in weeks multiplied by the number of areas with ice along the coasts). (From Schell, 1974.)
C. Variation in annual runoff in the United States as a whole. (From Leopold *et al.* 1964: 62.) The dotted line represents a generalized trend.
D. 10-year running means centred at date given of number of days for season with ice in the Baltic at Stugsund. (From Davis 1972.)
E. Final date of spring at Oxford, England, represented by a 10-year running mean. (From Davis 1972.)
F. Temperature changes of the northern hemisphere, shown by pentad means expressed as deviations from the 1880–4 pentad. (From Kalnicky 1974.)
G. 10-year moving mean of snowfall (in mm) at the Blue Hill Observatory, Mass., USA. (From data in Conover, 1967.)

cooled (Brown 1953). Off the Kola peninsula the water temperatures in the early twenties were 1.9 °C higher than they had been twenty years earlier. Similarly, between 1912 and 1931 the sea-water temperatures off north-western Spitzbergen rose by 1.5 °C. The sea around Iceland has shown a fairly continuous upward temperature trend in most, but not all sectors, amounting to about 1.5 °C between 1925 and 1960. In general the majority of areas showed their greatest rise after 1916–20.

The colonization of the west Greenland continental shelf by the cod (*Gadus morhua*) from Iceland is the best example of the response of fish to the general warming trend. Before 1917, except probably for short periods during the nineteenth century, only small local fiord populations of cod occurred in Greenland. After 1917 large numbers of adult fish appeared off the south-west coast as far north as Frederikshaab (62° N) and they migrated north through 9° of latitude in twenty-seven years (Ahlmann 1948). As a result 10 000 tons of cod were landed in Greenland in 1948 compared with only 5 tons in 1913. The haddock (*Melanogrammus aeglefinus*) and the halibut (*Hippoglossus vulgaris*) showed a similar northward movement towards both Greenland and to Novaya Zemlya. Between 1924 and 1949 swordfish, pollack, twaite shads, and dragonets were recorded for the first time off Iceland. Amongst the species that appeared more frequently were mackerel, tunny, horse mackerel, conger, basking shark, thorn-back ray, mullet, fork-beard, saury pike, and rudderfish. The great silver smelt and the Greenland shark extended their range (Cushing 1976). On the other hand, there was a striking response of typically cold-water forms such as the white whale (*Delphinapterus leucas*) and the capel (*Mallotus villosus*). Their southerly limits have contracted (Beverton and Lee 1965).

The Baltic Sea also benefited from the climatic amelioration. Its salinity increased as a result of increased frequency of south-easterly winds which tended to increase the outflow of brackish surface water from the Baltic and brought about a corresponding increase in the compensating inflow of saline North Sea water along the sea-bottom. Over the period 1933–9 salinities were up to 1.7 per cent higher than during the period 1923–32. High salinity levels improve spawning conditions for cod, and this led to an enormous, perhaps twentyfold increase in the abundance of the Baltic cod, which then supported a major fishery (Beverton and Lee 1965). A rise in salinity of 0.1 per cent also occurred in the north-east Atlantic (1919–38 compared with 1902–17) (Weyl 1968).

The results of the amelioration may also have included the dramatic decline of the Plymouth herring fisheries of the English Channel, the herring fisheries of the Firth of Forth, and the haddock fisheries of the North Sea. The Channel fisheries were partially replaced, after 1935, by warmer-water forms, especially the pilchard (*Sardina pilchardus*) and the cuttlefish (*Sepia officinalis*). A decrease in the amount of zooplankton and of nutrient salts in sea-water, especially in the winter months, was recorded in the Plymouth area during the warming period. In general, however, the results of the increased temperatures were beneficial for the north European fishing industry. A reversal to conditions comparable to those at the beginning of the century is now taking place, and the west Greenland cod fishery has now nearly disappeared. Already cod, ling, and haddock have returned to the Plymouth area of south-western England and the boreal barnacle (*Balanus balanoides*) has become greatly prominent along the shoreline. Many changes in fish numbers must also be put down to overfishing, but the role of climate has been an important one (Russell *et al.* 1971).

Land flora and fauna of Northern Europe have also shown changes in their distribution, though the biological consequences of changes in sea-temperature are likely to be more definite than the corresponding changes on land. The sea is more uniform so that it might be expected that temperature and salinity would offer the main restrictions to the spread of oceanic species. The influence of man also tends to be less direct, though of course, as just noted, overfishing has had near catastrophic effects on the distribution of certain coastal fisheries.

In Finland the polecat (*Mustela putorius*) began spreading into the country about 1810 and by the late 1930s had occupied the whole south Finnish interior to about 63°N (Kalela 1952). The colder the winter and the greater its snowfall the more difficult it is for this mammal to find its natural food of small rodents, frogs, and the like. In north-east Greenland the musk-ox had plenty of food from 1910 onwards, and its numbers increased markedly (Vibe 1967). Likewise, the roe-deer, common in south and central Scandinavia in the years before the Little Ice Age, was almost extinct there by the early nineteenth century, only to reappear and spread strongly northwards after 1870. Also in Finland some permanently resident birds such as the partridge, to which severe snow is inimical, the tawny owl, and many species of tit, extended northwards (Crisp 1959).

In Iceland and Greenland the distributions of birds give an equally clear indication of the effects of the amelioration (Harris 1964). For example, the fieldfare (*Turdus pilanis*) was unknown in Greenland and Jan Mayen before 1937, but breeding is now established there. Starlings (*Sturnus vulgaris*) arrived in Iceland in 1935 and became permanent in 1941. Swallows of one species (*Hirunda rustica*) appeared in the Faeroes and Iceland in the 1930s. The white-fronted goose and the long-billed marsh wren have begun to breed in Greenland, and some species like the mallard and long-tailed duck, which were summer visitors, now remain throughout the year. However, the reduction in numbers of the little auk provides a clear example of how an improving climate may have a directly adverse effect on a species. It feeds on small crustacea such as mysides, which are particularly plentiful in the surface water near the sea-ice front. As the sea-ice front retreated northwards from Iceland the birds in the Icelandic colonies had to fly over larger distances for food, and the colonies were thereby gradually deserted (Crisp 1959).

These environmental changes have also not been without their economic implications. The rise in temperature has increased the growing season of crops. In Finland, for instance, Helsinki showed, for the period 1934–8, twenty-three more days per year without frost compared to the mean for 1901–30. Over the same period there were also twenty-two more growing days (days when the average temperature persists above 5 °C) (Keranen 1952). The data for Sweden show a similar trend (Table 5.4(a)). Trees grew at a greater rate in Arctic Finland, and the Scandinavian countries experienced an extension of rye-, barley-, and oat-growing which was not occasioned solely by the breeding of more tolerant strains.

In various parts of North America, many sub-alpine meadows, which lie at high altitudes between closed forest and treeless alpine tundra, have experienced a massive invasion by trees. Although it might be argued that changes in humanly induced fires, or in grazing by domestic animals, might be contributing factors, it is probable that the bulk of the invasion by trees can be attributed to climate. Tree-ring studies indicate that the most intense phase of invasion coincided with the temperature peak from the early 1920s to the late 1940s. The increased temperatures and diminished depth and duration of winter snow have increased the growing season to the benefit of the trees (Heikkinen 1984).

A cooling episode at mid-century

When one compares temperature conditions of 1900–19 with those of 1920–39 one finds that about 85 per cent of the earth's surface experienced warming trends in mean annual temperature, whereas when one looks at temperature data for the period between 1940 and 1960 about 80 per cent of the total earth surface was probably involved in a net annual cooling (Mitchell 1963). Only a few areas, such as the western United States, New Zealand, south-east Canada, eastern Europe, the Pacific

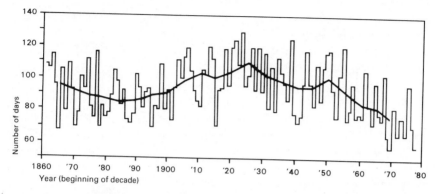

FIG. 5.3 Number of days with general westerly winds over the British Isles from 1861 to 1976. The bold line shows the 10-year averages at 5-year intervals (after Lamb and Morth 1978, fig. 11).

coast of Asia, the Brazilian plateau, and various portions of the western Indian Ocean continued to show a net warming. Parker and Folland (1988) suggest the cooling affected the northern hemisphere from about 1945 to 1970 over land, and from about 1955 to 1975 over the ocean.

The atmospheric conditions which gave the period of warming over Britain and Scandinavia in the first half of this century were highly zonal in type, with westerly winds being dominant. Thus the mean yearly number of days with general westerly winds over the British Isles (Fig. 5.3) was 109 in the 1920s and 99 in the 1930s. In more recent years (Lamb and Morth 1978) their frequency has been less (averaging 74 westerly days per year between 1970 and 1976), and since the mid-1950s the frequency of blocking anticyclones appears to have increased.

The cooling in the mid-years of this century had a series of consequences including the development of snow banks and glacierets in the Canadian Arctic (Bradley and Miller 1972); an increase in snowfall frequencies and quantities in New England (Fig. 5.2,G); a decrease in the length of the growing season in Oxford (Davis 1972) (Fig. 5.2,E); and an increase in sea-ice cover in the Baltic (Fig. 5.2,D).

The warming episode of the late twentieth century

Some parts of the world seem never to have experienced the brief cooling episode of the mid-twentieth century. Warming continued without interruption in New Zealand (Salinger and Gunn 1975) (Fig. 5.4) and over much of Australia (Tucker 1975).

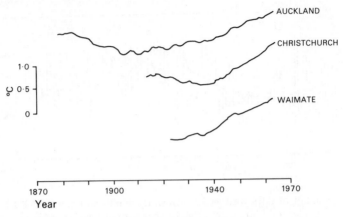

FIG. 5.4 20-year running mean of annual temperature for typical New Zealand stations (after Salinger and Gunn 1975).

In the last two decades, however, there is mounting evidence that warming is once again a rather general feature of the world's climate. This is shown in Fig. 5.5 and discussed by Jones *et al.* (1988). The climb since the 1970s has not yet been as long continued as the climb in temperatures that took place in the first four decades of the century, but there is no doubt that the 1980s was a particularly warm decade in many areas. Indeed, the average temperature for the 1980s was about 0.2 °C above the mean for the period 1950–79.

Precipitation changes during the period of instrumental record

Bradley *et al.* (1987) have attempted to provide a general picture of precipitation changes over northern-hemisphere land areas since the middle of the last century, using available instrumental records. They describe the general patterns of change for a series of major areas.

In Europe they found that annual precipitation totals have increased steadily since the middle of the nineteenth century, with well above average precipitation since a dry spell in the 1940s. Most of the upward trend was evident in the winter precipitation, with lesser amounts in autumn and spring. Summer rainfall, on the other hand, has shown a slight decline.

In the USSR they found that rainfall had increased dramatically since the 1880s, with most of the change occurring before 1900 and after 1940. The increase in annual totals is mainly accounted for by increases in

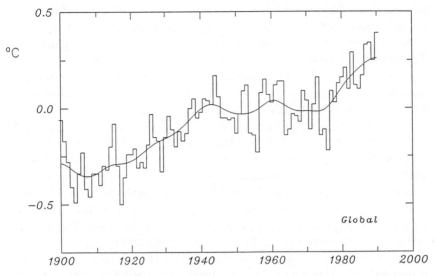

F<small>IG</small>. 5.5 Global average surface air temperature relative to the 1950–1979 average (based on Jones *et al.* 1988). Reprinted by permission from *Nature*, 322: 430–4. Copyright © 1986, Macmillan Magazines Ltd.

autumn, winter and spring. Summer totals have displayed very little trend.

In the USA precipitation totals declined from around 1880, reaching a low in the 1930s, and generally increasing thereafter. Precipitation has increased markedly in the last thirty years, principally as a result of autumn through to spring precipitation increases.

In North Africa and the Middle East very little trend was evident until the 1950s, when, after a relatively wet episode, precipitation declined drastically, especially in summer.

In south-east Asia a relatively wet episode in the 1920s and early 1930s separated two dry periods: the former centred on 1900 and the latter from the mid-1960s through to the present. The general trend for the last 40 years has been one of decline. Summer rainfall in the area shows virtually no trend since the 1870s.

We will now consider the more detailed picture from the British Isles, low latitudes, and desert margins.

Changes in rainfall in Britain

The changes in rainfall that have taken place during the period of instrumental records are as difficult to generalize about as are temperatures.

However, the changes have been considerable. Some of the most detailed studies of rainfall changes have been made in Britain and these show up both the quantitative variability in trends between different regions and the temporal variability in maxima and minima (Gregory 1956).

Those parts of Britain exposed to westerly influences showed a sequence with falling totals from 1881 until 1892–1901; increases until 1909–18; stability until 1922–31 at this high level; and falling totals until 1950. In general most British stations showed a rise in rainfall until the early 1920s, and then a fall, the fall generally starting in 1923–32, but there were regional differences in the beginning of the rise, the rate and mode of the rise, and the date at which most maximum values were recorded. However, taking the period 1900–59 in northern England, a contrast arises between those areas of rapid orographic uplift of prevailing westerly air-masses and juxtaposed areas on lee sides. The former shared a significantly large increase in rainfall amounting to 15 per cent in the Manchester lowland and the Lake District and 10 per cent over Rossendale, the Bowland Fells, and the head of the Lune Valley. Actual decreases occurred on the lee of the Pennines in the Eden Valley and in the Slaithwaite area. Slaithwaite had its maximum from 1910 to 1919 while the far west near the coast had a maximum from 1923 to 1932 (Barrett 1966).

Observations such as these have recently been updated by Wigley and Jones (1987). After analysis of precipitation data back to the 1760s (Fig. 5.6) they concluded (p. 245):

In spite of popular perceptions that precipitation values are changing, none of the regional, annual or seasonal time series show trends on time scales of 30 years or more. There is, however, considerable variability on time scales of the order of decades, more so in spring and summer than in autumn and winter. Spring precipitation showed a strong and steady increase over the period 1956–1969, especially in the NW and NE regions. In recent years, spring precipitation has been highly variable, with remarkably wet seasons occurring in 1979, 1981 and 1983. In contrast, spring 1984 was unusually dry in the NW region. Summer precipitation showed a decline from the late 1960s to the mid 1970s, with no trend since then. There has, however, been an unusual number of dry summers in the past 10 years (1976–1985).

One of the most important controls of agricultural activity is the incidence of drought conditions, and the particularly severe drought of 1976 caused a reassessment of the significance and frequency of these phenomena in Britain. Droughts are partially caused by low precipitation, and partly by high rates of evapotranspiration associated with high temperature levels. A combination of these two factors will cause a soil

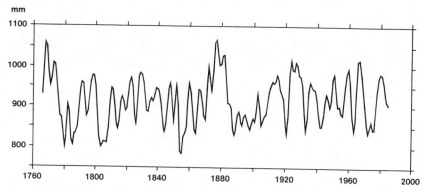

FIG. 5.6 England and Wales area-average precipitation. The data are annual totals, smoothed with a five-term binomial filter. In order to give filtered values for the end years, the series was padded at each end with two extra years using the mean values for 1766/7 and 1984/5. (From Wigley and Jones 1987, fig. 2.) Reproduced with the permission of the Royal Meteorological Society.

moisture deficit that can lead to stunted growth or even wilting of crop plants.

The incidence of major soil moisture deficits during the growing season of British crops (May to August inclusive) has been plotted from data for Kew (London) that go back to the end of the seventeenth century (Wigley and Atkinson 1977). From these data (Fig. 5.7,A) it can be seen that there have been fluctuations of some importance, and this is brought out further in Fig. 5.7,B. In particular it is noteworthy that of the 14 worst soil moisture deficit years, 8 have occurred since 1900. Moreover, more serious droughts occurred in the period 1960–76 than in any preceding 20-year period. It is likely that the temperature increase that has prevailed from 1880 has caused some of the drier soil conditions of this century.

Another important aspect of rainfall variability is the incidence of rainfall-related flooding. There is some evidence that serious flood frequencies have shown increases in recent decades. Howe *et al.* (1966) analysed flood data for the River Severn (at Shrewsbury) and the Wye (at Hereford). They found that during the period 1911–40 a flood height of 5.1 m was to be expected at Shrewsbury only once in 25 years; during the period 1940–64 the Severn reached this height once every 4 years. Fig. 5.8,A shows the trend for the Wye, and demonstrates the great clustering of high flood events since about 1930–40. The reasons for these increasing flood levels are complex, but probably include deliberate peat drainage in the Welsh uplands and an increase in the frequency of daily

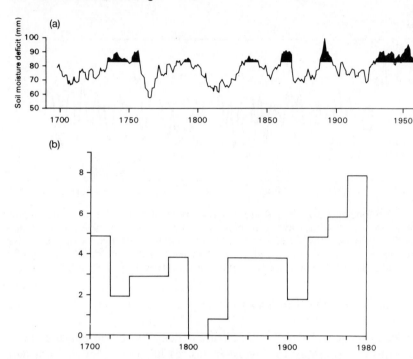

F IG. 5.7 Soil moisture deficits at Kew, London (after Wigley and Atkinson 1977). Reprinted by permission from *Nature*, 265: 431–4. Copyright © 1977, Macmillan Magazines Ltd.
A. 10-year running means of soil moisture deficit averaged over the growing season at Kew.
   An arbitrary datum level of 84 mm has been shown to accentuate the periods of higher deficits.
B. Number of times the soil moisture deficit averaged over the growing season exceeded 100 mm in 20-year intervals. Note the low incidence in the early nineteenth century and the very high incidence in the latest period.

rainfalls greater than 63.5 mm since 1940 (Fig. 5.8,B). Rodda's work (1970: Fig. 5.9) showed that an increase in intense rainstorms had also occurred in central England (as represented by climatic data from the Radcliffe Meteorological Station, Oxford). For the period 1881–1905 the return period for a storm of just over 50 mm was about 30 years, but for the period 1941–65 the return period of the same size of daily fall had dropped to little more than 5 years.

This climatic explanation for increasing flood levels has been confirmed for South Wales by Walsh *et al.* (1982). In the case of the Tawe valley near Swansea, of 17 major floods since 1875, 14 occurred from 1929–81

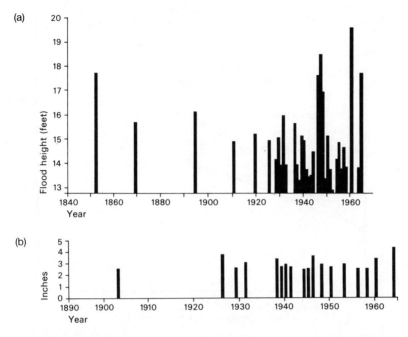

FIG. 5.8 Changes in flood levels and daily rainfall levels for the River Wye and Lake Vyrnwy, western Britain.
A. River Wye: recorded flood levels at Wye Bridge, Hereford.
B. Lake Vyrnwy: frequency of daily rainfalls of at least 2.5 inches (63.5 mm).

and only 3 during the 1875–1928 period. Significantly, of 22 notable widespread heavy rainfalls in the Tawe catchment since 1875, only 2 occurred from 1875 to 1928, but 20 from 1929 to 1981. However, Lawler (1987) has sounded a note of caution about over-generalizing about this particular climatic trend and argues not only that no simple synchroneity across the country exists in any changes identified but that since 1968 there has in some areas been a reversal in the trend of increasing storm rainfalls and associated floods.

Rainfall changes in the nineteenth and twentieth centuries in the low latitudes

Although Nicholson (1981) has surveyed various types of data to construct a history of droughts in Africa before the availability of direct meteorological observations, and has succeeded in demonstrating the existence of a widespread period of low rainfall over much of the con-

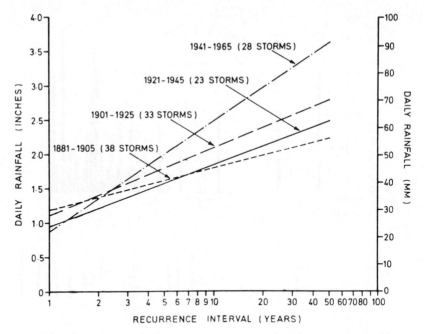

F<small>IG</small>. 5.9 Changes in the magnitude–frequency relations for daily rainfall amounts at Oxford exceeding 25 mm (after Rodda 1970, fig. 8).

tinent in the 1820s and 1830, it is not until the last decades of the nineteenth century that fully reliable data became available.

In many parts of the tropics and subtropics the period correspond-ing to the warming phase of the early twentieth century was a time of decreased precipitation. This is, for instance, shown by the data for eastern Australia presented in Table 5.5(b) and (c). In all about 2.5 m km² of Australia showed significantly decreased precipitation for the period 1911–40 compared with 1881–1910. Only 0.25 m km² showed an increase. The decreases were especially severe in the semi-arid area near Bourke (Fig. 5.10) where a decrease of 75 mm in mean annual rainfall took place at this time. This was equivalent to a regression of some 100 km in the isohyets (Fig. 5.11) (Gentilli 1971).

In the dry zone of south Asia, another region where any deterioration in rainfall would have severe human implications, there was a noticeable change in rainfall around 1890 to 1895. Conditions had been relatively wet in the 1880s and 1890s, but then there followed a period of low precipitation, with precipitation in the driest decadal period being gen-erally only between 52 and 69 per cent of that for the wettest decade of

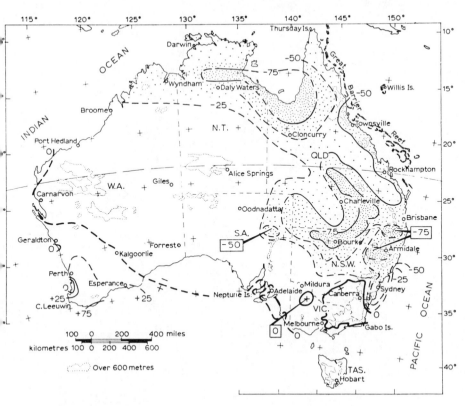

FIG. 5.10 Climatic changes in Australia since 1881 (after Gentilli 1971). Isopleths of the yearly rainfall difference (1881–1910 minus 1911–40).

this century. This change in regime is well illustrated by the graphs of the ten-year moving means of precipitation at Lahore and Karachi (Pakistan), Jaipur (Rajasthan), and Agra (United Provinces) (Fig. 5.12). After about 1940 or 1945 there seems to have been a return to more positive rainfall conditions.

This is mirrored in the record for central and southern Africa, where after relatively moist conditions in the pre-Boer War period, an abrupt change in rainfall conditions came in the mid-1890s which lasted until the 1930s.

Work elsewhere in the tropics, as Table 5.3(a) illustrates, suggests a similar rainfall decline in the first quarter of this century for a wide range of tropical locations. This decline resulted from a shortening of the wet season and a narrowing of the rainfall belts. Locations on desert margins

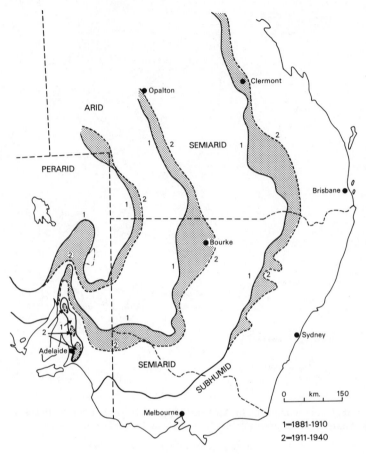

F ig. 5.11 Climatic changes in Australia since 1881 (after Gentilli 1971). Shifts in the climatic belt boundaries (1 is 1881–1910, and 2 is 1911–40).

showed a relatively greater decline. In West Africa, for example, Banjul in the Gambia showed a more substantial change than Freetown in Sierra Leone.

The data from Table 5.3(b) from the eastern parts of the USA and Australia further illustrate the decline in totals since the 1880s to a minimum in the period 1900–40. They also indicate that some rise has taken place since 1941.

However, in the 1960s a particularly sharp increase in rainfall took place in the equatorial parts of East Africa. The rainfall figures for the 36-

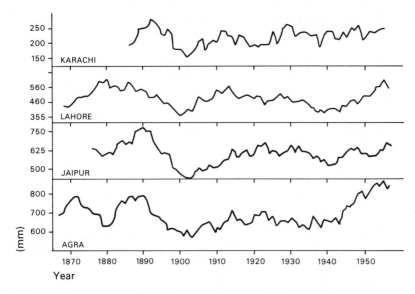

F<small>IG</small>. 5.12 10-year moving means of annual precipitation at Indian dry zone and Pakistan desert stations (from Goudie 1972).

month period up to mid-1964 were, according to Lamb (1966*b*), 130 or 140 per cent of the 1931–60 averages, and in places were over 250 per cent. This had very serious consequences in terms of changed discharges for the Nile and higher levels for the East African lakes. The Nile, for instance, at its outlet from Lake Victoria, had a mean discharge for the 63-year period before 1962 of around 600 m^3/sec. After 1963 this has more than doubled to a figure of around 1200–1300 m^3/sec.

However, this marked rainfall increase did not occur universally in the tropics, and the zones between 10–15° and 30° north and south of the Equator were drier than usual, with catastrophic drought conditions being experienced in, for example, Botswana, where during parts of the mid-1960s, the bulk of the population were forced to subsist on famine relief. Similarly, the Dead Sea, unlike the equatorial lakes, had a very high level between 1898 and 1932 of −393.1 to −395.3 m and then dropped very rapidly between 1957 and 1963 to as little as −398.8 m (Klein 1965) as a result of both increased irrigation in the Jordan valley and low rainfall.

In southern Africa, analyses of long-term precipitation changes have been made by Tyson (1986). He could find no statistical evidence that southern Africa is becoming progressively drier (as has often been popularly maintained), but he did find evidence for a number of quasi-periodic

Table 5.3 Rainfall fluctuations in the tropics, USA, and Australia in the late nineteenth century and in the first half of the twentieth century

(a) Tropical Rainfall

	Percentage deviations from 1881 normal of mean annual rainfall	
	1874–98	1907–11
Barbados	+14	− 9
Bogota	+10	− 3
Colombo	+ 4	− 8
Freetown (1875–99)	+11	−12
Georgetown (Queensland)	+20	− 8
Havana	+10	− 5
Honolulu	+13	−12
Recife	+42	−11
Townsville	+17	− 4
Trinidad	+10	− 7
Vizagpatan	+10	0

(b) Average rainfall as per cent of 1881–1940

	1861–80	1881–1900	1901–20	1921–40	1941–60
E USA (7 stations) (30–43°N)	125	109	91	99	108
E Australia (4 stations) (19–38°N)	113	111	96	93	108

(c) Mean rainfall in Queensland (cm/year)

	Georgetown	Townsville	Gilbert River
1872–96	95.25	137.2	97.3
1911–40	73.2	90.7	68.3

(d) USA rainfall changes (percentage deviations from the mean)

	Mean 1881–1940 (cm)	1861–1900	1901–40
Charleston	112.3	+21.4	−7.8
Washington	103.9	+ 5.8	+1.0
New York	108.7	+ 4.3	−2.9
Albany	83.1	+17.1	−2.9
Boston	100.8	+12.3	−2.3

(After Kraus, 1954, 1955 (a), 1955 (b))

rainfall oscillations. He reported (p. 197) that the most noteworthy of these 'is that with an average period of about 18 years. Since the turn of the century eight approximately 9-year spells of either predominantly wet years, which average to show above-normal rainfall during the spell, or predominantly dry years, showing below-normal rainfall, have occurred.'

Analysis of long-term precipitation data for the western Mediterranean area (Maheras 1988) enable an assessment of fluctuations between 1891 and 1985 (p. 187): 'On an annual basis, two principal moist periods occur (one from 1901 to 1921, another from 1930 to 1941) and also two secondary ones; between, we have the dry periods, primary and secondary, the second being the most marked (from 1942 to 1954). Another important feature is the decrease in precipitation beginning in 1980 and lasting up to 1985.' Maheras believes that annual precipitation in the western Mediterranean over the past century has shown an approximate periodicity of 20 years.

Further east, in the Balkans the 1930s also emerge as a wet decade (Maheras and Kolyva-Machera 1990) and the early 1980s as a very dry phase.

One of the most interesting, and, it has to be said, least comprehended features of climatic change in low latitudes in the twentieth century, has been the change that has taken place in the frequency of hurricanes. For example, in the American tropics their frequencies have increased so that while there were fifty in the period from 1911–20, there were over one hundred in the period 1950–60 (Dunn and Miller 1960). In view of the damage inflicted by hurricanes this is probably economically and socially highly significant. Also, since the start of this century, the location of hurricane tracks has undergone some change which seems to be .correlated with changes in sea-water temperatures (Riehl 1956). In the early years of the century most recurvatures in the hurricane tracks took place to the east of Florida. They then shifted westwards to the Gulf between 1910 and 1920 (a period of relatively cool sea-water temperatures); later (after 1920) they returned at first to Florida and adjoining waters, and finally (in the 30s and 40s) to the west Atlantic. In all, the shift in the average longitude of hurricane track recurvature near latitude 20° N was no less than 20°. In general, when sea-water temperatures decreased the hurricane tracks migrated westward, and when temperatures increased they returned eastwards.

The increase in hurricane frequencies in the Americas appears to have been followed in the Indian Ocean, off Australia (Milton 1974), and in Japan (Fujita 1973). Moreover, some investigators (e.g. Grant 1981) believe that there appears to be no clear sign that this activity is declining. Certainly, Fig. 5.13 shows a generally increasing trend for much of the period since the 1920s. Indeed, Nunn (1990) reports a great increase in tropical cyclones with hurricane-force winds in the South Pacific. Only 12 such events affected Fiji between 1941 and 1980 (3.1 per decade on average), yet 10 occurred between 1981 and 1989 (11.4 per decade on average). It is tempting to see such reported increases in hurricane frequencies over the last century or so as a result of a general warming trend. However, the Intergovernmental Panel on Climate Change (Houghton *et al.* 1990: 232–3) was wary of such an interpretation:

Current evidence does not support this idea, perhaps because the warming is not yet large enough to make its impact felt. In the North Indian Ocean the frequency of tropical storms has noticeably decreased since 1970 . . . There is little trend in

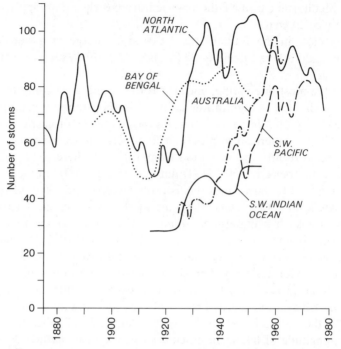

FIG. 5.13 Cyclone frequencies, Bay of Bengal, North Atlantic, south-west Pacific, Australian, and south-west Indian Ocean regions, 1880–1980 (partly after Milton 1974, in Spencer and Douglas 1985, fig. 2.3). Reproduced by permission of Routledge.

the Atlantic . . . There have been increases in the recorded frequency of tropical cyclones in the eastern North Pacific, the south-west Indian Ocean, and the Australian region since the late 1950s. However, these increases are thought to be predominantly artificial and to result from the introduction of better monitoring procedures.

Further information on changing hurricane frequencies is given in Spencer and Douglas (1985), and in Reading (1990).

Changing rainfall patterns on desert margins

In recent decades great concern has been expressed that climatic deterioration may be contributing to the phenomenon of desertification. The situation appears to be complex. When one examines recent rainfall data for the arid areas of the Sudan–Sahel in Africa, central Australia, north-west India and Arizona, USA (Fig. 5.14), it is apparent that some areas

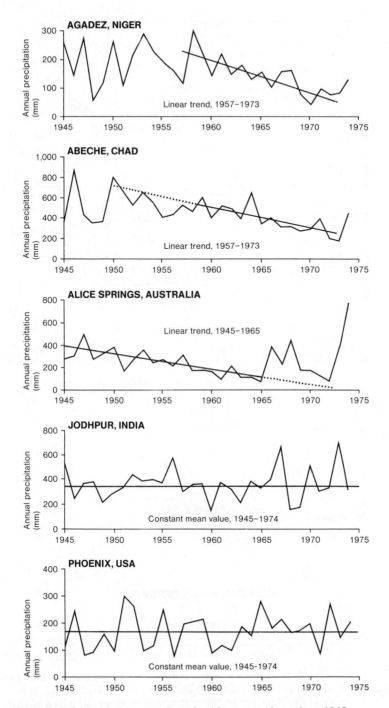

FIG. 5.14 Rainfall variations at selected arid-zone stations since 1945.

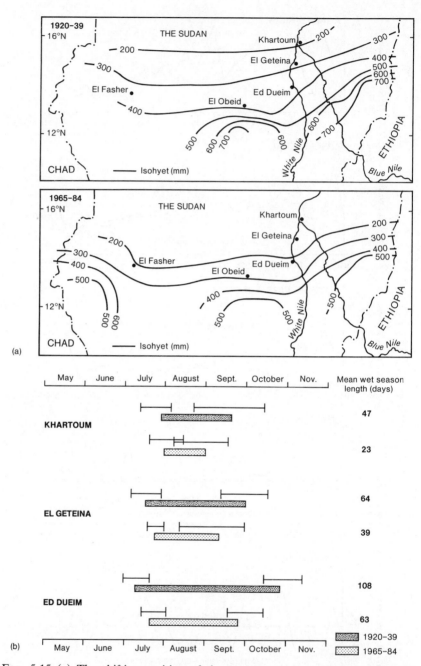

FIG. 5.15 (a) The shifting position of the mean annual rainfall in the Central Sudan between 1920 and 1939, and 1965 and 1984. (From Walsh *et al.* 1988, fig. 4.) (b) Changes in the duration of the wet season at three locations in Central Sudan over the same period. The bars indicate the interquartile ranges of wet season onset and termination dates. (From Walsh *et al.* 1988, fig. 5.)

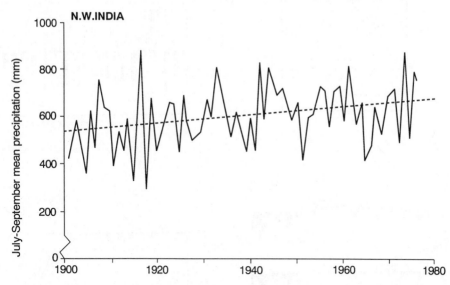

FIG. 5.16 The trend of July–September mean rainfall for north-west India (modified after Pant and Hingane 1988).

show relatively little evidence of a downward trend in the last three or four decades, whereas others do.

Recent climatic deterioration has certainly been severe in the Sudan (Fig. 5.15) where, in White Nile Province, annual rainfall in 1965–84 was 40 per cent below 1920–39 levels, and wet season length contracted by 39–51 per cent (Walsh *et al.* 1988). The dry epoch which started in the mid-1960s has continued and intensified in the 1980s. Further west, the dry epoch has had dramatic effects on Lake Chad. Its area declined from 23 500 km^2 in 1963 to about 2000 km^2 in 1985, which is probably the lowest level of the century (Rasmusson 1987: 156). A recent appraisal of rainfall trends in the Sahel is provided by Dennett *et al.* (1985) and for the western Sudan by Eldredge *et al.* (1988). Both studies agree on the existence of a clear downward tendency since the mid-1960s.

In the Rajasthan Desert of north-west India the trend of rainfall in the twentieth century appears to be very different from that in the Sahel. The latest analyses of monsoonal summer rainfall for the Rajasthan desert (Pant and Hingane 1988) indicate that there has been a modest upward trend in precipitation levels between 1901 and 1982 (Fig. 5.16).

In the drought-prone region of north-east Brazil, Hastenrath *et al.* (1984) have undertaken an analysis of rainfall records since 1912 (Fig.

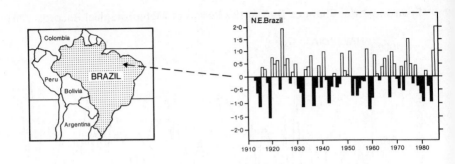

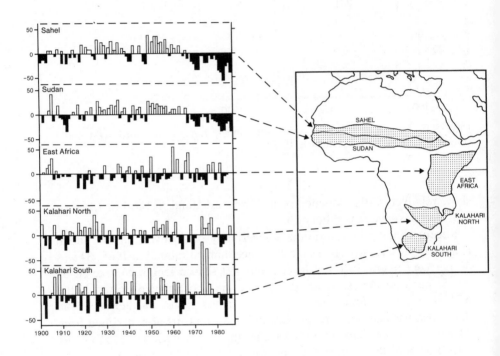

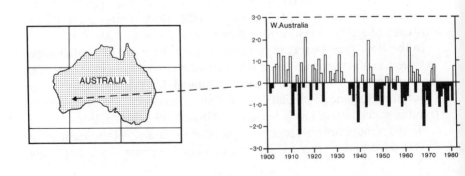

5.17A). The incidence of runs of dry years is thereby highlighted, but there is no very conspicuous evidence of any long-term trend, either upwards or downwards.

In Australia, Hobbs (1988) provides an up-to-date analysis of rainfall trends, and data for Western Australia are presented in Fig. 5.16C. There is no clear-cut trend comparable to that found in the Sudan and Sahel belts of Africa. However, some trends are evident, though they vary in direction across the continent. Hobbs concludes (p. 295):

The picture of variability for Australia is complex in both time and space, but this is not unexpected in view of the size of the continent. The mediterranean climatic regions [South Australia and Western Australia] . . . both show considerable rainfall variability on apparently irregular time scales. The major variations in the two regions have been out of phase with each other. . . . The evidence for any sustained long-term climatic changes, at least as far as rainfall is concerned, is unclear.

River discharge fluctuations

The publication by UNESCO in 1971 of many carefully selected discharge records for some of the world's major rivers enables one to see the way in which river discharges have fluctuated in response to the changes in both temperatures and precipitation. The rivers selected by UNESCO are ones for which there are long, reliable records and where the direct effects of human interference (such as irrigation, diversion of drainage, and the like) are not too significant. The data have been analysed by the author to obtain ten-year moving means of the mean monthly annual discharges in m^3/s (Goudie 1972). Thirty rivers from the northern hemisphere were selected for this study on the basis of the length and continuity of their records.

The graphs of the variability in ten-year moving means, some of which

←_____

FIG. 5.17 The trend of annual rainfall in selected dryland areas in the twentieth century.
A. N. E. Brazil from 1913–85 showing the average annual rainfall expressed as the normalized departure (σ) from the mean.
B. Africa from 1901–87 showing rainfall expressed as a percent departure from the long-term mean.
C. Western Australia showing yearly winter rainfall expressed as the normalized departure (σ) from the mean.

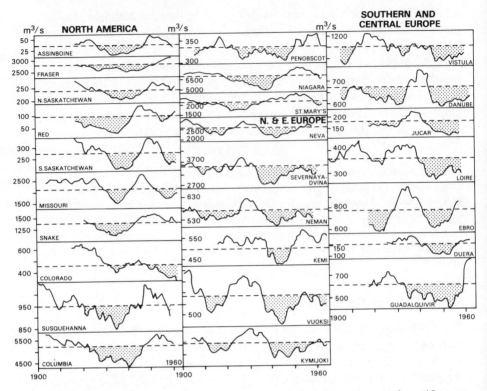

F<small>IG</small>. 5.18 Data on river discharge for selected stations expressed as 10-year moving means of the mean annual monthly discharge. Periods of low discharge are shaded (after Goudie 1972).

are reproduced in Fig. 5.18, show that considerable fluctuations have taken place, and a better impression of this can be gained by examining the ratios of the maximum to minimum decadal discharges for the period of observation. The mean ratio for the 30 rivers was 1.78, though there was a range from 1.19 to 6.49. This is equivalent to having had minimum mean ten-year periods with discharges only a little over 50 per cent of the maximum mean ten-year periods.

Analyses of the dates of the maximum and minimum mean ten-year periods do not suggest any general progressive decline in discharges as some of the proponents of the concept of progressive desiccation might have hoped. However, of the 30 rivers considered, no fewer than 17 showed their minima between 1935/6 and 1945/6 (mid-years of the ten-year periods). The maxima are much less clustered, though 9 rivers showed maxima between 1948/9 and 1958/9, and many others showed

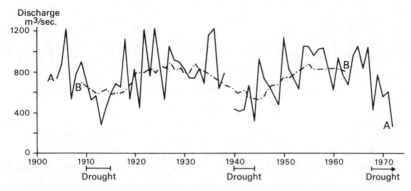

FIG. 5.19 The discharge of the Senegal River at Bakel since 1900.
A. The record of mean annual discharge (after Sircoulon, in Rapp 1974). Three drought periods of 1910–14, and 1968–72 are evident. Data for 1939 are missing.
B. 10-year moving mean of discharge (after Goudie 1972).

something of an upward trend in this period after low levels in the 1930s and 1940s.

It will be noticed how the North American rivers, of which 13 are used in Fig. 5.18, showed in almost all cases a general decrease during the first three or four decades of this century, with the lowest discharges being attained during the so-called 'dust bowl' years of the 1930s. This was a time of higher than average temperatures and lower than average precipitation over much of North America (see also Fig. 5.2,C).

The Russian rivers, Neman, Neva, and Severnaya-Dvina also show their minimum flows in the early 1940s and in this way they compare closely with the Finnish rivers, Kymijoki, Vuoksi, and Kemi. However, some of the rivers from further south or west in Europe (Labe, Danube, Duera, Guadalquivir, Jucar, and Ebro) show either their minima or a discharge trough in the late 1940s or early 1950s. The Russian and Finnish discharges also exhibit a secondary trough between 1910 and 1920.

In this respect they are comparable to the two rivers considered from West Africa, the Niger and the Senegal. They show low discharges in the period 1910 to 1920, and again, after higher discharges, low discharges in the early 1940s. These periods of low flow were comparable to those of the late 1960s and early 1970s (Fig. 5.19).

The discharge of the White Nile has also fluctuated markedly. The minimum ten-year period of discharge as determined at Lake Albert occurred in 1926–7, a time when the lake levels in East and Central Africa were also low. For the decade centred on 1926/7 the discharge was

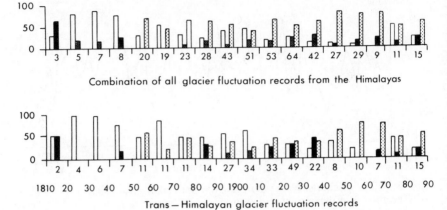

FIG. 5.20 Glacier fluctuations in the Himalayan region (after Goudie *et al*. 1983).

only 19.2 milliards of cubic metres per year while for the wet decade centred on 1915/16 it was up to 28.0.

Glacial fluctuations in the twentieth century

The last Little Ice Age, as already noted, had largely ended by the end of the nineteenth century. Since then many mountain glaciers have been in retreat, often at very fast rates. Some have disappeared altogether, and in the tropics, for example, where recent glacial histories seem to be comparable to those of higher latitudes, six glaciers have disappeared from Ruwenzori since they were first described by explorers in the middle of the last century, and at the present rate Mount Speke will be completely deglaciated within less than four decades (Whittow *et al*. 1963). The Elena Glacier on Ruwenzori averaged a retreat of 5 m per year between 1900 and 1952, but this accelerated between 1952 and 1960 to a rate that reached between 6.5 and 25 m per year. In the Himalayas most glaciers are also retreating (see Fig. 5.20). When first visited in the early nineteenth century most of them were either advancing or at a standstill. By the 1860s, however, retreat became evident in many cases, and with the exception of a phase of advance that affected some of the great Karakoram Glaciers, at around the turn of the present century, retreat has persisted up until the present (see Goudie *et al*. 1983).

Similar rates of recession have occurred in high latitudes. Between 1899–1901 and 1936, for example, the Lady Franklin Glacier of Svalbard retreated about 2.5 km (Thorarinsson 1940). The Jostedals Glaciers of Norway on average retreated about 160 m between 1910 and 1921. They then advanced 60 m until 1930 and retreated at a steadily increasing rate so that 58 m of recession was recorded by 1946 (Ahlmann 1948). Equally, the Svartisen Ice Cap was reduced in area from 469.1 km^2 (1894–1905) to 400 km^2 (1965) (Theakstone 1965). The Jostedals advances were paralleled in the Alps where retreat was arrested about 1906, and advance developed, culminating between 1916 and 1920, with almost 75 per cent of Alpine Glaciers participating. However, after 1926 the bulk of the Alpine glaciers were in retreat (Fig. 5.21), largely because of a reduction of cyclonic conditions compared to the decades 1886–95 and 1906–15 which were followed by glacier advances (Hoinkes, 1968). However, the advances of 1915–25 were generally inadequate to bring their snouts back to the 1895 positions.

The loss of glacial area in these regions was considerable. By 1925 or thereabouts, for instance, about a 25 per cent reduction in area of Swiss glaciers had taken place compared to the position in 1875. Similarly, in Italy, of the 239 Lombardy glaciers at the turn of the century, no fewer than 66 had disappeared completely between 1905 and 1953.

In spite of several decades of decreased temperatures in Europe most Alpine glaciers continued their retreat into the 1960s but by the late 1960s there were signs of strong advance in the Swiss, Italian, and Austrian Alps.

The Alaska record shows the same basic features. Since their Little Ice Age maximum, for example, Herbert and Eliot Glaciers had receded 3.2 km and 610 m respectively. The Muir ice-front has been particularly conspicuous by its retreat. Its area has been reduced by 450 km^2 between the 1890s and 1940s; during the period 1899–1913 alone it receded about 12.9 km, and the Muir Glacier of the 1890s had been dismembered into twelve separate glaciers. The Lemon Creek Glacier recession since 1759 has been well documented (Heusser and Marcus 1964), and it is interesting to see that the rate of retreat, which had been very rapid since 1891, was somewhat reduced in the period 1902–29, which seems to correspond partially with the period of partial advance in the European Alps.

Not all glaciers, however, have followed the general pattern outlined above. In Alaska, for example, the Taku Glacier, in the Juneau ice-field, advanced 5.6 km in forty-eight years while the other glaciers were retreating (Lawrence 1950). Equally the Crillon Glacier in south-east Alaska advanced at a rate of 28 m per year between 1894 and 1933, and showed

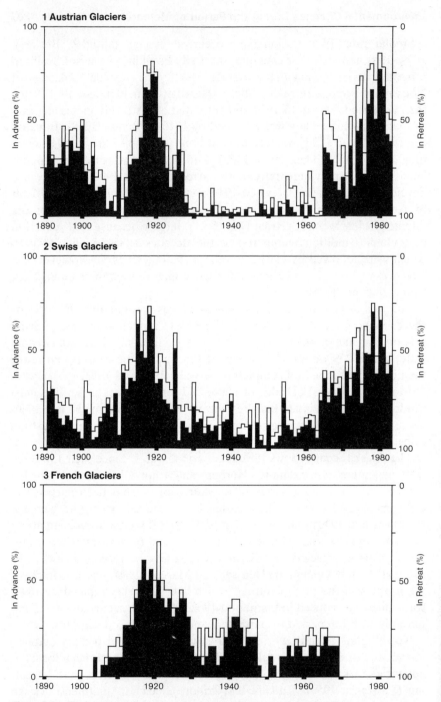

FIG. 5.21 The behaviour of glaciers in Austria, Switzerland, and France since 1890, in terms of the percentage of those observed each year which were found to be advancing, retreating and stationary. Notice the great concentration of retreat round about 1920 and again in 1980 (after Grove 1988, fig. 6.7).

an overall advance of 4.5 km since 1768 (Goldthwait *et al.* 1963). In the European Alps the Brenva Glacier showed an advance between 1925 and 1940 when all its neighbours were in retreat.

The reasons for such anomalous movements are many. In the case of the Brenva it was due to the effects of a spread of protective avalanche debris which descended on to the Glacier in 1920 from Mont Blanc de Courmayeur (Grove 1966). The Taku and Crillon Glaciers of Alaska owe their anomalous advance to the fact that they are nourished from a higher source area. Taku's source begins at 1800 m, for example, whereas nearly all the other ice streams begin at 1200–1500 m (Heusser *et al.* 1954). The advance of the Jan Mayen glaciers since the early 1950s is due to a greatly increased cyclonic precipitation in these high latitudes which has more than outweighed any increased temperatures associated with the cyclonic activity (Lamb *et al.* 1962). It has been estimated that precipitation in the 1950s had almost doubled in comparison with that of the 1920s. Indeed, conditions which cause one glacier to advance may cause another to retreat. In western Norway, for example, advection of warm moist cyclonic air generally plays the greater part in ablation, whereas in Sweden, radiation, which is often reduced during cyclonic conditions, appears to be even more important.

Ground laid bare by the retreat of ice-caps and glaciers under the pressure of the relatively warm conditions of the first half of the twentieth century, became colonized by vegetation in stages. In Alaska three main stages have been found in the succession (Lawrence 1958). The first, or pioneer stage sees initial colonization by hardy *Rhacomitrium* mosses, a gradual increase in perennial herbs, particularly the broad-leaved willow herb and horsetail, and finally the establishment of *Dryas drummondii*, the latter being a low-growing evergreen under-shrub.

The second stage, the thicket stage, witnesses the appearance of dwarf, creeping willows which together with the *Dryas* lead to an increasing degree of shade which leads to the gradual suppression of the shade-intolerant mosses and herbs of the pioneer stage. Towards the end of this stage there is a dominance of shrubs of willow (*Salix* spp.) and alder (*Alnus* spp.). Alder is important because it is an important source of soil nitrogen enrichment, and thereby provides improved soil conditions for the next stage in the succession.

The third stage in the succession is the forest stage when there is a dominance of Sitka spruce (*Picea sitchensis*) and later a mixture of spruce and hemlock (*Tsuga* spp.).

This sequence whilst only strictly applicable to the Glacier Bay area of Alaska nevertheless gives a broad view of the situation that probably existed in most areas during deglaciation.

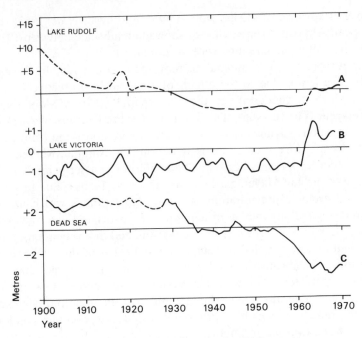

FIG. 5.22 Annual changes in lake levels in the twentieth century.
A. Lake Rudolf (Turkana), East Africa, showing the low levels from 1930 to 1960, and the relatively high levels around 1900 and since 1960.
B. Lake Victoria, East Africa, showing the marked 'stepwise' rise in level since 1960.
C. Dead Sea, showing the generally positive hydrological budget from the start of the century to around 1930, and the highly negative budget during the 1960s, resulting in part from the diversion of Jordan waters for irrigation. (From Butzer 1971, figs. 5.5 and 5.4.)

Changing lake levels

One of the most interesting examples of environmental change in the present century has been the fluctuating level of lakes in the tropics. In particular many equatorial lakes in Africa showed a dramatic increase in level in the early 1960s, which led to the flooding of port installations, deltaic farming land, and the like (Butzer 1971). This rise contrasted sharply with the frequently low levels encountered in the previous decades. Lake Malawi (Nyasa) seems to have reached a minimum around 1927–9, Lake Tanganyika was very low in the 1920s and again between 1948 and 1956, Lake Victoria's lowest level was reached in 1922, while Lake Naivasha showed a very sharp progressive fall after 1938. The relatively dry phase which seems to have been the dominant reason for

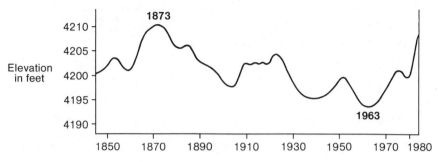

FIG. 5.23 The changing level of the Great Salt Lake, USA (1851–1984) (after Stockton 1990).

these low levels in the 20s and following decades started to develop in the 1880s in the basins of Lakes Nyasa, Tanganyika, and Victoria but rather later (around 1898) in the Turkana and Stefanie (Chew Bahir) basins. The lakes studied by the colonial scientists were thus very different from those described by the great explorers and may have contributed towards the concept of progressive desiccation and desertification which concerned foresters and others in Africa between the wars.

In the 1960s the level of the lakes of East and Central Africa rose sharply, Lake Tanganyika stood about 3 m higher in 1964 than it had in 1960, Lake Victoria rose by 1.5–2.2 m (Fig. 5.22), and rises of 2.3 m have been recorded for Lakes Baringo, Nakuru, and Manyara. Lake Turkana began to rise 4 m in late 1961 and submerged over 300 km² of the Omo Delta.

The Great Salt Lake in Utah, USA, has an especially impressive record of fluctuating levels dating back to the middle of the last century. As Fig. 5.23 shows, the lake rose from an elevation of about 4200 feet in 1851 to a peak of around 4210 feet in 1873. Thereafter it declined very markedly, reaching the lowest recorded level in 1963. Since then it has risen from around 4194 feet in elevation back to the sort of elevation achieved in the 1870s. There has thus been a total fluctuation of this great water body of the order of 5 m over that time period (Stockton 1990). The latest rise is a consequence of a run of very wet years since the mid-1970s.

One demonstration of the complex relationship between changes resulting from the activity of man, and changes resulting from natural causes can be seen in the recent history of the Caspian Sea. Since 1929 this large inland sea has fallen by 3 m in level (Fig. 5.24). This has resulted largely because of a reduction in river inflow from the Volga, which, on average, contributes 80 per cent of the overall surface dis-

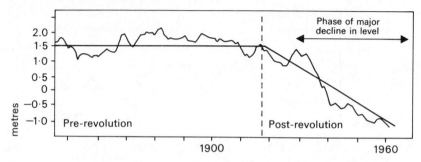

FIG. 5.24 Annual fluctuations of level (in m) of the Caspian Sea, illustrating the rapid post-revolutionary trend.

charge to it. Thus over the period from 1929 to 1965 the annual loss of water exceeded the annual gain by around $26 \, km^3$, though the deficit was at its greatest between 1930 and 1945 when there was an average deficit of $50 \, km^3$. The economic consequences of this change in inflow and in the level were marked and generally adverse. The Caspian fishing industry diminished as a result of the increasing salinity and the reduction in the numbers of shallows, and fish production declined from about half a million tons per year between 1925 and 1935 to only 82 000 metric tons in the period 1965–8. Moreover, the quality of the fish also declined, with the proportion represented by prized types such as sturgeon, white fish, salmon, and herring dwindling particularly strongly (Micklin 1972).

The causes of this phenomenon are both natural and man-induced. The climatic factor may well have been predominant, particularly in the earlier phases. In the years before 1929 the airflow over the European USSR was predominantly westerly but this changed to a chiefly merid-ional and easterly pattern during the 1930s and 1940s. As a result the number of depressions penetrating from the Atlantic dropped, whereas the frequency of dry anticyclones from the Arctic and Siberia went up fairly substantially during the winter season. Lake and river (see p. 205) discharges declined. However, especially over the last four decades, reservoir formation, irrigation, and municipal and industrial withdrawals have been of major importance.

Changing dust-storm frequencies

The changes in temperature and precipitation conditions in the twentieth century have had an influence on one further element of the environ-ment: the development of dust storms. These are events in which visi-

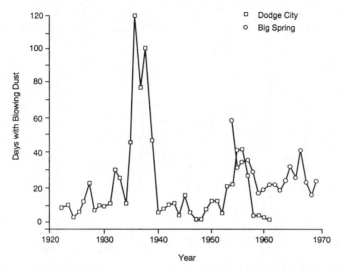

F<small>IG</small>. 5.25 Frequency of dust-storm days at Dodge City, Kansas (1922–61) and for Big Spring, Texas (1953–70) (after Gillette and Hanson 1989).

bility is reduced to less than one kilometre as a result of particulate matter, such as topsoil, being entrained by wind. This is a process that is most likely to happen under conditions of high winds and large soil-moisture deficits. Dust storms have many environmental consequences, including a potential influence on atmospheric temperatures and on convectional activity (see p. 262).

As with lake-level fluctuations it is not always easy to separate the importance of climatic change and human influence (especially vegetation removal and surface disturbance) in affecting the incidence of dust storms. Probably the greatest incidence of dust storms occurs when climatic conditions and human pressures combine to make surfaces susceptible to wind attack.

Two twentieth-century examples serve to show how dust-storm incidence has been affected to a very marked degree. The first is from the High Plains of the USA in the 'Dust Bowl' years of the 1930s when 'black blizzards' caused great distress (Fig. 5.25). There was a great spike in the curve of dust-storm days at Dodge City, with over 100 a year occurring in the mid-1930s. Dust was transported over huge areas and distances, and was noted in New England. This dust-storm era was caused by an era of sod-busting, caused by the increasing use of mechanized agriculture in response to increasing wheat prices after the First World War, and to the

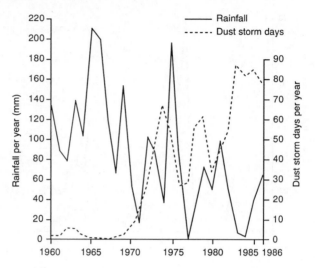

FIG. 5.26 Annual frequency of dust-storm days and annual rainfall for Nouakchott, Mauritania, 1960–1986.

increasing availability of the internal combustion engine to power trucks and tractors.

The second example comes from the Sahel belt in West Africa since the late 1960s. Because of the great Sahel drought (see p. 198) and the ever-increasing population levels in the area, dust-storm incidence showed a massive increase in a great belt from Mauritania in the west to the Horn of Africa in the east. Many stations show a great increase in the 1970s and 1980s.

The variation in frequency of annual dust-storm days and annual-rainfall totals for Nouakchott (Mauritania) is shown in Fig. 5.26. The increase in dust-storm days after 1968 is dramatic. Low-rainfall totals of 48.1 mm in 1970 and 17.9 mm in 1971 represented just 32 and 12 per cent respectively of the 1949–67 average, and can be seen as the main onset of the drought. The number of dust-storm days increased markedly from 6 in 1970 to 65 in 1974 before a reasonably high annual rainfall of 190.6 mm in 1975; dust-storm activity declined to 25 days in 1976 and 27 days in 1977. In 1977, however, the rainy season brought just 2.7 mm of precipitation, making it the driest year since records began in 1931, and dust-storm activity rose to 55 and 61 days in 1978 and 1979 respectively. The total dropped to 33 dust-storm days in 1980 after a relatively heavy

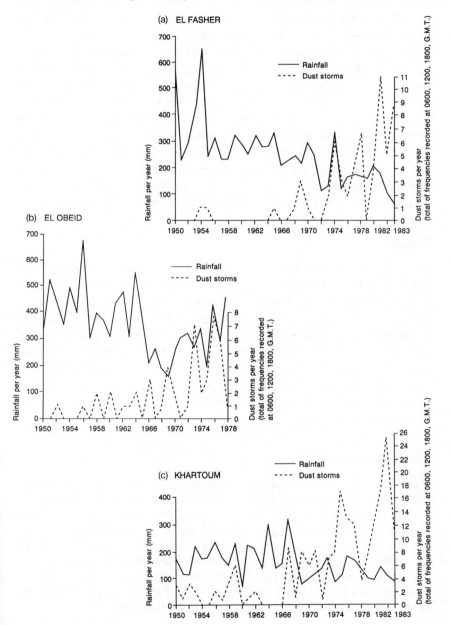

FIG. 5.27 Annual rainfall totals and dust-storm frequencies for Sudan.
(A) El Fasher (1950–83)
(B) El Obeid (1950–78)
(C) Khartoum (1950–83)

rainfall in 1979, but rose to an unprecedented 85 days in 1983 and remained at around 80 days per year until 1986.

The annual dust-storm frequency and annual-rainfall totals for El Fasher, El Obeid, and Khartoum (Fig. 5.27) show a marked rise in dust-storm activity dating from the late 1960s/early 1970s. Particularly low rainfall in 1972 and 1973 at El Fasher, for example, was followed by a distinct rise in dust-storm frequency, peaking in 1974, falling in 1975 and 1976 after high rainfall in 1974, but remaining at increasing levels after that year as annual rainfall remained for the most part below 200 mm. The zero dust-storm reading for 1979 followed the wettest year in the central Sudan (1978) in the last 20–5 years (Trilsbach and Hulme 1984), although particularly high rainfall was not evident at El Fasher itself.

Conclusion

The great weight of data now available on the question of trends and fluctuations in the twentieth century has led to a great change in attitudes to climatology. As Lamb remarked (1966a), 'Not so very long ago . . . climate was widely considered as something static, except on the geological time-scale, and authoritative works on the climates of various regions were written without allusion to the possibility of change, sometimes without mention of the period to which the quoted observation referred.' As Lamb and many others have now shown this static attitude of the 'old climatology' must now be replaced by the dynamic attitude of the 'new climatology'.

Selected reading

Climatic changes during the period of meteorological observations (and earlier) have been discussed by H. H. Lamb (1977), *Climate: present, past and future*, and (1982), *Climate, history and the modern world*. The British picture has been analysed by P. B. Wright in T. J. Chandler and S. Gregory (1976) (eds.), *The climate of the British Isles*. A valuable series of essays, which takes a world-wide perspective, is S. Gregory (1988) (ed.), *Recent climatic change*. However, possibly the most authoritative treatment of the question of climatic changes of the last century or so is provided by J. T. Houghton, G. J. Jenkins, and J. J. Ephraums (1990) (eds.), *Climate Change: the IPCC Scientific Assessment*.

6

Sea-Level Changes of the Quaternary

The importance of sea-level changes

The climato-vegetational changes of the Quaternary era were only equalled in importance by the relative sea-level changes that took place, though these themselves were partially caused by climatic factors. Other contributing factors included tectonic and orogenic forces, local compaction of sediments, and loading of sediments into coastal basins of sedimentation. It is also possible to categorize the changes according to whether they were of a world-wide nature and involving sea-level changes (eustatic changes) or of a local nature, and involving changes in landlevels (tectonic changes).

The effects of such changes can be seen along most shorelines. Where there are stranded-beach deposits, marine-shell beds, and platforms backed by steep cliff-like slopes one has evidence of emerged shorelines. One also often has evidence of submerged coastal features such as the drowned mouths of river valleys (rias), submerged dune-chains, notches and benches in submarine topography, and remnants of forests or peat layers at or below present sea-level. Many coasts show evidence of both emergent and submergent phases in their history and numerous techniques are available to assist in their interpretation (Rose 1990).

Table 6.1 attempts to categorize and list the various causes of sea-level change according to whether they are dominantly world-wide or local. The eustatic types of sea-level change will be discussed first in that they have general significance, while the 'anomalies' on this general pattern caused by local factors, such as isostasy, orogeny, and epeirogeny, will be discussed second.

Table 6.1 Factors in sea-level change

Eustatic (World-wide)	Local
Glacio-eustasy	Glacio-isostasy
Infilling of basins	Hydro-isostasy
Orogenic-eustasy	Erosional and depositional isostasy
Decantation	Compaction of sediments (autocompaction)
Transfer from lakes to oceans	Orogeny
Expansion or contraction of water volume	Epeirogeny
because of temperature change	Ice-water gravitational attraction
Juvenile water	
Geoidal changes	

Eustatic factors

Although glacio-eustasy is the most important of the eustatic factors that have affected world sea-levels during the course of the Quaternary, it is worthwhile to look at some of the other eustatic factors which play a role, especially over the long term. Two very minor factors are the addition of juvenile water from the Earth's interior and the variation of water-level according to temperature. The latter could raise the level of the sea by about 60 cm for each 1 °C in temperature of the sea-water, while the former could probably add about 1 m of water in a million years. The evaporation and desiccation of pluvial lakes, some of which had large dimensions, would be unimportant in affecting world sea-levels, adding a maximum of about 10 cm to the level of the sea, were they all to be evaporated to dryness at the same time (Bloom 1971).

Another cause of eustatic changes of sea-level, especially in the Holocene, is the process called 'isostatic decantation'. Isostatic uplift in the neighbourhood of the Baltic basin and of Hudson Bay has led to a reduction in the volume of these seas, and the water from them has thus been decanted into the oceans to affect world-wide sea-levels. A comparison of the area and volume of the late-glacial precursor of Hudson Bay with Hudson Bay itself suggests that the volume of water decanted could only be sufficient to cause a rise in world sea-level of about 0.63 m. The contribution of the Baltic Sea would be even less. This factor can thus be largely ignored.

These minor factors are of a very limited degree of importance in terms of a Quaternary time-scale, especially when they are compared with the changes brought about by glacio-eustasy.

Rather more important may have been the infilling of the ocean basins by sediment, which would tend to lead to a sea-level rise. With current rates of denudation Higgins (1965) has estimated that this could lead to a rise of 4 mm/100 years. This is equivalent to a rise of 40 m in a million years.

Glacio-eustasy

During the first decades of the twentieth century, following on in part from the work of Suess, a number of workers, including De Lamothe, Deperet, Baulig, and Daly, proposed that most sea-level oscillations and strandlines of the Quaternary were glacio-eustatic (see Guilcher 1969). They believed, correctly, that sea-level oscillated in response to the quantity of water stored in ice-caps during glaciations and deglaciations.

They proposed that there was a suite of characteristic levels in Morocco and elsewhere around the Mediterranean which could be related to different glacial events:

Sicilian (80–100 m)
Milazzian (55–60 m)—between the Gunz and the Mindel
Tyrrhenian (30–35 m)—between the Mindel and the Riss
Monastirian (15–20 and 0–7 m)—between Riss and Würm
Flandrian—the present post-Würm transgression.

The transgressions of the interglacials were succeeded by regressions during glacials and the height of the various stages declined during the course of the Pleistocene (Fig. 6.1). Total melting of the two main current ice-caps—Greenland (2480 km^3) and Antarctica (22 100 km^3)—would raise sea-level a further 66 m if they both melted. Deep-sea core evidence, however, does not suggest that in previous interglacials of the Pleistocene these two ice-caps did disappear, and without a general melting of them, sea-level would only have been a few metres higher than now in the interglacials. This fact does not tie in too happily with the simple glacio-eustatic theory of progressive sea-level decline during the Pleistocene. Some factors other than glacio-eustasy must be responsible for the proposed high sea-levels of early Pleistocene times if indeed they are a reality at a global scale. Furthermore, because other factors have played a role, some local, few people now seriously believe that through height alone can one correlate shorelines over wide areas on the basis of a common interglacial age.

Nevertheless, low Quaternary sea-levels brought about by the ponding up of water in the ice-caps were quantitatively extremely important. Donn *et al.* (1962), on the basis of theoretical considerations from known ice volumes, reckon that in the Riss, possibly the most extensive of the glaciations, sea-levels might have been lowered by 137 to 159 m below

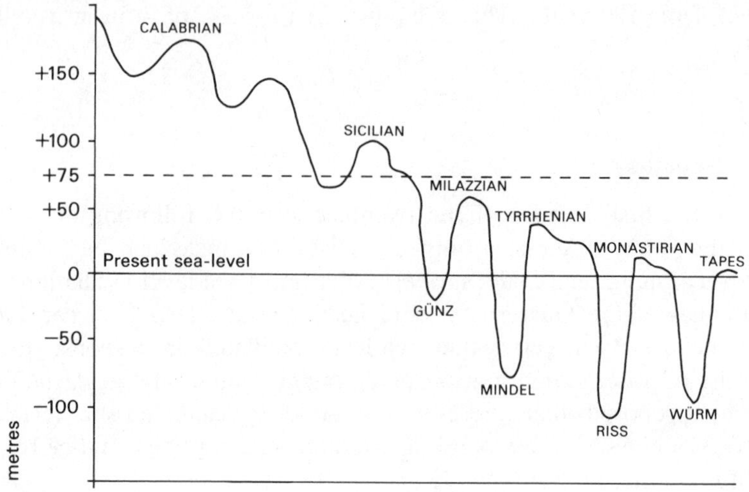

FIG. 6.1 The classic sequence of Pleistocene sea-levels, showing the downward trend in the elevation of raised beaches through time. The dashed line represents the approximate elevation of sea-level if the ice in Greenland and Antarctica were melted (after Frenzel 1973, fig. 92).

current sea-level. In the Last Glacial (Würm-Wisconsin-Weichsel) they give a figure for lowering of rather less—105 to 123 m.

There is a growing body of geomorphological and sedimentary evidence to support this assessment. For example, the bottom of the Iroise (a body of water off Western Britanny) is covered down to 100 m below sea-level with solifluxion-derived materials which are only very slightly reworked by the action of the sea. Off New England, McMaster and Garrison (1966) claim changes down to 144 m as a result of detecting old shorelines. However, on the basis of dates for coral and associated material in the Great Barrier Reef (Australia), California, and south-east Caribbean sea areas, Veeh and Veevers (1970) favoured the conclusion that 13 600 to 17 000 years ago, that is towards the end of the Last Glaciation, sea-level dropped universally to at least −175 m, some 45 m deeper than hitherto suspected.

There is in fact some range of opinion as to the precise figure for glacio-eustatic lowering at the time of the last glacial maximum. Bloom (1983), in reviewing the evidence, suggests that opinions range within 60 m either way of 120 m. Part of this uncertainty arises from debates about the areal and volumetric extents of the great ice-caps and ice-shelves.

The consequences of this low sea-level included the linking of Britain

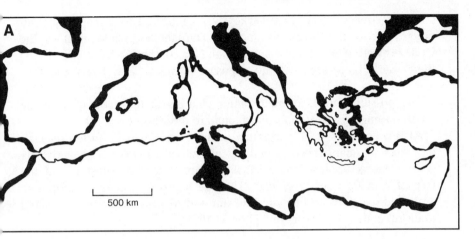

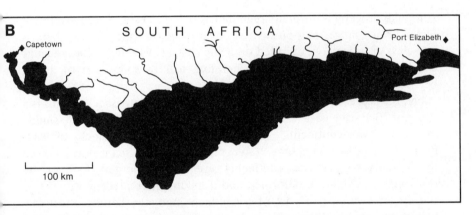

FIG. 6.2 Changes in coastal configuration at the time of the last glacial maximum *c.*18000 years BP. The area in black shows the area of the continental shelf that would have been dry land were sea-level *c.*120 m lower than at the present. (A) The Mediterranean basin, (B) The southern coast of South Africa (modified after Van Andel 1989, figs. 3 and 6).

to the continent of Europe, the linking of Ireland to Britain (Whittow 1973), of Australia to New Guinea, and of Japan to China (Emery *et al.* 1971). The floors of the Red Sea (Olausson and Olsson 1969) and the Persian Gulf (Sarnthein 1972) were also dry land.

   In the Mediterranean (Fig. 6.2,A) large plains existed off the coast of Tunisia and fringed most of Italy, southern France, eastern Spain, and much of Greece. Anatolian Turkey was connected to Europe by land-bridges across the Bosporus and the Dardanelles, while most of the Cyclades were merged into a single island. In southern Africa (Fig. 6.2,B) there was a large area of land exposed between Cape Town and Port Elizabeth.

The possible effects of such major changes in the geography of the Earth on migrations of flora and fauna are discussed on p. 92.

At certain times, as for example in the early stages of an interglacial, the rates of change occasioned by glacio-eustasy may be surprisingly great. This is because some ice-sheets are potentially unstable in the face of a modest sea-level rise. Through a process called 'decoupling', grounded ice starts to become buoyant and to float, whereupon it starts to disintegrate (Anderson and Thomas 1991).

Orogenic eustasy

Although orogeny is normally regarded as being an essentially local factor of sea-level change, and eustasy as being of world-wide nature, there is one class of process, here called orogenic eustasy, whereby a local change can have world-wide effects. It therefore acts as some sort of a link between these two main types of change.

Fig. 6.3 represents the sort of picture that one can envisage and this is a situation that can easily be represented in the laboratory with simple materials. Two 'continents', represented by rectangular blocks of lead, float on a mantle of mercury. Water, representing the sea is poured on to the mercury so that the 'continents' are just submerged. One of the continents can then be deformed and a mountain created by the simple process of turning one part of a continent upright, thereby effectively halving the width and doubling the thickness of the 'mobile belt'. The deformed 'continent' will displace the same amount of mercury as the undeformed continent, although the mountain will have a deeper root. Thus the level of the mercury will remain the same, but the water now has a larger area to spread over, so that it will spread out reducing the depth of the water. The stable 'continent' will thus emerge from the sea. In effect one is producing a world-wide regression of the sea by means of a local orogenic event (Grasty 1967). It has been calculated that an increase of only 1 per cent in the area of the oceans would lower the sea-level by about 40 m, assuming the average depth of the oceans to be 4 km. Over a long time-period this process could be significant, though it prob-ably cannot explain the shorter amplitude sea-level fluctuations of the Pleistocene. On the other hand, the gradual fall of interglacial sea-levels

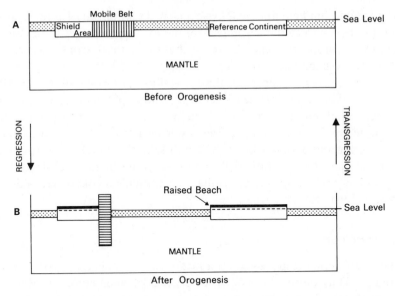

FIG. 6.3 The production of world-wide (eustatic) changes of sea-level as a result of orogenesis (after Grasty 1967).

during this epoch which has frequently been proposed (see p. 219) could have been caused partly by this mechanism.

Bloom (1971: 355), on the basis of information derived from studies of global tectonics, estimated that as ocean basins are spreading at rates of up to 16 cm per year, 'The spreading of the ocean basins since the Last interglacial could accommodate about 6 per cent of the returned melt-water, and the post-glacial shorelines would be almost 8 m lower than the interglacial shorelines of 100 000 years ago.'

Geoidal eustasy

In recent years the importance of a third type of eustatic change has been identified, notably by Mörner (1980). This is termed geoidal eustasy. The shape of the Earth is not regular, and at the present time the geoid (due to the Earth's irregular distribution of mass) has a difference between lows and highs of as much as 180 m. The ocean surface reflects this irregularity in the geoid surface. The geoid surface varies according to various forces of attraction (gravity) and rotation (centrifugal), and will respond by deformation to a change in these controlling forces. The possible nature of such changes is still imperfectly comprehended, but they include fundamental geophysical changes within the Earth, changes in tilt in response to the asymmetry of the ice-caps, changes in the rate of rotation of the Earth, and the redistribution of the Earth's mass caused by ice-cap waxing and waning.

Two major consequences of the recognition of this cause of variability in sea-level are that both sea-level change and uplift/subsidence can only be measured relative to a fixed point—the Earth's centre—and that apparent uplift could occur as the result of a lateral (tangential) movement of a geoid anomaly or of the land itself. Nunn (1986) has suggested that this may be a better explanation for some of the high-level fossil coral reefs of the tropical oceans than the normally invoked mechanism of local tectonic uplift.

Isostasy

The earth's crust responds when a load is either applied to it or removed from it. Thus during the course of a cycle of denudation as perceived under the Davisian model, erosion would remove a considerable mass of rock from on top of underlying strata and a certain amount of compensatory uplift would delay the point at which base level would be attained. However, it is unlikely that this isostatic effect would be of more than local significance during the relatively short time-span of the Quaternary.

In areas of intense volcanic activity the loading of volcanic sediment on to the earth's crust might cause some local crustal depression which might in turn be compensated for by uplift at some critical distance away from the eruption (McNutt and Menard 1978) (Fig. 6.4).

However, isostasy would be important on a broader scale in two main ways: by the application and removal of large masses of ice to certain parts of the earth's crust; and by the application and removal of large bodies of sea-water, and, occasionally, lake water, from the continental shelves, and from lake basins.

The latter mechanism has recently been called hydroisostasy and is probably of less importance than glacio-isostasy, but its effects have often been forgotten. The hydroisostasy theory can be summarized thus. Eustatic sea-level changes brought about by the melting and freezing of the ice-caps would alternately add and take away water to and from the ocean basins. This water would thereby add or remove a load from the ocean-floor, and if one assumes that the density in the sub-crustal zone is nowhere less than 3.00 or 4.00 then total isostatic adjustments to the water loads resulting in depression of the sea-bed would be expected to

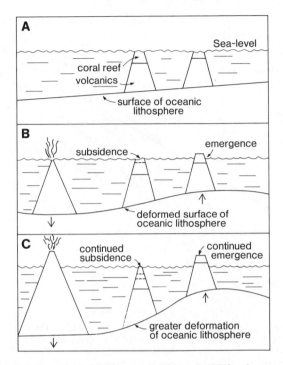

FIG. 6.4 A model (based on McNutt and Menard 1978) of apparent sea-level change brought about in the vicinity of coral reefs by volcanic loading on to the elastic lithosphere. (A) At around one million years ago atolls developed on two extinct volcanoes are in equilibrium with sea-level on an oceanic lithosphere surface that is slowly sinking as it cools (right to left). (B) By c.0.3 million years ago, active volcanism loads the lithosphere and causes subsidence in a near-field 'moat' and emergence at a greater distance (typically c.200 km). In the moat region new reef formation is induced by the subsidence and reefs become emerged at a greater distance. (C) At the present time, continued volcanic activity produces still more volcanic material so that subsidence and uplift continue.

range from one-third to one-quarter of the effective depth of the water (Higgins 1969). In fact, however, the rate and amount of hydroisostatic deformation would vary from place to place according to various factors. Coasts with nearby ocean water more than 100 m deep had the load of water from the post-glacial eustatic rise of sea-level added early and close to the shore, whilst coasts which now border shallow seas had the load added late and, generally, far offshore. One would expect that the amount of submergence would be roughly proportional to the proximity

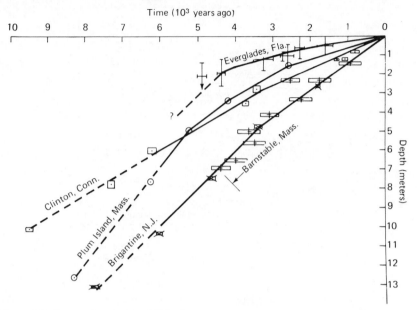

Fig. 6.5 Comparison of published eustatic sea-level curves from the eastern seaboard of the USA to illustrate the effects of hydro-isostasy (after Bloom 1967).

of deep water. This has tentatively been confirmed by recent studies on the coast of America. In this interpretation, it is the shallow offshore depth of the Everglades area of Florida with causes that region to be relatively stable (Fig. 6.5) in comparison with New England (Bloom 1967).

Other factors which might affect the degree of submergence consequent on the sea-level rise would be the local sub-crustal density, its dynamic viscosity, and the degree of isostatic adjustment achieved before loading or unloading began. Estimation of the general effects of this mechanism, however, suggests that the melting of all the present Antarctic ice would raise sea-level eustatically by around 60 m, but that compensatory hydro-isostatic sinking of the ocean floors would reduce the effective sea-level rise to about 40 m, that is to around two-thirds.

The effects of hydro-isostasy, however, can also be seen in the case of pluvial Lake Bonneville, which, as already stated, attained a great depth and area during the pluvial phases of the Pleistocene. It attained a depth of some 335 m. The shorelines around it, which are well known following the studies of G. K. Gilbert and later workers, are all warped according to their position relative to the area of maximum water depth. The

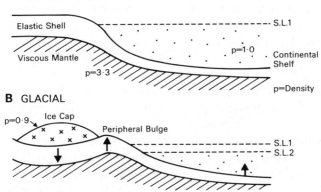

A   INTERGLACIAL

B   GLACIAL

FIG. 6.6 A simplified model of the effects of glaciation on local sea-levels. 'A' represents the interglacial or pre-glacial situation. 'B' represents the situation when an ice-sheet has developed. Depression takes place beneath the ice-sheet as a result of the transference of mass from the oceans, but the land rises as a bulge at some distance from the ice-cap. The continental-shelf rises as a result of the removal of a mass of water as sea-level falls.

Bonneville high shoreline is at least 64 m higher on islands in the ancient lake than it is at the periphery (Crittenden 1963). Pluvial Lake Lahontan also shows warping but only of 6 to 9 m. More recent evidence of this phenomenon can be seen at the man-made Lake Mead in the USA, where the building of the Hoover Dam has led to the impounding of 40 000 million tons of water over 600 km², creating a usual load of about 140 pounds per square inch. This has caused subsidence of the order of 170 mm over the centre of the lake between 1935 and 1950. The degree of subsidence decreases progressively from the lake centre (Longwell 1960).

The principle of glacio-isostasy can be summarized thus: during glacial phases, water loads were transferred from the oceanic 70 per cent of the Earth's surface to the glaciated 5 per cent. This led to depression of the crust, whilst the release of the weight of the ice resulting from melting leads to uplift (Fig. 6.6).

Relatively little is known about the nature of the depression sequence, but as the emergence sequence is both evident in raised beaches and measurable at the present time, more is known about this (Andrews 1970). It is possible to identify three main stages of isostatic response: the period of restrained rebound as the ice-sheet begins to lose mass; the period of post-glacial uplift during which ice-loss takes place at an accelerating rate giving a smooth acceleration of uplift; and, the period of residual glacio-eustatic recovery when some coastal uplift still goes on in spite of total ice removal. In most areas this is the position that we are in today. Because of this sequence the gradients of tilt lessen on younger and younger shorelines in a way that is clearly related to the exponential form of post-glacial isostatic uplift. Some quantitative data are available from Scotland which suggest that in the eastern part of the country, a degree of tilt of 18 mm/km/1000 yr. was established between 9500 and 5500 BP, but that this rate decreases to a tilt of 10.9 mm/km/1000 yr. between 5000 BP and the present.

In areas peripheral to the ice-sheets, such as parts of the eastern coast of the United States, the Baltic, and North Sea, there are indications that zones not subject to an ice-load bulged up during glacial phases (Newman *et al.* 1971), perhaps because of volumetric displacement of the upper mantle's low velocity layer, but that they have collapsed peripherally in post-glacial times, giving greater submergence than could be explained by the eustatic Flandrian (Holocene) transgression alone. This is thought to be the result of some compensatory transference of sub-crustal material.

In the areas that were covered by ice, however, the degree of maximum isostatic uplift has been considerable: around 300 m in North America and 307 m in Fennoscandia, but less in Great Britain (Fig. 6.7,A, B, and C). The ice-caps of Greenland and Antarctica, moreover, are still exerting sufficient weight to create a considerable degree of isostatic depression (see Fig. 6.7,D). Much of the bedrock surface of interior northern Greenland is currently at or below present sea-level. Gravity readings and ice-thickness determinations obtained by trans-Greenland expeditions suggest that before the ice-sheet formed, the presently low-lying bedrock areas of northern Greenland, some of which extend below sea-level, were in the form of a plateau about 1000 m above sea-level. If the ice were to be removed, the bedrock surface would slowly rise up to this height once again.

The effect of glacio-isostasy on the lakes of Finland is also of interest. It is known by tradition that the woodland and meadows on the low southern shores of the Great Finnish lakes had a tendency to become more swampy and marshy with time, whereas the northern shores tended to become drier. This is a consequence of the post-glacial tilting which has sometimes had spectacular results: fjords opening towards the north, to the Gulf of Bothnia, have been converted into lakes by the development of a threshold. Equally, many of the large lakes used to have their outlets on their northern sides, but they too have them diverted to the south shores by the same process. The rivers draining the lakes of the southern sides have not had time to fully adjust their profiles to this

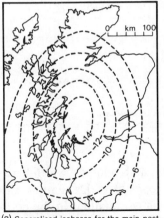

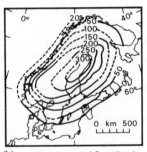

(b) Isostatic recovery of Scandinavia in the last 10,000 years (m)

(a) Generalised isobases for the main post-glacial raised shoreline in Scotland (m)

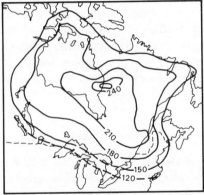

(c) Maximum postglacial rebound of northeastern North America, in meters

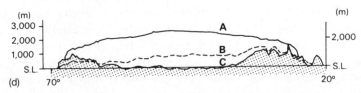

(d)

FIG. 6.7 The effects of glacio-isostasy on different land areas:
A. Generalized isobases for the main post-glacial shoreline in Scotland (m) (after Sissons 1976).
B. The degree of isostatic recovery (m) of Scandinavia in the last 10 000 years.
C. Maximum post-glacial rebound of north-eastern North America (m).
D. Cross-section of north Greenland showing the present shape of the ice-sheet, A, the present level of the bedrock surface, C, and the estimated profile of the landmass with no ice-load (after Hamilton 1958).

Table 6.2 Areas of Pleistocene vulcanicity and tectonic uplift

Europe
  Greece, the Aegean, Vesuvius, Etna, Sardinia, Catalonia, Massif Central, Northern Bohemia,
  Romania, Silesia, Eifel area, Spitzbergen, Iceland.
Asia
  Armenia, Asia Minor, Caucasus, Iraq, North Palestine, Transjordan, Arabia, Dead Sea and
  Galilee, Northern Siberia, Mongolia, Manchuria, Korea, China, Sea of Okhotsk, Japan, Kuriles,
  Java, Sumatra, and various Pacific islands.
America
  Alaska, Sierra Nevada, West Indies, Central America, the Andes.

(After Charlesworth, 1957)

phenomenon and as a result they show rapids and falls which have proved to be popular sites for electricity development in this century. Elsewhere some of the lakes have dried up through decantation. In the north of the country the uplift of the land has also caused difficulties for port authorities who have to deal with a progressive shallowing of their harbours. On the bonus side the uplift has provided further usable land for the nation. It was in fact the need to allot newly-emerged land to owners around the Gulf of Bothnia which led the director of the cadastral survey of Finland, Efraim Otto Runeberg (1722–70), to postulate in 1765 that small movements of the Earth's crust were responsible for many of the gains in land (Wegmann 1969).

Miscellaneous causes of local changes in sea-level

One of the prime causes of the observed changes of the land relative to the sea is orogenic activity, the process by which mountains are built. Signs of Pleistocene vulcanicity and earth movements are visible in many parts of the world (Table 6.2) and Charlesworth (1957) has written that the Pleistocene 'witnessed earth-movements of a considerable, even catastrophic scale. . . . The Pleistocene indeed represents one of the crescendi in the earth's tectonic history.' This is not, however, a universally accepted point of view. In place of this 'crescendo' belief, others would maintain that the Quaternary has witnessed mountain-building on a scale that is not dissimilar to that of previous eras, whilst others believe that the Quaternary was a new and distinct phase of activity which replaced supposed stability of the Middle Tertiary. This last hypothesis, the neo-tectonics hypothesis, has much support at this time, but all three models involve a recognition of the considerable extent of orogenic movements in the Quaternary (King 1965).

The zones where orogenic activity has been most intense in the Pleistocene have been recognized over the last twenty years. Seismic activity, vulcanism, and mountain-building occur for example in a well-defined series of narrow belts (Fig. 6.8), with that surrounding the Pacific Ocean being especially notable.

The most important type of plate-boundary for mountain building (orogenesis) is the convergent plate boundary where plates converge at a rate of several centimetres per year and subduction takes place. Mountains are thought to be created along convergent plate boundaries in at least four major ways: where a continent collides with an island arc; where a continent collides with another continent; where the oceanic lithosphere underthrusts a continental margin; and where the oceanic lithosphere underthrusts island areas.

Plate-tectonic theory also helps to explain some zones of subsidence in the oceans. Flat-topped submarine mountains called *guyots* appear at some point in the past to have been volcanic islands above sea-level. They formed at zones of sea-floor spreading, but as the spreading process continued they ceased to be active, became eroded by wave action, and gradually suffered subsidence, which brought the formerly wave-eroded surface to as much as 900–1200 m below sea-level. If the subsidence rate has been sufficiently gradual the guyots may be capped by coral atolls.

On the other hand, some areas, the continental platforms, located away from plate margins, have suffered from relatively little mountain-building during the Pleistocene. They stand in contrast to the areas of new fold mountains, some of which may have been uplifted as much as 2000 m in the last few million years.

In some localities, and on a much more limited scale, the flattening of sediments by the weight of overlying material (autocompaction) can lead to subsidence of some consequence. It is often apparent in peat and other such materials which have a very high porosity and a weak skeletal framework of vegetable fibres. Salt-marsh peats, for example, which make up a large part of many transgressive sedimentary deposits, have an 80 per cent porosity, and frequently in section one sees logs that have become flattened from their original round shape to a more oval form (Kaye and Barghoorn 1964). The subsidence caused by compaction of Holocene beds in Holland is estimated at 2.5 cm/100 years (Veenstra 1970).

In some localities man may have caused some fall in land-levels relative to the sea. One of the clearest indications of this is provided by the example of Venice. Currently there is an increasing flood risk which is leading to the frequent inundation of St. Mark's Square and other parts of the city. Although currently rising sea-level of eustatic origin and more long-continued subsidence in the area both play a role, one of the

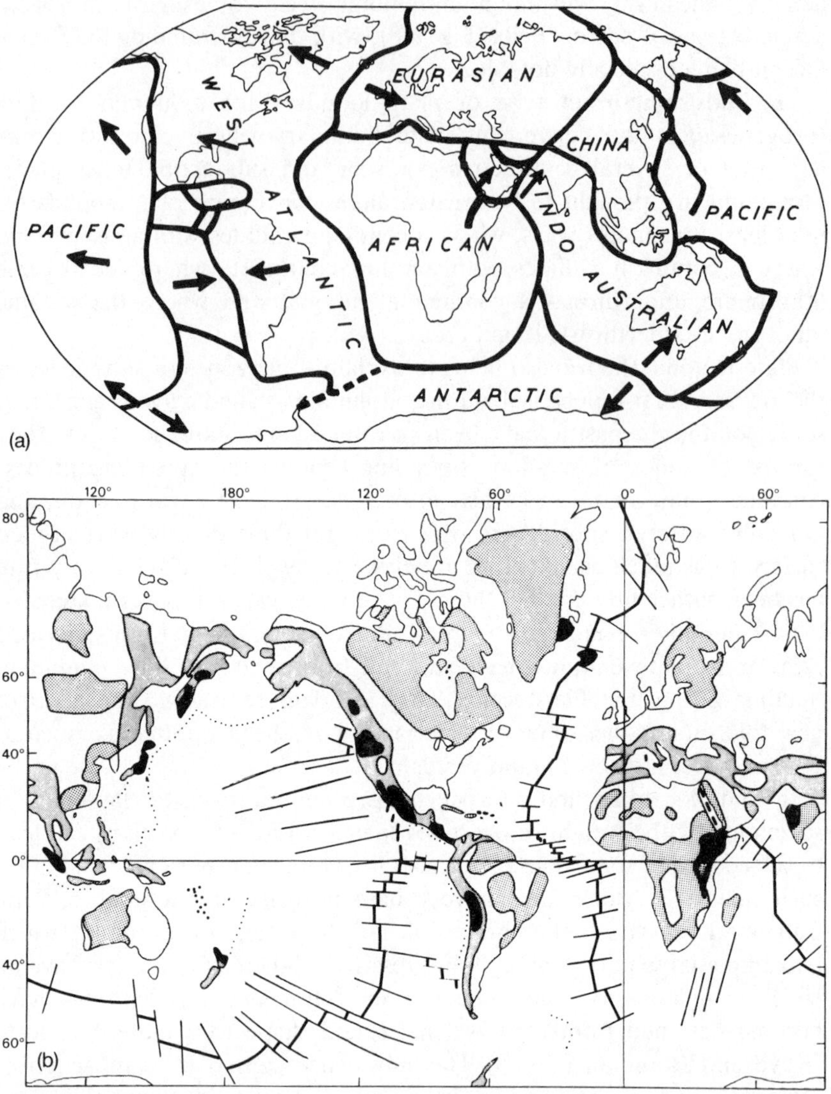

FIG. 6.8
A. Suggested plates of the Earth's crust (after Tarling and Tarling 1971: 95, fig. 40b).
B. The major tectonic features of the world (after P. J. Wyllie and Smith 1973: 9, fig. 1.3.6).
   Light shading = continental platforms; irregular shading = continental shields; dark shading = Tertiary folded mountain chains; black areas = Cainozoic volcanic regions; dotted lines = oceanic trenches; heavy lines = active rift systems of oceanic ridges; light lines = oceanic faults.

prime causes of the problem is the abstraction of groundwater by large new industrial complexes on the other side of the Venetian lagoon. This abstraction has caused subsidence to take place (see Fontes and Bortolami 1973, and Gambolati *et al.* 1974).

Subsidence produced by oil abstraction is an increasing problem in some parts of the world, the classic area being Los Angeles, where 9.3 m of subsidence occurred as a result of the exploitation of the Wilmington oilfield between 1928 and 1971. The Venezuelan and Russian fields have also been affected, though to a less extreme degree.

Operating on a broader scale than orogenic movements are epeirogenic movements. These do not involve complicated deformation with folding, faulting, tilting, and warping, but involve large elevations and depressions of continents and ocean basins, with warping restricted to a so-called marginal hinge line. On the landward side there would tend to be elevation, producing stairways of terraces, many of which have been attributed to eustasy, whereas there would tend to be subsidence on the seaward side. Broad-scale epeirogenic movements have marked the evolution of the African continent, where broad planation surfaces initiated during phases of tectonic stability are separated by massive scarps initiated during each uplift episode (King 1962).

Clark (1976) has postulated another possible mechanism of local sea-level change: that the growth of massive Pleistocene ice-sheets, such as that of Canada, would cause sufficient gravitational attraction for sea-level to rise locally relative to the land. As the ice-sheets melted and lost mass, the sea-level would also fall in response to the reduced gravitational attraction. He suggests that this ice-water gravitational effect alone could cause raised beaches to occur 85 m above present sea-level in Hudson Bay.

The nature of pre-Holocene sea-levels

Although, as noted, the classic sequence of Pleistocene sea-levels involves a gradual reduction in the height of interglacial sea-levels during the course of the Pleistocene, there is relatively little accurate dating of the sequence of sea-level change. Recently, however, new dating techniques have enabled some generalizations to be put forward, though far more dates are required for any degree of certainty to be reached.

Changes of sea-level due to glaciation and deglaciation can in large measure now be correlated by the oxygen isotope curves in deep-sea core sediments, for these directly reflect changing ice-volumes on the continents (see pp. 19–21), and through them of the quantities of water in

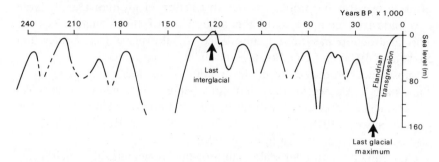

FIG. 6.9 Sea-level curve for the last quarter of a million years based on work in New Guinea (modified from Chappell and Thom 1977).

the oceans. This record plainly implies that sea-level has changed frequently, rapidly, and somewhat repetitively during the Pleistocene. A sea-level curve derived from Core V28–238 correlates extremely well with a curve derived from uranium series dates on Barbados emerged coral terraces (Mesolella *et al.* 1969), and both the sea-levels also appear to correlate closely with the maxima on theoretical insolation curves (Veeh and Chappell 1970; see also pp. 263–9).

Oxygen isotope studies can also be used to ascertain whether or not glacio-eustasy caused sea-levels to be higher than now in earlier interglacials. Shackleton (1987) doubts that sea-levels were appreciably higher due to this mechanism (say by more than a few metres) during any interglacial of the past 2.5 million years.

Although Fig. 6.9 shows a sea-level higher than the present until 120 000 years ago, some earlier studies suggested that sea-level was close to that of the present during the mid-Würm Interstadial, say 30 000 years ago (Thom 1973). There are a substantial number of apparently confirmatory dates from raised beach deposits from Tanzania, Aldabra (in the Indian Ocean), the Atlantic, and the Red Sea. It needs to be pointed out, however, that dates of this order are close to the limit of C14 dating, and even slight contamination could produce misleading dates. Indeed, for Red Sea and Aldabran samples discrepancies have been revealed between C14 dates and those obtained by other radiometric methods of the Uranium Series (Thomson and Walton 1972). The difficulty of radiocarbon dating of shells and corals of this age can be highlighted by considering the way in which contamination by a certain percentage of modern calcium carbonate can affect the apparent C14 age of a carbonate which is in reality 100 000 years old. The following situation has been found, and indicated that slight contamination of a sample of interglacial age could appear to give a date of the interstadial of the mid-Würm (Newell 1961).

Per cent Contamination by modern $CaCO_2$	Apparent C14 years
50	5 600
10	19 000
5	24 500
1	37 000
0.01	74 000

Nevertheless, if a mid-Würm Interstadial was a reality as seems likely (see p. 77) it would be expected to have caused a relative sea-level rise, though on the basis of such climatic information as is available one might expect eustatically controlled sea-levels during these relatively warm interstadials to be at −40 to −50 m. Temperature conditions in the interstadials were only sufficient to cause partial melting of the big ice-caps. The Laurentide and Scandinavian Ice-Caps persisted during the last Glacial, colder-than-present vegetation was present in Western Europe, and oxygen-isotope records from deep-sea and Greenland ice-cores show values not equivalent to interglacial or present conditions.

Thus the concept of an interstadial sea-level higher than that of the present interglacial is now treated with scepticism. Indeed, Bloom (1983: 218) has said,

The Middle Wisconsin high sea-level alleged for the south-eastern coast of the United States and elsewhere is invalid; the hypothesis is based on contaminated or misinterpreted radio-carbon dates and is not supported by other dating methods, by documented ice-marginal positions, or by the deep-sea oxygen-isotope record, or by the radiometric dating of emerged coral reefs.

The post-glacial rise in sea-level or Flandrian transgression

The low sea-levels of glacial times were succeeded as the ice melted by a large rise in sea-level: the Flandrian transgression. At times the rate of rise was of a remarkable magnitude, perhaps because of the near catastrophic collapse of ice-shelves and ice-caps. In Lancashire, north-west England, sea-level may have risen by as much as 7 m in 200 years round about 8000 years BP, possibly because of the rapid final collapse of the massive Laurentide Ice-Sheet of North America (Tooley 1978).

One of the best available records of the progress of the Flandrian Transgression is provided by the study of radio-carbon-dated samples

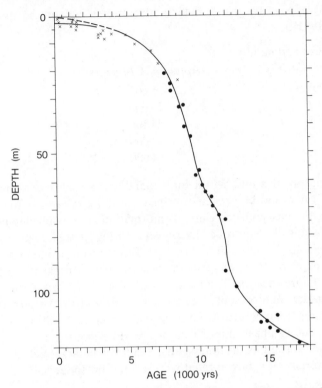

FIG. 6.10 The sea-level curve for Barbados, based on radio-carbon dated corals (*Acropora palmata*). The filled circles are data from Barbados corrected for estimated uplift. The crosses are data for four other islands in the Caribbean. The drawn curve reflects the depth range of live *Acropora palmata*; the dashed line is adjusted to sea-level (after Fairbanks 1989). Reprinted by permission from *Nature*, 342: 637–42. Copyright © 1989, Macmillan Magazines Ltd.

drilled from submerged coral reefs on the Barbados shelf in the Atlantic (Fairbanks 1989) (Fig. 6.10). The coral that was involved, *Acropora palmata*, grows within a few metres of sea-level, and thus permits a relatively accurate determination of past water-depths. The oldest and deepest of the samples indicates that at the time of the last glacial maximum the maximum depth of the shoreline was at −120 m. A gradual rise started shortly afterwards, and accelerated to a rate of 24 m in less than 1000 years beginning in 12 500 BP. The rate of rise decreased between 11 000 and 10 500 BP (the Younger Dryas, see p. 134) and then accelerated until 8000 BP, when the shoreline was at a depth of around −25 m. The curve of sea-level rise flattens out 5000 years ago at −7 m.

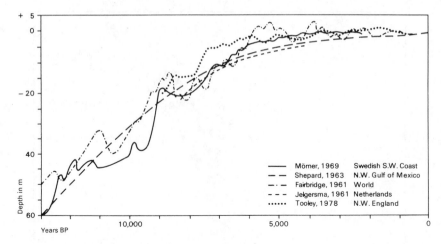

FIG. 6.11 Sea-level curves for the Holocene.

Fig. 6.11 presents detailed sea-level curves for the world as a whole, and for north-west England, the Gulf of Mexico, the Netherlands, and for the west coast of Sweden.

The importance of this event has been summarized thus by Newell (1961: 87):

Flooding of the continental shelves by the sea, was certainly the most important geological event of recent time. It initiated the building of modern deltas, many coral reefs, alluviation of river valleys, and the formation of existing beaches and barrier islands. Doubtless, it also had far-reaching effects on climate and the migrations of marine and terrestrial organisms, including man.

On morphological grounds, including the presence of steps on coastal shelves, it seems reasonable to postulate that the rapid transgression suffered some stillstands, and even slight regressions. This point of view has been put forward as a result of submarine notch and terrace studies for numerous areas in recent years including the Persian Gulf (61–64 m, 40–53 m, and 30 m below present sea-level), the Bass Strait (60 m below sea-level), the Gulf Coast of the United States (60 m, 32 m, and 20 m) and the Mediterranean (5, 10, 27, 55 and 96 m) (Ballard and Uchupi 1970, Flemming 1972).

Whether such minor stillstands took place or not, the general Holocene trend until about 6000 BP was one of rapidly rising sea-levels. Over low-angle shelves this rapid rate of rise means that the sea must have advanced laterally at a fast rate. In the Persian Gulf region, for instance,

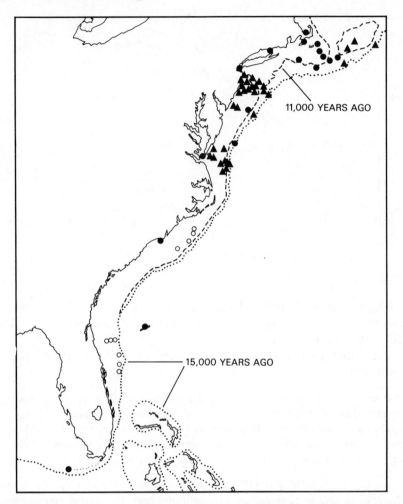

Fɪɢ. 6.12 A comparison of the Atlantic coast shoreline of the United States at 15 000 years ago, 11 000 years ago, and at the present. Confirmation that the continental-shelf was once laid bare is found in discoveries of elephant teeth (triangles), fresh-water peat (dots), and the shallow-water formations called oolites (circles) (after Emery 1969).

there was a shoreline displacement of around 500 km in only 4000–5000 years, a rate of no less than 100–120 m per year. This must have had profoundly disturbing results for inhabitants of the coastal plains. Comparable rates of transgression for the North Sea are 60 km/1000 years, for the Bristol Channel 15 km/1000 years, and for the Gulf of Mexico 18–32 km/1000 years (Evans 1979).

Table 6.3 The stages of world sea-level change since 13 000 BP according to different sources

Years BP	Shepard (1963)	Schofield (1960)	Fairbridge (1961)	Godwin *et al.* (1958)
1 000	−0.5	+1	+1	
2 000	−1	+2	−2	
3 000	−2	+3	−3	
4 000	−3	+5	+2	
5 000	−4	−2	+3	0
6 000	−7	−0.5	0	−4
7 000	−10	−4	−6	−9
8 000	−16	−19	−16	−17
9 000	−22	−33	−14	−28
10 000	−31	−36	−32	−35
11 000	−40			−44
12 000	−48			−52
13 000	−58			−62

(Levels in m)

Fig. 6.12 illustrates the way in which the shorelines of 11 000 and 15 000 years ago compare to those of North America today. The shoreline at 15 000 BP, being at a lower level, was a considerable distance across the continental shelf, and the various groups of islands of the east coast of Florida were linked up to form much larger land areas.

The main problem that arises with the interpretation of this Holocene transgression lies in what happened after about 6000 BP. There are three fairly distinct schools of thought on this, though it has to be stated that the arguments are about possible changes of the order of only a couple of metres (Table 6.3). It is generally accepted that in the last six millennia the rate of sea-level rise, if present, has been far less than it was in the Early Holocene. One point of view has it that there has been a continuously rising sea-level to the present time, though the rate of rise has diminished with time (Shepard's hypothesis (1963)). Godwin *et al.* (1958) on the other hand hypothesize that sea-level rose steadily until about 3600 BP, since when it has more or less remained constant. Fairbridge (1958) and others (Mörner 1971*a*) have maintained, in contrast to these other two ideas, that Late Holocene sea-level oscillated to positions both above and below the present level. He suggested that the sea was at levels 1–4 m above the present about five times between 6000 BP and the Middle Ages (Table 6.4). A considerable amount of evidence has been raised against Fairbridge's concept. Some people have reworked or reinvestigated some of the sites claimed by Fairbridge to illustrate high Holocene stillstands and they give a Pleistocene rather than Holocene age for the raised beaches and terraces. Jelgersma (1966) has said that if the high sea-levels

Table 6.4 Fairbridge's Holocene oscillatory sequence

Transgression	Emergence	sea-level (m)	Date (BP)
Older Peron		+3 or 4	5000
	Bahama	−3	4300
Younger Peron		+3	3900–3400
	Crane Key	−2	3300
	Pelham Bay	−3	2400–2800
Abrolhos		+1½ − 2	2300
	Florida	−3	2000
Rottnest		+½ to 1	1200–1000
	Paria		700

(Fairbridge, 1958)

had taken place one would expect that coastal plains would have been inundated on a very large scale. She says that data from the Gulf of Mexico, Florida, and the Netherlands fail to reveal such a degree of transgression. Very detailed archaeological researches in relatively stable parts of the Mediterranean, using diving techniques, have convinced Flemming (1969) that to within an accuracy of ±0.5 m there has been no eustatic change of sea-level in the Mediterranean in the last 2000 years. Dating of freshwater peats in Australia, one of Fairbridge's field areas, by Thom *et al.* (1969) fails to indicate that sea-level rose above its present position between 2985 BP and 9000 BP. Likewise, dating of Chenier ridges in Queensland leads to a broadly comparable conclusion (Cook and Polach 1973). Similarly, after an expedition around some Pacific atolls, Newell and Bloom (1970) said 'We found no unequivocal evidence for recent higher sea-level, and abundant evidence that the characteristic morphology of the Indo-Pacific coral reefs is most probably in adjustment to the slow rise in sea-level that has characterized the last 6000 years.'

One of the main lines of evidence that has been used to substantiate the high Holocene sea-level concept is the presence of small raised terraces ('Daly levels') in many parts of the tropics. Radio-carbon dates for slightly elevated reefs of Holocene age, which cluster around about 4000 years BP as reviewed by Stoddart (1969) do indicate the possibility of a slight Holocene transgression, but, he says, there are so few that they could be due to local emergence.

However, many of the terraces, from which samples of *in situ* coral have been dated by various isotopic means, indicate that a large number of the 'Daly levels' may be the product of earlier high stands of sea-level. Higher sea-levels than the present have been suggested for the middle of the Würm (around 30 000 BP), and for the pre-Würmian times, notably 70 000 to 180 000 BP, and to a lesser extent 190 000–240 000 BP. In general, as explained on p. 234, it seems probable that sea-level stopped at or slightly above its present level for much of the period $70-190 \times 10^3$ years ago. This would give a much longer time for the terraces to develop than the limited number of years available during supposed Holocene oscillations. Recent work on rates of erosion on coral islands suggests that time would have been inadequate in the Holocene to produce the degree of planation recognized (Stoddart 1971).

Behind the arguments as to whether global sea-levels have been steadily rising, steady, or oscillatory, was a desire to try and achieve a world-wide eustatic curve. More recently, however, the search for a eustatic sea-level curve having global significance has ended (Kidson 1986). Studies of the importance of geoidal eustasy (see pp. 223–4) and recognition that the geoid has not remained stable over time, have resulted in the recognition that there must have been regional differences in response to post-glacial sea-level rise. Moreover, as part of this re-evaluation there has been a growing appreciation that crustal isostatic response to the removal of the weight of ice-sheets has been accompanied by a consequential hydro-isostatic response (see pp. 224–7), especially in the areas of the shelf seas. It has also been recognized that there are very few, if any, areas that have been sufficiently stable to use as 'eustatic dip-sticks'. Furthermore, the interpretation of past sea-levels may have been complicated by changes in tidal levels through time as a response to changes in coastline position and topography. Kidson (1986: 54–5) has summarized the need for caution in the light of these arguments:

No clear view on higher eustatic SLs in the Holocene has emerged. The differences in view in the 1980s seem as wide as they were in the 1950s and 1960s. . . . The difficulties in separating the eustatic, tectonic and isostatic components of SL change are so complex that subjective analysis will always be possible. For the moment the only possible course is to keep on open mind about outstanding issues such as smooth or spasmodic change of level and higher than present stands of the sea. Only detailed and rigorous studies in individual localities may provide an adequate accumulation of data which may make an eventual solution of these outstanding problems possible.

In the light of considerations relating to the importance of changes involving hydro-isostasy and glacio-isostasy in determining the response of sea-levels to glacio-eustasy, Clark and Lingle (1979) identified six different zones with certain distinctive sea-level change characteristics (Fig. 6.13). Zone I consists of those regions beneath the ice-sheets at the time of the glacial maximum. These would have curves that were charac-

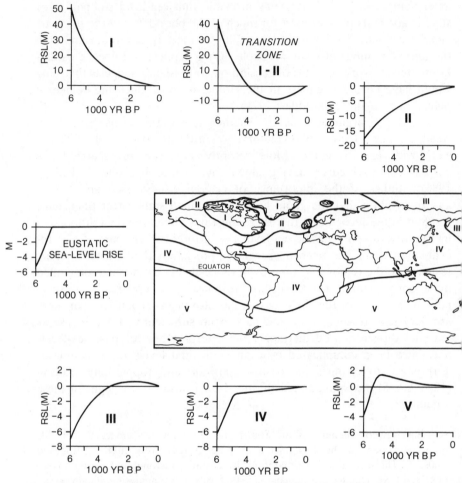

FIG. 6.13 Typical Holocene relative sea-level (RSL) curves predicted for each of six global zones. Predictions are based on northern-hemisphere deglaciation and a resulting eustatic sea-level rise of 75.6 m between 17 000 and 5000 years ago. No eustatic change is assumed during the last 5000 years. Significant Holocene deglaciation in Antarctica would modify the boundaries of the six zones. (From Clark and Lingle 1979, fig. 1.)

terized by immediate emergence during the shrinkage or disappearance of the ice, due to the elastic uplift of the earth and a reduction in the gravitational attraction exerted by the ice-mass on the surrounding ocean. Emergence would continue as viscous flow within the mantle caused the land formerly depressed beneath the ice to be displaced upwards. A transitional zone between Zone I and Zone II would be characterized by

Table 6.5 Current rates of sea-level rise

Source	Rate (cm/100 yr)	Date
Gutenberg (1941)	19.4	1880–1942
Fairbridge and Krebs (1962)	12.0	1900–50
Fairbridge and Krebs (1962)	55.0	1946–56
Scholl (1964)	18	1914–64
Scholl (1964)	12	1940–64
Donn and Shaw (1963)	42	1890–1940
Donn and Shaw (1963)	24	1940–60
Hawkins (1971)	25	1916–62
Pirazzoli (1989)	4–6	c1889–1989

emergence after ice-sheet thinning, but at a later time submergence followed by more gradual submergence would begin as a collapsing pro-glacial forebulge migrated towards the reduced ice-sheets. Zone II would be characterized by submergence following ice-sheet thinning. This would be caused by a flow of the mantle material into the uplifting areas from peripheral regions resulting in collapse of the pro-glacial forebulge. Zone III would be characterized by rapid submergence followed by more gradual submergence until a few thousand years ago, then slight emergence of less than 0.75 m began. Zone IV would be characterized by continuous submergence of 1–2 m during the last 5000 years, despite an assumed constant ocean volume during that period. Zone V is a region where initial submergence would be followed by slight emergence beginning when water was no longer added to the oceans. A Holocene emergence of up to 2 m is predicted. Finally, there is Zone VI, which consists of all continental margins except those adjacent to Zone II. Their emergence is due to the addition of meltwater to the oceans causing depression of the ocean floor. Material within the mantle then flowed from beneath the ocean to beneath the continents, causing upward displacement of the continents and tilting of the crust along coasts.

There is certain evidence from present tide-gauge records to suggest that following on from the current post-neoglacial (Little Ice Age) amelioration in climate there has been a corresponding rise of sea-level. It is of course difficult to separate current tectonic submergences and other factors from the eustatic effect of current glacial melting, and Pirazzoli (1989) believes that some previous estimates may have been excessive as a result.

Some of the data are summarized in Table 6.5. It is interesting from the glacio-eustatic viewpoint that the tendency towards a reduced rate of glacial retreat and of a reduction or reversal in the climatic amelioration of the twentieth century (see p. 184) has, according to the studies of

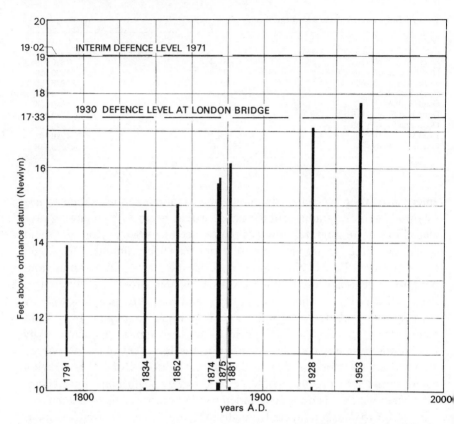

FIG. 6.14 Increasing high-tide levels at London Bridge (after Horner 1972, fig. 1).

Scholl (1964) and Donn and Shaw (1963), led to a reduction in the rate of eustatic rise in the last two or three decades. Similarly Binns (1972) claims to have found evidence of a number of shorelines related to a fall in sea-level during a cold neoglacial phase (see p. 162) around 2500–2400 years BP.

In some localities present-day subsidence, combined with eustatic rise, is of a sufficient magnitude to present a threat to low-lying settlement concentrations. In the case of London, for instance, historical records show that high-tide levels and surge levels relative to Newlyn Ordnance Datum are becoming progressively higher (Fig. 6.14), with an increase of the order of 1.3 m between 1791 and 1953. This increases flood risk, but it is not clear how much this is due to subsidence and eustatic rise alone, and to what extent embanking by man, changes in water temperatures affecting tides through changing viscosity of the water, and changes in climatic conditions (including rainfall and wind directions), have played a significant role (Bowen 1972, Horner 1972).

A comparable picture is evident at Venice where storm surges (locally called 'acqua alta') have reached new levels of intensity and frequency (Berghinz 1976). Out of 58 surges recorded in the past hundred years, 48 occurred in the last 35 years, and 30 in the last 10 years. In other words in the first 65 years there was one 'acqua alta' every 5 years, in the following 25 years almost one per year, and in the last ten years, three per year. Eustasy plays a role, but subsidence appears to be the major problem, with the rate increasing fivefold from 0.9 mm/year for the period 1908–25, to 5 mm/year for the period 1953–61.

Elsewhere in the world, the relative stability of sea-level in the last few thousand years may be one contributing factor to the widespread sand loss from many beaches, with consequent erosion and threats to human activities (Russell 1967). As long as the Flandrian transgression was taking place at an appreciable rate, new areas of coastal plain were being inundated and fresh supplies of sediment were encountered, the coarser components of which were driven forward to produce beaches. An increase in beach volume took place so long as the rise of sea-level continued, and surplus sand was blown downwind to form extensive coastal-dune tracts. The beach-dune system probably reached its greatest volume as the stillstand was approached. However, once that level was attained, new sediment supplies were no longer encountered, and marine processes brought about a net loss to the system in many parts of the world. There is a pressing need for work to be undertaken on the relative importance of this factor compared to changes in wind conditions or storminess, and to humanly induced changes in sediment budgets.

Post-glacial sea-level changes in northern Europe: the combined effects of eustasy and isostasy

Thus far we have been concerned solely with the operation of the various individual factors that have influenced sea-level changes in the Quaternary Era. In reality, of course, the sea-level fluctuations at any one point involve a combination of several factors. This is especially clearly illustrated in the case of the countries of north-western Europe, where both glacio-eustasy on a world-wide basis and glacio-isostasy on a more local basis have acted together in post-glacial times to give the observed sequences of sea-level changes. In essence the post-glacial eustatic changes have tended to lead to marine transgressions, whilst isostatic rebound from the weight of the ice-caps has tended to lead to regression. In north-western Europe one can see how, through time, the rates of these two contrary processes have led to distinctive patterns.

In Scotland the rate and degree of isostatic rise was relatively small compared with parts of Scandinavia and Canada (see p. 228), so that at times when the eustatic sea-level rise was going ahead with great speed (e.g. 14 000 to 6000 BP) it overtook the regression of the sea, interrupting the emergence of the coast. Thus in certain parts of Scotland, notably the Firth–Clyde lowlands, there are alternating deposits of marine and freshwater sediments that comprise a valuable record of land and sea-level changes, and detailed survey, radio-carbon dating, and pollen analysis allow a comprehensive picture to be presented (Donner 1970, Walton 1966).

Because of the weight of the ice the Late Weichselian situation was one of relatively high sea-levels and submergence. It was during this phase that many of the high level-platforms were cut by the sea. The highest altitude of the Late Weichselian submergence is shown by raised beaches at 30–35 m above Ordnance Datum in Central Scotland. The Flandrian eustatic transgression, however, reached its maximum between 8000 and 6000 BP after a low position of sea-level between 10 000 and 8000 BP. During that regressive phase of low sea-level some distinctive peats were formed and these now often underlie marine deposits of the Flandrian transgression, such as the Carse Clay, which produced another suite of raised beaches at a level generally lower than those formed during the maximum of the Late-Weichselian. This post-glacial shoreline reaches an altitude of 15 m in central Scotland, but because of the varied nature of the isostatic effect it declines in all directions from this peak.

In Norway the pattern of sea-level change in post-glacial times is essentially similar with high Late Glacial shorelines resulting from the iceload effect being followed by a low stand of sea-level between about 12 000 and 10 000 BP, and a transgression, often called the 'Tapes' transgression in Norway and the 'Nucella' transgression in Iceland.

Another situation where the effects of the post-glacial isostatic and eustatic sea-level changes can be well seen is in the development of the Baltic Sea. At the maximum of the Weichselian Glaciation, 18 000 to 20 000 years ago, the Baltic area was covered by a great ice-body which deposited the Brandenburg and later moraines over the North European Plain. As this great ice-sheet spasmodically retreated, a series of small ice-dammed lakes developed in the southern part of what is now the Baltic. The coalescence and expansion of these lakes produced the first major stage of the post-glacial evolution of the Baltic—the Baltic Ice Lake, dated at about 11 000 BP. This lake was not totally landlocked and managed to overflow to the low Early Holocene ocean level by a series of channels, including one into the White Sea. The location of the overflow channels varied according to deglaciation, glacial re-advances, and isostatic updoming.

In due course the deglaciation of southern Sweden enabled the sea to open the Baltic basin, and thus the Yoldia Sea was established at about 10 300 BP, and lasted for less than 1000 years. However, the link between the Yoldia Sea and the ocean via southern Sweden was severed as isostatic updoming in that area raised the connection above sea-level. This thus led to the formation once again, after 9800 BP of a lake—the Ancylus Lake—which lasted for more than 2000 years, and its outlet was eventually through the Oresund, the channel that still separates Sweden from Denmark. Finally, the general world-wide rise in sea-level exceeded the rate of isostatic uplift and the Oresund was submerged, thereby changing the Ancylus Lake into the so-called 'Littorina Sea' from about 7000 BP onwards. This sea was characterized by a relatively warm-water fauna, of a salt-water type, and was probably contemporaneous with the Tapes Sea recognized by Scandinavian workers outside the Baltic Basin. After around 3000 years (at about 4000 BP) the 'Littorina' Sea merged into the current Baltic Sea, though faunal evidence suggests that as a result of the reduction in its outlet by updoming, the waters of the Baltic became progressively more brackish, with the associated constriction of the oceanic connection. Given present rates of uplift, however, and assuming no general rise of sea-level, the Baltic, currently linked to the ocean by channels only 7–11 m deep, could become a lake again in 8000–10 000 years.

Likewise the Black Sea was until around 9300 BP an oxygenated freshwater lake separated from the Mediterranean by a sill in the vicinity of the present Bosporus, before being converted into the anoxic marine habitat of today (Degens and Hecky 1974).

A combination of eustatic rise and subsidence in the southern North Sea led to the development of the North Sea as we know it now. Fig. 6.15 illustrates how around 9500 BP the North Sea was largely dry land, and that the only sizeable body of water was the course to the south-west of the Rhine–Meuse system. However, by 8300 BP the situation had changed very radically, and the North Sea was then linked through to the English Channel. Compared to the present shorelines there remained extensive areas of low-lying ground at the mouths of the major estuaries, but these too were flooded as the Holocene progressed. The gradual sequence of

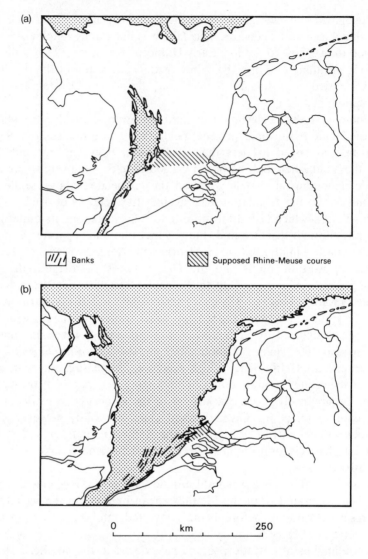

FIG. 6.15 The extension of the southern North Sea during the early Holocene around:
A. 9300 BP
B. 8300 BP
(after de Jong 1967, figs. 25 and 26).

inundation can be seen from an examination of the history of the British coastal lowlands.

Holocene sea-level movements in southern Britain

The various alternations of sea-level and climate in the Holocene had very marked effects, notably on the Dutch coast, the Fens of East Anglia, and the Somerset Levels. The sediment of these areas show the alternations that took place between marine, brackish, and freshwater sedimentation according to the degree of sea-level rise, sediment delivery by streams, and climatic conditions. These alternations had important consequences for human settlement, whilst the present-day arrangements of clays, silts, and sands, resulting from the complex Holocene history are reflected in current land utilization.

The rapid rise of sea-level in the Holocene caused the reworking of sediments deposited on the present continental shelf by the rivers of glacial times. These sediments were combed-up as the transgression progressed to create some major sedimentary features such as Chesil Beach, Dungeness, Orford Ness, and Dawlish Warren. In addition, the sea flooded low-lying areas, including the lower reaches of river valleys, to create rias (especially in south-west England and Pembrokeshire) and embayments (e.g. the present sites of Romney Marsh, the Pevensey Levels, the Broads, the Fens, and the Somerset Levels). These have subsequently become filled with alluvial sediments or cut off from the sea by the growth of spits (e.g Poole Harbour). The rising sea flooded the floor of the North Sea and the English Channel (c.9600 BP), achieving the final separation from the Continent in about 8600 BP (Jones 1981) and the almost total inundation of the Thames Estuary by 8000 BP.

In very general terms the temporal pattern of Holocene (Flandrian) transgression is comparable in different areas of England and Wales. Fig. 6.16 illustrates how in the last 10 000 years there has been a phase of very rapid sea-level rise to around 6500 years BP, followed by a rather gentle rise, or even stability, thereafter. However, in detail the pattern is more complex (as revealed by the stratigraphy of areas like the Fens) and there are also differences between areas, caused by subsidence (in the south-east) and by isostatic uplift (in the north).

The sedimentary record in lowland areas is frequently complex. This is exemplified in the Fens (Shennan 1986) where there have been multiple sequences representing either positive tendencies of sea-level (described as WASH) or negative (described as FENLAND). These are shown in Table 6.6.

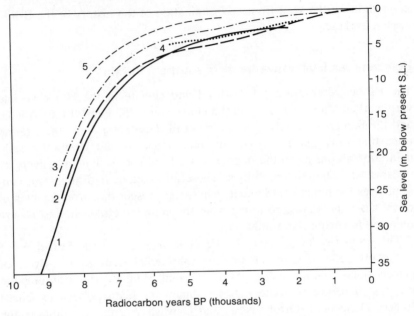

F<small>IG</small>. 6.16 Curves of sea-level change since 10 000 years BP for different parts of England and Wales: 1 = Bristol Channel; 2 = English Channel; 3 = Cardigan Bay; 4 = Somerset Levels; 5 = North Wales. The datum level is MHWST (mean high water spring tide) (after Shennan 1983, fig. 8).

The sediment in the Broadland valleys of eastern Norfolk preserves evidence of the main phases of Holocene marine transgression, represented by two layers of estuarine clay sandwiched between layers of freshwater peats (Coles and Funnell 1981). The lower clay is 12–16 m thick at its seaward limit, extends 20 km inland, at its upper surface (dated at c.4500 BP) and is at present at −5.5 to −6.5 m below Ordnance Datum. The upper clay, which is 6 m thick at the coast and extends 23 km inland, dates to around 2000 BP, while the upper surface dates to about 1500 BP.

In Somerset the sequence is not dissimilar. At the base, one has the valleys formed during Pleistocene low sea-levels. Borehole records show that these valleys are filled with a blue-green clay to the approximate level of the present sea, and on faunal grounds it seems likely that this clay was deposited under brackish water conditions associated with a rising sea-level. This first transgression, which occurred earlier than in the Fens (from about 5500 BP) was succeeded by a period of freshwater marsh development during which Sphagnum bogs were developed on the water-

Table 6.6 Holocene sea-level tendencies in the East
Anglian Fens

Period	Tendency
>6,300BP	WASH I
6,300–6,200	FENLAND I
6,200–5,600	WASH II
5,600–5,400	FENLAND II
5,400–4,500	WASH III/IV
4,500–4,200 (3,900)	FENLAND IV
4,200 (3,900)–3,300	WASH V
3,300–3,000	FENLAND V
3,000–1,900	WASH VI
1,900–1,550	FENLAND VI
1,550–1,150	WASH VII
1,150–1,000 (950)	FENLAND VII
950 onwards	WASH VIII

Source: Data in Shennan (1986).

logged surfaces of the underlying marine clay. The period of bog growth,
however, was not without its fluctuations, for in the Bronze Age (possibly
3000 to 2500 BP) an increase in wetness necessitated the construction of
many wooden track-ways (corduroy roads), since buried, which crossed
from one upstanding island (such as the Isle of Avalon or Brent Knoll) to
the next. It was, however, as in the Fens, inundated in Romano-British
times (about 2000 BP) and more clay was deposited in the valleys. To
what extent this marine inundation was due to an actual sea-level rise,
and to what extent it might have been due to such factors as occasional
high tides or storm surges, is a matter of some controversy, and has been
discussed by Kidson (1977).

Current rates of subsidence and uplift

Precise levelling, archaeological evidence, and tide-gauge records are
among the sources of information that can be employed to assess the
current rates of subsidence (Table 6.7,a) and uplift which complicate the
eustatic picture.

The main areas of subsidence are probably the deltaic basins of the
world's great rivers such as the Rhine, Mississippi, Rhone, and Narbada
(India). As with eustatic changes there is a considerable quantity of data
which illustrate the order of change which is progressing. Other areas
of change are the tectonically unstable volcanic areas, such as parts of
Japan and the East Indies. Also, as noted previously, parts of the world
peripheral to zones of current isostatic uplift may currently be showing
some isostatic depression.

Table 6.7 (a) Current or recent rates of subsidence

Source	Location	Rate cm/100 yr
Veenstra (1970)	NW Germany	2.5
Veenstra (1970)	S. Netherlands	10–20
Veenstra (1970)	S. Denmark	15
Tjia (1970)	Tokyo (Japan)	10
Tjia (1970)	Osaka (Japan)	12
Tjia (1970)	Alaska	100
Kvitkovic and Vanko (1971)	New Carpathians	up to 5
Tjia (1970)	Indonesia	30
Van Veen (1954)	Netherlands	25 (since 7200 BP)
Churchill (1965)	East Anglia and Kent	9
Coleman and Smith (1964)	South-central Louisiana	7.3

Table 6.7 (b) Current rates of vertical movement associated with seismic activity

Source	Specific events location	Displacement	Date
Plafker (1965)	Alaska	10 m–15 m	1964
Plafker (1965)	Yakutat Bay (Alaska)	14.3 m	1899
Twidale (1971)	New Zealand (Murchison quake)	5 m	1929
Twidale (1971)	Adelaide (Australia)	5–8 m	1954
Daly (1926)	California	7.01 m	1872
Daly (1926)	Sonora (Mexico)	6.10 m	1887
Daly (1926)	Japan	6.10 m	1891
Daly (1926)	India	10.67 m	1897
Daly (1926)	California	0.91 m	1906
Daly (1926)	Formosa	1.83 m	1906
Daly (1926)	Mexico	0.61 m	1912
Daly (1926)	Nevada	4.57 m	1915

One effect of coastal subsidence and landward uplift in deltaic areas is shown by the terraces of the Mississippi in Louisiana and Texas. These are progressively steeper in gradient as they become older. The Williana terrace has a gradient of 1.86 m/km, the Bentley 0.66–1.49, the Montgomery 0.47–0.94, the Prairie 0.20–0.45 and the present flood-plain only 0.002–0.26. As a result of the associated subsidence more than 3000 m of Quaternary beds have been deposited in this area.

Locally, major uplift may still be taking place at measurable rates because of earthquake and related activity. Some of the measured vertical movements, determined by accurate geodetic levelling and other methods have been quite considerable (Table 6.7,b). The movements take place both because of individual seismic events and because of a more general process of gentle seismic creep (Table 6.8).

Individual events may lead to as much as 15 m of uplift, as in the Alaska earthquake of 1964, while more gradual tectonic changes may lead to uplift that sometimes exceeds 10 m in 1000 years. There are very

Table 6.8 Gradual rates of uplift (metres per 1000 years)

Source	Location	Rate
Kvitovic and Vanko (1971)	Carpathians	up to 1
Kafri (1969)	Russia	up to 14.6
Kafri (1969)	Israel	60
Tjia (1970)	New Zealand	4.0
Tjia (1970)	Indonesia	0.2–1.5
Collins and Fraser (1971)	New Zealand	11–12
Collins and Fraser (1971)	Japan	0.7–1.6
Collins and Fraser (1971)	Garlock Fault (Calif.)	10
Schumm (1963)	California	4.0–12.8
Chappell and Veeh (1978)	Timor	0.47
Ollier (1981)	Costa Rica	2.5
Ollier (1981)	Crimea	2.0–4.0
Ollier (1981)	Caucasus	6.0–12.0
Ollier (1981)	Poland	5.0–10.0
Ollier (1981)	New Guinea	2.0–4.0
Ollier (1981)	New Hebrides	5.5–7.7
Ollier (1981)	Iran	7.4–12.0
Ollier (1981)	Swiss Alps	1.0

Table 6.9 Current rates of isostatic uplift as determined for Finland from tide-gauge data

Station	Mean rate (cm/10 yr)
Kemi	7.37
Oulu	6.53
Raahe	7.63
Pietarsaari	8.50
Vaasa	7.60
Kaskinen	7.03
Mantyluoto	6.20
Rauma	4.93
Turku	3.53
Degerby	4.20
Hanko	2.73
Helsinki	1.83
Hamina	1.80

(Determined by author from data in Lisitzin, 1964)

few parts of the world where rates of denudation can keep pace with such rates of uplift, the average rate for upland areas probably being around 0.5 m per thousand years (Ollier 1981: 225).

Most areas, of course, do not have such high rates of uplift, and it is possible to relate rates of recent vertical crustal movement (RVCM) to different geotectonic zones (Fairbridge, 1981) in the following way:

Shields and platforms	$1\,\text{mm/yr}^{-3}$
Cenozoic orogenic belts	up to $20\,\text{mm/yr}^{-1}$
Older Phanerozoic orogenic belts	up to $5\,\text{mm/yr}^{-1}$
Intra-orogenic Basin-Range belts (with regional block faulting)	up to $10\,\text{mm/yr}^{-1}$

The degree of isostatic adjustment taking place at the present time has also been estimated by analysis of tide-gauge data. Trends can be identified in rates, with minimum rates occurring away from the former centres of ice-cap growth. In Finland, for example, during the present century, rates in the south at stations like Helsinki and Hamina have been about four times lower than those at the head of the Gulf of Bothnia (Table 6.9) (Oulu, Kemi, etc.).

Selected reading

An enormous compilation edited by N. A. Mörner (1980), *Earth, rheology, isostasy and eustasy*, provides much useful data from many parts of the world, and explores the exciting new theme of the role of geoidal changes. The best book on the sea-level changes of the Holocene is that by M. J. Tooley (1978), *Sea-level changes: north-west England during the Flandrian stage*.

Techniques for sea-level studies are expertly analysed by J. Rose in A. S. Goudie (1990) (ed.), *Geomorphological Techniques*, and in O. van der Plassche (1986) (ed.), *Sea-level research: A manual for the collection and evaluation of data*.

D. E. Smith and A. G. Dawson (1983) (eds.) provide a good analysis of isostatic effects in *Shorelines and Isostasy*.

The association between sea-level changes and human activities are discussed in P. M. Masters and N. C. Flemming (1983) (eds.), *Quaternary coastlines and marine archaeology*.

Contemporary tectonic movements (neotectonics) are analysed by C. Vita-Finzi (1986), *Recent earth movements: An introduction to neotectonics*.

Finally, R. J. N. Devoy (1987) (ed.) provides a wide-ranging series of essays on many aspects of sea-level change studies in many parts of the world in *Sea-surface studies*.

7

The Causes of Climatic Change

Introduction

The climatic changes that have been established and described, and which formed the basis for associated environmental changes such as those of sea-level described in the last chapter, have caused a great deal of discussion about their causes. The purpose of this chapter is to summarize some of the main opinions that have been put forward, to stress the variety of factors involved, and to show the doubts still associated with the major hypotheses.

An indication of the complexity of factors that needs to be considered in any attempt to explain climatic change is given in Fig. 7.1. This flow diagram starts with the ways in which the input of solar radiation into the earth's atmosphere can fluctuate. For reasons such as varying tidal pull being exerted on the sun by the planets, the quality and quantity of outputs of solar radiation may change. The receipt of such radiation in the Earth's atmosphere will be affected by the position and configuration of the Earth and by such factors as the presence or absence of interstellar dust. Once the incoming radiation reaches the atmosphere its passage to the surface of the Earth is controlled by the gases, moisture, and particulate matter that are present. These materials may either be of natural or man-made origin. At the Earth's surface the incoming radiation may be absorbed or reflected according to the nature of the surface (the albedo). The effect of the received radiation on climate also depends on the distribution and altitude of land masses and oceans. These too are subject to change in a wide variety of ways—continents may move to or from areas where ice-caps might accumulate, mountain belts may grow or subside to affect world wind-belts and local climates, and the arrangement of the climatically highly important ocean currents may be controlled by changes in sill depths and the width of the seas, oceans, and

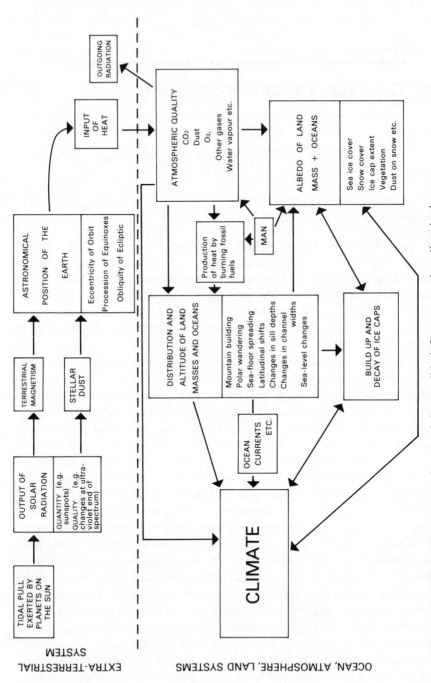

Fig. 7.1 A schematic representation of some of the possible influences causing climatic change.

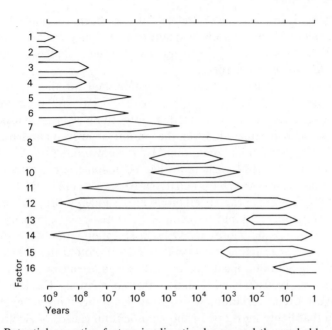

FIG. 7.2 Potential causative factors in climatic change and the probable range of time-scales of change attributable to each (after Mitchell 1968: 156).
(1) Evolution of sun; (2) Gravitational waves in the Universe; (3) Galactic dust; (4) Mass and composition of air (except CO_2, H_2O and O_3); (5) Polar wandering; (6) Continental uplift; (7) Orogeny and continental uplift; (8) CO_2 in the air; (9) Earth-orbital element; (10) Air, sea-ice, cap feedback; (11) Abyssal ocean circulation; (12) Solar variability; (13) CO_2 in the air (fossil fuel combustion); (14) Volcanic dust in the stratosphere; (15) Ocean–atmosphere autovariation; (16) Atmosphere autovariation.

channels. The situation is complicated, as the flow diagram suggests, by the existence of various feedback loops within and between the ocean, atmosphere and land systems.

In addition it needs to be remembered that the potential causative factors in climatic change operate over a very wide range of different time-scales, so that some factors will be more appropriate than others to account for a climatic fluctuation or change of a particular span of time. An attempt to show this diagrammatically is made in Fig. 7.2.

Solar radiation hypotheses

Following through the flow diagram, it is clear that changes in the output of solar radiation may lead to significant changes in the receipt of radiation at the Earth's surface. Indeed, it has been recognized that the sun's radiation changes both in quantity (through association with such phenomena as sunspots) and in quality (through changes in the ultraviolet range of the solar spectrum).

Cycles of solar activity have been established for the short term by many workers (see Meadows 1975), with eleven- and twenty-two-year cycles being ones particularly noted. Eighty- to ninety-year sunspot cycles have also been postulated. Over the period of instrumental record discussed in Chapter 5 it has been found by some workers (see, for example, Wood and Lovett 1974) that there has been some correlation between, for example, sunspot activity and East African rainfall and lake levels. Sometimes, however, the correlations may suddenly break down, while other correlations may not necessarily be statistically significant. Nevertheless, some of the more significant associations may have predictive value. Thus, for example, Strongfellow (1974) plotted the five-year moving mean of lightning incidence in Britain against mean annual sunspot numbers for 1930–73 and found a 0.8 correlation. He identified an eleven-year cyclic variation, with a trough for lightning being found in 1973. In that lightning is one of the main natural causes of electric power transmission failures in the United Kingdom such a relationship may assist the electricity authorities in planning maintenance services. At a less serious level a good correlation has now been found between sunspot activities and achievement in sport. King (1973: 445) found that:

Information contained in Wisden can be used to show that, of the twenty-eight occasions on which cricketers have scored 3000 runs in a season in England, sixteen have been in sunspot maximum and minimum years; the five years in which this rare phenomenon occurred more than once were all sunspot minimum or maximum years. Likewise, thirteen of the fifteen occasions in which a batsman has scored 13 or more centuries in a season took place in, or within a year of, a sunspot maximum, or minimum year.

Exceptional cricketing feats are produced at times of exceptional weather occurring at the extremes of the sunspot cycle.

Although the role of changes in solar activity has frequently been attacked, especially with regard to cycles, some striking correlations have been found between changes in solar activity and certain major characteristics of the general atmospheric circulation. Fig. 7.3, for example, shows a distinct similarity in trend between Baur's solar index and the yearly frequency of the general westerly zonal type over the northern hemisphere, with a general increase in both parameters until the 1930s from 1890, and with a rapid decrease in both into the 1960s. This suggests

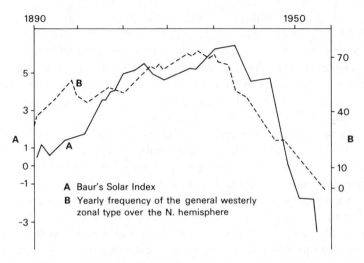

F IG . 7.3 Curves of Baur's solar index and the yearly frequency of general westerly weather type over the northern hemisphere (after Lamb 1969*b*).

that a portion of the observed climatic variation of the twentieth century may be attributable to a variation of the sun's energy output at source. Claims may also be made, however, for the importance of other factors.

The observation of sunspots in historical times also gives a measure of solar activity (Harvey, 1978). Telescopic observations on sunspots exist since AD 1610, while naked eye sunspot observations, while less reliable, exist from 527 BC. One very striking feature of the record is the near absence of sunspots between AD 1640 and AD 1710, a period sometimes called the Maunder Minimum. It is perhaps significant that this minimum occurred during some of the more extreme years of the Little Ice Age.

Over a longer time-scale it is far more difficult to show that the sun's output of radiation has changed sufficiently to affect the Earth's climate, since any substantial proof is lacking. Nevertheless, this is a possible hypothesis, and one which has received much support. Some evidence to support it comes from studies of the oscillation in the concentration of atmospheric C14, which in turn depends partly upon variation in the emission of solar radiation. C14 levels have fluctuated during the Holocene, and Denton and Karlén (1973) have argued that the major intervals of high atmospheric C14 activity coincide with periods of neo-glacial expansion, while the intervals of relatively low C14 activity co-incide with intervals of glacier contraction. Equally, Bray (1970) has suggested that Holocene glaciations show a periodicity of around 2600 years, and that an arithmetic progression starting with 22 years (the complete sunspot or 'Hale' cycle), and a first term of 4, results in a sequence of 88, 440, and 2640 years. Other workers, utilizing spectral analysis of an ice-core from Camp Century, Greenland, have claimed to find systematic long-term oscillations of a broadly comparable magnitude to those of Bray: 78, 181, 400, and 2400 years. These they also relate to varying solar activity (Johnson *et al.* 1970).

The causes of variations in solar activity are still imperfectly understood, but one possible cause of variability in receipt of insolation on the Earth's surface is the presence of clouds of fine interstellar matter (nebulae) through which the Earth might from time to time pass, or which might interpose themselves between the sun and the Earth. These would tend to lead to a reduction in the receipt of solar radiation. Similarly, the passage of the solar system through a dust lane bordering a spiral arm of the Milky Way galaxy might cause a temporary variation of the radiation output of the sun, and so lead to an ice-epoch on Earth (McCrea 1975).

Another possible cause of solar variations has been suggested by Opik (1958), but his model can neither be proved nor disproved at the present time. He proposed the following theoretical cycle of solar activity. Initially a 'normal' situation exists of the type responsible for relatively warm climates on Earth. With the passage of time metals that diffuse slowly are left behind as a result of the diffusion of hydrogen from the sun's mantle to its core. These metals accumulate to form a barrier to radiation from the core and in keeping with maintenance of a steady-state condition, the sun contracts. However, the metal barrier becomes hot, convection currents develop, and the core becomes very large. This increases the hydrogen content available for fuel and the energy output increases. The production of the heat is such that it cannot be adequately transported to the surface; thus, the sun expands. In expanding, energy is expended thereby reducing the heat and light output from the sun. This gives reduced radiation and cooling on Earth. However, expansion lowers the temperature of the core and the amount of energy it produces. Thus the core shrinks and eventually the sun returns to its 'normal' position, giving relative warmth on Earth.

Atmospheric transparency hypotheses

Even if one supposes that changes in receipt of insolation from the sun have not taken place to a sufficient degree markedly to change the

Earth's climate, the effects of incoming radiation from the sun may have been moderated by changes in the atmospheric composition of the Earth.

Particular importance has been attached to the role of volcanic dust emissions (see Chester 1988, for a review). The presence of elevated-dust levels in the atmosphere could increase the backscattering of incoming radiation (thus encouraging cooling). In addition volcanic dust might reduce sunshine totals further by promoting cloudiness, for dust particles, by acting as nuclei, can promote the formation of ice crystals in sub-freezing air saturated with water vapour.

Lamb (1970) has been a major exponent of the idea that climatic deterioration could be produced by the volcanic production of a dust veil in the lower stratosphere.

The ash emissions of Krakatoa in the 1880s and Katmai (1912) produced a global decrease in solar radiation of 10 to 20 per cent lasting 1 to 2 years, and the ash from Krakatoa was injected into the stratosphere, reaching a height of 32 km. It has recently been shown that many of the coldest and wettest summers in Britain, such as 1695, 1725, the 1760s, 1816, the 1840s, 1879, 1903, and 1912 occurred in conjunction with times of high volcanic dust inputs into the stratosphere and upper atmosphere (Lamb 1971). Moreover, the period of temperature warming in the northern hemisphere that was experienced in the 1920s, 1930s, and 1940s coincides with a period when there were no major volcanic eruptions in the northern hemisphere, suggesting the possibility that the absence of a volcanic-dust pall in those decades was one factor in the warming process. Similarly, the little climatic optimum and the Little Ice Age (Bray 1974) seem to correspond to periods of low and high volcanic activity respectively, as do the glacial fluctuations of the past 1000 years (Porter 1981, 1986).

During the Holocene as a whole Bray (1974) has suggested on the basis of examination of C14 dates that the major advances of alpine and polar glaciers were exactly contemporaneous with the major post-Wisconsin volcanic activity phases in New Zealand, Japan, and southern South America (4700–5450 BP, 2150–2850 BP, and 50–470 BP).

Bryson (1989) has found evidence from radio-carbon dates of volcanic deposits for a peak of volcanic activity at the time of the Allerød oscillation (c.11 900 BP) and has suggested that there may be some causal link. Going back further, a study of the Byrd ice-core from Antarctica has produced evidence of particularly heavy and frequent volcanic-dust falls at 20 000 to 16 000 years BP—the same time as the maximum cold of the Last Glacial (Gow and Williamson 1971). Additional support for the role of vulcanism comes from the analysis of coarse volcanic ash in 320 deep-sea cores. Kennett and Thunell (1975) find that such ash is very frequent in the Quaternary, being about four times the neogene average.

What is intriguing, therefore, is that there is a great deal of prima-facie evidence for periods of high volcanic activity being associated with periods of colder climate. Equally intriguing is that these periods cover a wide range of scales from the decadal fluctuations of the last century or so, through the neoglacial events of the Holocene, to the climatic deterioration of the Younger Dryas, to the last glacial maximum, and to the Pleistocene as a whole. Although correlations of time series do not necessarily imply causality, the implications are certainly strong. There is the possibility, however, that for some of the longer time scales (i.e. 10^3 to 10^6 years) the relationship may be reversed, with climatic changes creating changes in volcanic activity. For example, increased volcanism might result from the stresses imparted by glacio-isostatic and hydro-isostatic loading of the earth's crust associated with ice-cap growth and changes in the volume of water in the oceans.

Changing vulcanism is not the only way in which climatically important changes in atmospheric transparency might occur. For example, dust can be emplaced into the atmosphere by deflation of fine-grained surface materials by wind. There is plenty of evidence that aeolian-dust loadings in the atmosphere have changed through time (see Leinen and Sarnthein 1989) and that they were very high both in high latitudes (as demonstrated by the presence of dust particles in dated ice-core sequences), and lower latitudes (as demonstrated by the presence of loess accumulations and of dust in deep-sea core sediments in many of the world's oceans) around the last glacial maximum. Such dust may have contributed to cooling (by backscattering incoming solar radiation) and to aridification (by suppressing convective activity). In other words dust could have contributed to the accentuation of the dry, cold conditions of glacial times.

Carbon dioxide and other trace gases in the atmosphere can also affect climate by modifying the receipt of solar radiation. This is a matter that is discussed further on p. 272.

Climatic change and variations in terrestrial magnetism

In recent years a large quantity of work has been initiated into the relation between changes in the intensity of the Earth's magnetic field and changes in climate. The work is in its early stages, but already some strong relations have been established between temperatures, over time-scales of ten years to 1.2 million years, and magnetic intensity. For example, Wollin and his co-workers (1971, 1973) have found that over the period 1925–70 magnetic intensity has been decreasing at observatories in Mexico, Canada, and the United States at the same time as temperatures have been increasing. Equally, at observatories in Greenland, Scotland, Sweden, and Egypt, the intensity is increasing whereas the climate is getting colder. In other words there appears to be a close inverse correlation between changes of the Earth's magnetic field and climate (Fig. 7.4).

Why this situation should exist is not clear. It is possible that the Earth's magnetic field changes in response to changes in solar activity and that both climate and terrestrial magnetism are yoked together in their response to solar events (Wollin *et al.* 1974). If this is the case then magnetism does not in itself have a simple cause and effect relation with climate. On the other hand, it is possible that magnetism may be modulated to some degree by the ability of the Earth's magnetic field somehow to provide a shield against solar corpuscular radiation.

Thus the relationship of these two phenomena appears to be proven, but the reason for the relationship is still not clear.

Most palaeoclimatologists remain to be convinced about the importance of this particular mechanism, and this scepticism is well displayed by Bradley (1985: 98):

For the time being then, reasons for the observed correlations remain ambiguous and only further research will resolve the controversy. . . . There are many areas of controversy as yet unresolved, and some apparently promising lines of research may yet prove to be erroneous. Where correlations do exist, the explanation of cause and effect is often little more than guesswork.

Earth geometry theories—the Croll–Milankovitch hypothesis

Following through Fig. 7.1 it is reasonable to assume that if the position and configuration of the Earth as a planet in relation to the sun were to change, so might the receipt of insolation from the sun. Such changes do take place, and there are three main astronomical factors which have been identified as of probable importance, all three occurring in a cyclic manner (Fig. 7.5): changes in the eccentricity of the Earth's orbit (a 96 000 year cycle), the precession of the equinoxes (with a periodicity of 21 000 years), and changes in the obliquity of the ecliptic (the angle between the plane of the Earth's orbit and the plane of its rotational equator). This last has a periodicity of about 40 000 years.

The Earth's orbit round the sun is not a perfect circle but an ellipse. If

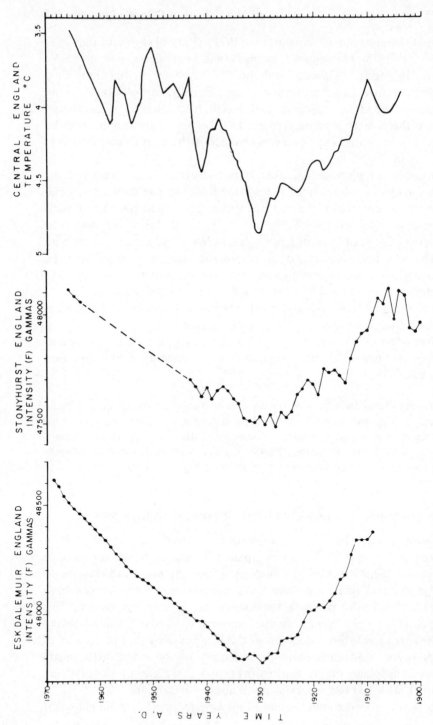

Fig. 7.4 Magnetic intensity curves based on annual means compared with 10-year running average of winter temperature for central England (1900–70). (From Wollin et al. 1974, fig. 6.)

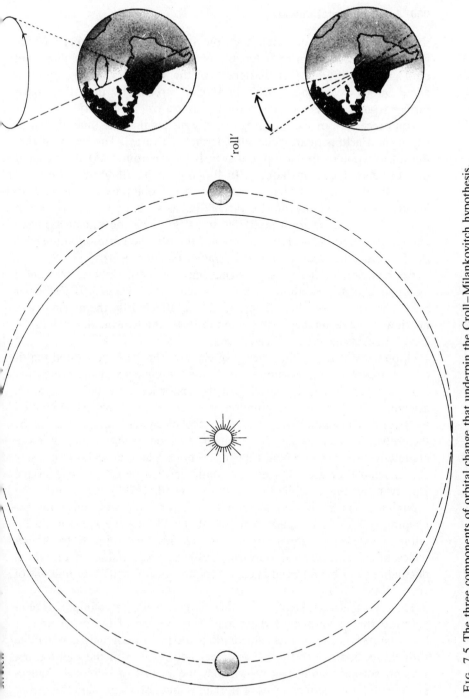

'roll'

FIG. 7.5 The three components of orbital change that underpin the Croll–Milankovitch hypothesis.

the orbit were a perfect circle then the summer and winter parts of the year would be equal in their length. With greater eccentricity there will be a greater difference in the length of the seasons. Over a period of about 96 000 years the Earth's orbit can 'stretch' by departing much further from a circle and then revert to almost true circularity.

The precession of the equinoxes simply means that the time of year at which the Earth is nearest the sun (perihelion) varies. The reason is that the Earth wobbles like a top and swivels its axis round. At the moment the perihelion comes in January. In 10 500 years it will occur in July.

The third cycle perturbation, change in the obliquity of the ecliptic, involves the variability of the tilt of the axis about which the Earth rotates. The values vary from 21° 39′ to 24° 36′. This movement has been likened to the roll of a ship. The greater the tilt, the more pronounced is the difference between winter and summer (Calder 1974).

Appreciation of the possible significance of these three astronomical fluctuations of the Earth goes back to at least 1842, when J. F. Adhemar made the suggestion that climate might be affected by them. However, his views were developed by Croll in the 1860s and by Milankovitch in the 1920s (Beckinsale 1965, Mitchell 1965).

Opponents of the orbital theory of climatic change have argued that it does not explain the medium- and short-term changes of the Holocene, that it cannot explain the spacing of the major ice ages, that it does not account for the timing and initiation of an ice age, and that the computed variations in insolation resulting from the mechanism are too small by themselves to caused significant change. In one sense many of these arguments are valid—orbital changes do not explain all scales of climatic change, and they may well need to be intensified by other mechanisms. However, as the compilation of Berger *et al.* (1984) has shown, the orbital mechanism is one of immense power and significance for explaining some of the major features of the Pleistocene climatic fluctuations. There is substantial evidence to link this mechanism to the longer scales of climatic change (Goreau 1980), and the influence of orbital fluctuations has been traced back in the geological record over millions of years (Kerr 1987).

The major attraction of these ideas is that while the amount of temperature change caused by them may well only be of the order of 1 or 2 °C, the periodicity of these fluctuations seems to be largely comparable with the periodicity of the ice advances and retreats of the Pleistocene. Recent isotopic dating has shown that the record of sea-level changes preserved in the coral terraces of Barbados and elsewhere, and the record of heating and cooling in deep-sea cores, correlates well with theoretical

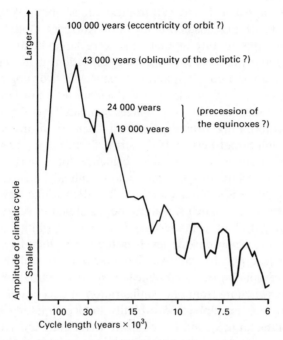

FIG. 7.6 The spectrum of climatic variation over the past half-million years. This graph shows the relative importance of different climatic cycles in the isotopic record of the Indian Ocean cores and relates them to the Earth's orbital variations (modified after Imbrie and Imbrie 1979, fig. 42).

insolation curves based on those of Milankovitch (Mesollela *et al.* 1969, Broecker *et al.* 1968).

Indeed the variations in the Earth's orbit have recently been seen as 'the pacemaker of the ice ages' (Imbrie and Imbrie 1979) for detailed statistical analysis of the ocean cores shows that they possess statistically significant wave-like fluctuations with amplitudes of the order of about 100 000 years, 43 000 years and 19 000–24 000 years (Fig. 7.6). The most important of these cycles is the longest one, corresponding to variations in eccentricity (Hays *et al.* 1976). This applies back to 900 000 years ago, but probably not further (Pisias and Moore 1981). Spectral analysis has shown broadly comparable wavelengths in Chinese loess profiles laid down over the last 700 000 years (Lu 1981), with spectral peaks at 41 700 and 25 000 seeming to correspond to the obliquity of the ecliptic and the precession of the equinoxes respectively.

They may also help to explain the very marked expansion of lakes that took place in tropical and sub-tropical areas about 9000 BP. Recent analyses of theoretical insolation levels by Kutzbach (1981) indicates that radiation receipts in July at that time were larger than now (by about 7 per cent) and that this led to an intensification of the monsoonal circulation and associated precipitation. Changes in hydrological conditions, as exemplified by studies of lake-level fluctuations in low latitudes, show a good correspondence with the changes simulated by general circulation models incorporating orbitally induced changes in insolation receipts (Kutzbach and Street-Perrott 1985, Street-Perrott *et al*. 1990).

Although there is now excellent evidence for a very close relationship between orbital forcing and major climatic changes during the Pleistocene, there is still a need to consider the various ways in which the effects of the Milankovitch signal can be magnified to cause the observed degree and speed of climatic changes. One possibility is the role of fluctuations in carbon dioxide levels in the atmosphere as an intensifying mechanism (see pp. 272–5). Another possibility is that the oceans play a role by responding in a non-linear way to an initial climatic perturbation, and switching from one configuration of circulation characteristics to another. For example, the orbitally induced melting of an ice-cap might liberate large quantities of fresh water into the North Atlantic, which might affect the density and temperature of that water, which might in turn affect the whole pattern of heat-exchange in the oceans of the world, with a whole series of concomitant climatic effects. This is the view put forward by Broecker and Denton (1989: 2465):

We propose that Quaternary glacial cycles were dominated by abrupt reorganizations of the ocean–atmosphere system driven by orbitally induced changes in fresh-water transports which impact salt structure in the sea. These reorganizations mark switches between stable modes of operation of the ocean–atmosphere system. Although we think that glacial cycles were driven by orbital change, we see no basis for rejecting the possibility that the mode changes are part of a self-sustained internal operation that would operate even in the absence of changes in the Earth's orbital parameters. If so . . . orbital cycles can merely modulate and pace a self-oscillating climate system.

One such model of coupled ocean–atmosphere change that has been invoked to account for rapid deglaciation is that put forward by Ruddiman and McIntyre (1981*a,b*). They proposed that when, as a result of orbital change, there was a large receipt of insolation at high latitudes in the northern hemisphere, this would cause a reduction in the volume of continental ice, and would be accompanied by iceberg calving into the oceans. This would cause sea-level to rise (which would make ice-shelves unstable so that accelerated iceberg calving occurred) and the ocean water in the North Atlantic to cool. Ocean cooling would lead to an increased area of winter sea-ice and decreased moisture provision to the atmosphere, which would in turn cause less nourishment of ice-caps, thereby promoting their demise.

Another coupled ocean–atmosphere model which has been proposed to account for the 'rapidity, near synchroneity, and global extent of the events associated with the termination of the last glacial cycle' is that of Broecker and Denton (1990). They maintain that towards the end of the last glacial cycle large masses of fresh water were delivered to the North Atlantic Ocean by rivers draining from the wasting Northern Hemisphere ice-caps. This in turn affected the salinity and density of the oceans which in turn caused a change in the pattern of ocean circulation. They conclude (p. 336),

The atmosphere and ocean are tied together in a non-linear manner, making the combined system susceptible to abrupt mode switches . . . We propose that mode switches occur when orbitally induced seasonality changes alter the flow of water vapour from one part of the ocean to another. The Atlantic conveyor belt appears to be the most vulnerable part of the system. Therefore, salinity changes in this part of the sea are most likely to create the instabilities that cause the ocean–atmosphere system to reorganize.

Hypotheses involving changes in terrestrial geography

Although some climatic changes take place over a short time-span, such as the Little Ice Age or the twentieth-century phase of warming up, some of the longer-term changes, including possibly the initiation of glaciation in certain parts of the world, may have resulted from changes in the positions of the continents, shifts in the position of the polar axis, and uplift of continents, by orogeny. Of these the first two are probably relatively unimportant in terms of the Pleistocene, in that the rates of change involved are very slow and over the time-span concerned rather slight. The rate of polar wandering, for example, has been estimated as 3×10^{-7} degrees yr^{-1}, and this would be insufficient to affect the pattern of Pleistocene glaciation (Cox 1968). The rate of continental drift is a little higher, with a mean rate of around 1.0×10^{-7} degrees yr^{-1}, which is equal to a displacement of only 1 during the last 10^7 years (possibly only 0.2 since the start of the classic glaciations). Even with a maximum postulated rate of 6.0×10^{-7} degrees yr^{-1}, the displacement is not of very great significance. However, Ewing (1971) has suggested that if sea-floor spreading operated at a rate of about 2 cm/yr. the width

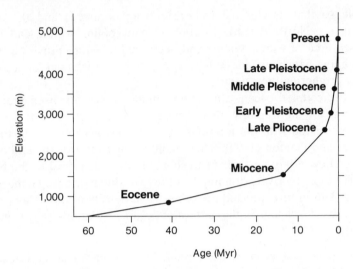

Fig. 7.7 The uplift history of the Tibetan Plateau and the Himalayas over the last sixty million years. The curve is inferred from palaeobotanical data (based on work by Xu Ren in Molnar and England 1990, fig. 4). Reprinted by permission from *Nature*, 346. Copyright © 1990, Macmillan Magazines Ltd.

of a rift such as that between Spitzbergen and Greenland would increase by about 200 km in 10 m.y., sufficient to affect the entry of ocean currents into the Arctic and thereby the climate of the surrounding areas. Nevertheless, many workers consider that terrestrial causes of climatic change can be narrowed down to vertical uplift of mountains through orogeny so that their summits become sufficiently high and cold for snow and ice to accumulate. This could well have important effects, for as pointed out on p. 251 there has been considerable tectonic movement in many areas in the Pleistocene and Late Tertiary.

If one assumes a rate of uplift in a tectonically highly active area of 10 m per 1000 years it would only require about 10 000 years to obtain a mean temperature fall of 0.65 °C, since temperature falls at a rate of about 0.65 °C with every 100 m rise in altitude. Thus in the course of the Pleistocene it would be quite possible for a mountain to develop sufficiently quickly to cause a marked temperature depression at its summit. Also, total precipitation is known to show a general increase with altitude, at least up to about 3000 m, so that the overall increase in height of mountains would serve to produce veritable snow traps.

Some support for this hypothesis is given by the fact that not all areas appear to have undergone multiple glaciation, and in many cases it seems possible or probable that uplift in the mid and late Pleistocene brought the mountains of some areas into a position where ice could accumulate. Thus, for example, Mauna Kea (Hawaii), Tasmania, and the Pyrenees, have been cited as areas which anomalously only suffered one major glacial phase, and that late in the Pleistocene.

However, orogenic effects could also have had a global dimension, and the role of the mountains and plateau of High Asia and the south-west USA have been accorded particular importance in this respect (Ruddiman *et al.* 1989).

Tibet and the high Himalayas (Fig. 7.7) have undergone a net uplift of 2–3 km during the past 3 million years, so that 'As much as 75 per cent of the net elevation of Tibet (an area approximately one third of the size of the contiguous United States) may have been attained during the time of Northern Hemisphere glacial inception and intensification.' (Ruddiman and Raymo 1988: 420). Significant Plio-Pleistocene uplift has also occurred in several regions of the south-west United States.

Such uplift has had two effects on northern hemisphere climate that could be relevant to glacial inception and intensification. One of these is to enhance albedo-temperature feedback on a globally significant scale (Ruddiman and Raymo 1988: 420–1):

The cold air over the large, newly created region of high-standing topography at middle latitudes provides a temperature discontinuity that accelerates the normal southward movement of the snow-line in autumn–winter. This enlarged high-albedo surface in turn leads to global cooling during the transitional and winter seasons.

The second of these mechanisms is a change in the form of the major waves in the general atmospheric circulation. It is proposed that the increased elevation of both the Tibetan-Himalayan and south-western USA regions would have altered the planetary wave structure in such a way as to cool the North American and European land masses and increase their susceptibility to orbitally driven insolation changes (the Milankovitch mechanism). Ruddiman and Raymo (1988: 425) suggest: 'Before late-Pliocene mountain uplift, early Pliocene cold air masses were probably more confined to the northernmost parts of North America and Europe . . . We suggest that the degree of southward penetration of the waves over North America and Europe gradually deepened during the late Pliocene and Pleistocene . . .'.

Feedback (autovariation) hypotheses

Thus far in our consideration of the possible causes of climatic change we have suggested that it has been caused by significant changes in the output of solar radiation, the position and configuration of the Earth in relation to other heavenly bodies, the quality of the atmosphere, and the arrangement of land masses, oceans, and mountains. However, there has been a variety of hypotheses in which it is envisaged that the atmosphere possesses a degree of internal instability which might furnish a built-in mechanism of change. It is conceivable that some small change might, through positive feedback, have extensive and long-term effects. As Mitchell (1968: 158) put it, 'minor environmental disturbances may have sufficed to "flip" the atmospheric circulation and climate from one state to another, and to "flop" it back again'. Some selected examples will give an indication of the importance of the hypotheses involving feedback relations.

Wilson's (1964) hypothesis

At a time when the total thickness of the ice-cap of the Antarctic is less than a critical value, the rate of thickening produced by the accumulation of precipitation exceeds the rate of subsidence produced by plastic flow of the ice-cap and by mass loss through calving at the perimeter. When, however, the ice thickness reaches a critical threshold value, the transverse shearing stress near the base of the ice-cap becomes so large that the ice flow abruptly accelerates. This produces heat by friction, so that flow accelerates further until the whole ice-cap subsides at a more or less catastrophic rate. This fills the world oceans with cold ice and thereby leads to a reduction in world temperatures, which promotes glaciation in certain other parts of the world (Hollin 1965, Selby 1973).

In addition, as a result of the surge of the ice-cap, possibly one-third of the ice-sheet is transferred to the continental shelf to form a huge ice-shelf. This shelf would increase the surface albedo of $25 \times 10^6 \, \text{km}^2$ of ocean from 8 per cent to 80 per cent, adding to the cooling effect by decreasing the heat input to the Earth as a whole by 4 per cent.

The Plass hypothesis

The CO_2 levels in the atmosphere have an effect on the world's heat balance because CO_2 is virtually transparent to incoming solar radiation but absorbs outgoing terrestrial infra-red radiation—radiation that would otherwise escape to space and result in heat loss from the lower atmosphere. In general, through the mechanism of this so-called 'greenhouse effect' low levels of CO_2 in the atmosphere would be expected to lead to cooling. Under the Plass hypothesis (1956) an unspecified cause reduces the carbon dioxide content of the atmosphere by several per cent. This would lower atmospheric temperatures, and after 50 000 years or so the oceans would cool by a similar amount and come to a new CO_2 equilibrium with the atmosphere. The lowered temperature would promote continental glaciation, which would in turn lower sea-levels, and thus cause a new imbalance of carbon dioxide with the atmosphere by increasing its concentration in the oceans. This increased CO_2 content of the atmosphere leads to atmospheric warming, leading to glacier melting, and a restoration of the oceans to their original volume.

Until recently there was no method by which this and related hypotheses could be tested. However, it has now proved possible to retrieve carbon dioxide from bubbles in layers of ice of known age in deep-ice cores. Analyses of changes in carbon dioxide concentrations in these cores have provided truly remarkable results. The work of Delmas *et al.* (1980) on the Dome Core in Antarctica showed for the first time that round about the last glacial maximum (*c.*20 000 years ago) the level of atmospheric carbon dioxide was only about 50 per cent that of the present. This seeming coincidence of cold temperatures and low CO_2 levels lended some prima-facie support to the Plass hypothesis. Subsequent analyses have confirmed the findings of the Delmas group, and have demonstrated that carbon dioxide changes and climatic changes have progressed in approximate synchroneity over the last 160 000 years. Thus the last interglacial was a time of high CO_2 levels, the last glacial maximum of low CO_2 levels, and the early Holocene a time of very rapid rise in CO_2 levels (Fig. 7.8).

The explanation for this remarkable coincidence is still the matter of intense speculation and debate, and a large number of models has been produced to explain the causes of the fluctuating atmospheric CO_2 levels. Because the oceans are such a large store of CO_2 in comparison with terrestrial sinks, the explanation almost certainly lies in the oceans and changes in their circulation and turnover, or in changes in the productivity of the various organisms that live in them (Sarnthein *et al.* 1988). An initial hypothesis was put forward by Broecker (1981). He hypothesized that a possible cause of the high level of CO_2 during the interglacials may be a loss of phosphorus to continental shelf sediments during transgressions of the oceans. This would, he believes, reduce the amount of plant matter formed in the sea per unit area of upwelled water and would thereby increase the CO_2 pressure in surface water and in the atmosphere. Regressions during glacials would tend to reverse this trend. Given that transgressions and regressions of the oceans are tied up so closely with the decay and growth of ice-caps, these CO_2 changes could

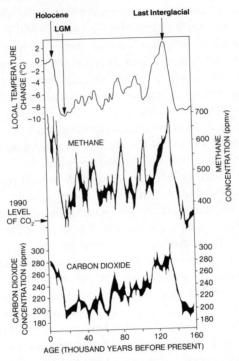

FIG. 7.8 Analysis of air trapped in Antarctic ice-cores showing that methane and carbon dioxide concentrations were closely correlated with the local temperature over the last 160 000 years. Present day concentrations of carbon dioxide are indicated. The temperatures were derived by measuring the proportion of deuterium in the ice. Interglacial CO_2 levels have been 70 to 80 ppm higher than those of the last glacial maximum. (From J. T. Houghton, G. J. Jenkins, and J. J. Ephraums (1990) (eds.), fig. 2.)

amplify the effects of orbital variations of the type postulated by the Croll–Milankovitch theory.

Another ingenious explanation for decreased CO_2 levels in glacials was provided by Martin (1990). He suggested that phytoplankton consume carbon dioxide, and that phytoplankton require iron. Combining these two assumptions he suggested that at the last glacial maximum the world was very much dustier than today (because of deflation from expanded deserts and from outwash plains, etc.), that the dust was iron rich, and thus when it entered the oceans it caused enhanced phytoplankton growth which caused the phytoplankton to consume carbon dioxide, which led to lowered atmospheric carbon dioxide levels. Changes in biological productivity could also be achieved by changes in processes such as low-latitude upwelling (Sarnthein *et al.* 1988) or in the temperatures of high-latitude surface waters (Sarmiento and Toggweiler 1984). The role of changes in ocean circulation patterns are discussed in Siegenthaler and Wenk (1984) and Broecker and Denton (1989).

It is also possible that methane (CH_4) may have played a similar role in intensifying orbital effects, for molecule for molecule it is a highly effective greenhouse gas. As with CO_2, CH_4 levels appear to have been greater in their atmospheric concentrations during interglacials than in glacials (Raynaud *et al.* 1988). As Fig. 7.8 indicates, CH_4 concentrations have tended to be less than 400 ppbv in glacials and over 600 ppbv in interglacial times. Nisbet (1990) has suggested that during the glacials, many high-latitude bogs and wetlands (which are major sources of the gas) would have been covered by ice or rendered less productive by extensive permafrost distribution. Moreover, many natural gas-fields, which give out methane to the atmosphere, would have been sealed off by ice-sheets and permafrost. By contrast, in an interglacial wetlands and bogs would produce large amounts of methane, and at the transition between glacial and interglacial conditions the methane that had been trapped in and beneath ice would be released into the atmosphere as the ice cover decreased. The rapid release of the gas would lead to atmospheric warming which would in turn lead to more melting, and more gas release.

The Ewing–Donn hypothesis (1956, 1958)

The cycle of events begins with high interglacial sea-levels, and with a flow of warm water into the Arctic ocean, which both keeps that ocean ice-free and favours the accumulation of precipitation in the form of snow on the surrounding land masses. This lowers the sea-levels so that a submarine ridge between Iceland and the Faeroes begins to block the further flow of warm water into the Arctic. The increased area of the growing ice-cap would also lead to a greater reflection of solar-radiation. This would accelerate the rate of cooling. Such a tendency would be further reinforced by the formation of an anticyclone over the ice, with outblowing winds repelling the moderating influence of Atlantic conditions. Thus the Arctic freezes over and prevents any continuing replenishment of the ice-sheets which thus gradually waste away. Sea-level thus rises and warm water flows in again, starting the cycle off once more.

There is some degree of argument as to whether the Arctic has in reality been ice-free in the Pleistocene. On the one hand Larsen and Barry (1974: 258), basing their work on the study of ocean-core sediments, suggest that the Arctic Ocean was never ice-free in the Pleistocene and so could not have been a factor in the growth or melting of Pleistocene continental glaciers. On the other hand Worsley and Herman (1980) suggest after a review of the evidence that in Pliocene and Pleistocene times variations in the nature and number of calcareous nannofossils indicate that one is dealing with an ocean that was episodically ice-free. The controversies surrounding the age of the Arctic ice and its behaviour in the Pleistocene, and the light this sheds on the Ewing–Donn model have recently been well reviewed by Barry (1989).

Albedo-based hypotheses

One factor which controls heating levels in the earth-atmosphere system is the degree to which incoming solar radiation is reflected or absorbed by the Earth's surface. Changes in the albedo of that surface, brought about perhaps by seemingly minor events, might be able to cause major changes in climate. For example, the deposition of a layer of relatively dark volcanic dust over the ice-caps, as a result of a chance volcanic explosion, might lead to the melting of that ice-cap which might in turn set a train of events in motion. Similarly, the presence of an unusually widespread and persistent snow cover over northern Canada as a result of a chance association of snowy winters and cool summers, could either help to trigger off climatic change directly, or could play a role as part of a feedback reaction (Williams 1975). Such a snow cover, persisting through all or most of the summer and autumn, would reflect the sun and further chill the air (Calder 1974). This in itself might increase the likelihood of snow the next winter. The snow gradually accumulates, leading to a great ice-sheet. A popular name for this mechanism is 'Snowblitz'.

The human impact on climate

The miscellaneous hypotheses discussed so far have been applied with varying degrees of success to a variety of different time-spans. When, however, one considers the immediate past and contemplates the near future, the role of man takes on a position of probable importance, for, although as explained in Chapter 5, the twentieth-century climatic changes have greatly affected man, it is more than likely that man has himself been partially responsible for some of the observed changes, particularly because of his effect on atmospheric quality. As yet, because of the complexity of the atmospheric system and the large number of possible causes, it is difficult fully to assess and quantify the role that man

Table 7.1 Possible ways in which human activities may cause climatic change

Gas emissions
CO_2—industrial and agricultural
methane
chlorofluororcarbons (CFCs)
nitrous oxide
krypton 85
water vapour
miscellaneous trace gases
Aerosol generation
Thermal pollution
Albedo change
dust addition to ice caps
deforestation
over-grazing
Extension of irrigation
Alteration of ocean currents by constricting straits
Diversion of fresh waters into oceans

has played, though certain mechanisms of man-induced climatic changes on a global (as opposed to a microclimatic scale) can be recognized (Table 7.1).

However, there is no space in this volume to pursue this important matter further. It is a subject that has been reviewed elsewhere (Goudie 1990).

Conclusion

Our knowledge of the causes of climatic changes is still highly imperfect. Moreover, as Sparks and West (1972: 26) put it, 'It is not a field in which many people can dwell comfortably for a long time because it is almost entirely speculative.' No completely acceptable explanation of climatic change has ever been presented, and it is also clear that no one process acting alone can explain all scales of climatic changes. Some coincidence or combination of processes in time is probably required, such as Flint's 'solar-topographic' theory (1971), which depends mainly on variations in the intensity of solar variation and of mountain building. Moreover, numerous feedback loops may exist and quite small triggers may set off a chain of events which lead to substantial change. Some hypotheses appear plausible to explain variations over a long period (for example the Croll–Milankovitch hypothesis is being seen as more and more applicable to the glacial–interglacial cycles of the last half-million years), while others may appear more plausible for short-term fluctuations (changes in sunspots may be a hypothesis relevant at a scale of a decade or more). Two other basic problems exist. One is that to test certain hypotheses we need precise knowledge of the exact pattern and dating of past fluctuations—this we seldom have. The second is that we are dealing with an immensely complex series of interrelated systems; the solar system, the atmosphere, the oceans, and the land. It is thus unlikely that any simple hypothesis or model of climatic change will have very wide applicability.

Given these considerations it is clearly impossible at the present state of knowledge to make any safe prognosis of the climatic developments of the future.

Equally, as the debate intensifies about the potential role of human activities in causing climatic changes over the coming decades and centuries, it will be important to remember the scale and range of past natural environmental changes. Moreover, the knowledge we have gained about how the Earth has responded to past climatic changes will be invaluable for an understanding of what the future has in store.

Selected reading

There are three excellent surveys of theories on climatic change, which although now old, give a good survey of the development of ideas.

In *Essays in Geography for Austin Miller* (1965) (ed. J. B. Whittow and P. D. Wood), R. P. Beckinsale contributes 'Climatic change: a critique of modern theories', pp. 1–38. Another review, of a similar type but with greater concentration on the auto-variation ideas is J. M. Mitchell 'Theoretical palaeclimatology' in H. E. Wright and D. G. Frey (1965) (eds.), The *Quaternary of the U.S.A.*, pp. 881–901. J. Mitchell (1968) has also edited *Causes of climatic change*, Meteorological Monographs, 8. These three sources give lengthy bibliographies and summaries of the major theories.

The Croll–Milankovitch ideas have been discussed in their historical context by Imbrie and Imbrie (1979), *Ice Ages*. They are also discussed in a series of essays in A. Berger *et al.* (1984) (eds.), *Milankovitch and climate*.

The most comprehensive assessment of the role of volcanic eruptions is by H. H. Lamb (1970), 'Volcanic dust in the atmosphere: with a chronology and assessment of its meteorological significance', *Philosophical Transactions Royal Society of London*, 266 A, 425–533. A brief review of the role of man in causing climatic change is given in Goudie (1990), *The Human Impact*.

References

ADAMSON, D. A., GASSE, F., STREET, F. A., and WILLIAMS, M. A. J. (1980), 'Late Quaternary history of the Nile', *Nature*, 287: 50–5.

AGRAWAL, D. P., DODIA, R., and SETH, M. (1990), 'South Asian climate and environment at *c*.18 000 BP', in O. Soffer and C. Gamble (eds.), *The World at 18 000 BP*, vol. 2 (London).

AHLMANN, H. W. (1948), 'The present climatic fluctuation', *Geographical Journal*, 112: 165–95.

——(1953), 'Glacier variations and climatic fluctuations', *American Geographical Society* (New York).

AHMAD, N. and SAXENA, H. B. (1963), 'Glaciations of the Pindar river valley, southern Himalayas', *Journal of Glaciology*, 4: 471–6.

AITKEN, M. J. (1989), 'Luminescence dating: a guide for non-specialists', *Archaeometry*, 31: 147–59.

ALLCHIN, B. (1978), *Prehistory and Palaeogeography of the Great Indian Desert* (London).

ALPERS, C. N. and BRIMHALL, G. H. (1988), 'Middle Miocene climatic change in the Atacama Desert, northern Chile: Evidence for supergene mineralization at La Escondida', *Bulletin, Geological Society of America*, 100: 1640–56.

ANANOVA, E. N. (1967), 'Palynological correlation of the flora and vegetation of the Likhvin-Mazovian l-Holstein-Neede Interglacial', *Review of Palaeobotany and Palynology*, 4: 175–86.

ANDERSON, J. B. and THOMAS, M. A. (1991), 'Marine ice-sheet decoupling as a mechanism, for rapid episodic sea-level change: the record of such events and their influence on sedimentation', *Sedimentary Geology*, 70: 87–104.

ANDREWS, J. T. (1970), 'A geomorphological study of Post-Glacial uplift with particular reference to Arctic Canada', *Institute of British Geographers Special Publication*, 2.

——(1975), *Glacial systems—an approach to glaciers and their environments* (North Scituate, Mass.).

ATKINSON, T. C., HARMON, R. S., SMART, P. L., and WALTHAM, R. C. (1978), 'Palaeoclimatic and geomorphic implications of $^{230}Th/^{234}U$ dates on speleothems from Britain', *Nature*, 272: 24–8.

BAKER, C. A. and JONES, D. K. C. (1980), 'Glaciation of the London Basin and its influence on the drainage pattern', in D. K. C. Jones (ed.), *The Shaping of Southern England* (London): 131–75.

BALLARD, R. D. and UCHUPI, E. (1970), 'Morphology and Quaternary history of the continental shelf of the Gulf Coast of the United States', *Bulletin of Marine Science*, 20: 547–59.

BANDY, O. L. (1968), 'Changes in Neogene paleo-oceanography and eustatic changes', *Palaeogeography, Palaeoclimatology, Palaeoecology*, 5: 63–75.

—— and MARICOVICH, L. (1973), 'Rates of Late Cenozoic uplift, Baldwin Hills, Los Angeles, California', *Science*, 181: 653–4.

BARD, E., LABERYRUE, L. D., PICHON, J. J., LABRACHERIE, M., ARNOLD, M., DUPRAT, J., MOYES, J., and DUPLESSY, J. C. (1990), 'The last deglaciation in the southern and northern hemispheres', in V. Bleil and J. Thiede (eds.), *Geological history of the polar oceans: Arctic versus Antarctic* (Dordrecht): 405–15.

BARNOLA, J. M., RAYNAUD, D., KOROTKEVICH, Y. S., and LORIUS, C. (1987), 'Vostok ice-core provides 160 000-year record of atmospheric CO_2', *Nature*, 329: 408–14.

BARRETT, E. C. (1966), 'Regional variations of rainfall trends in northern England, 1900–1959', *Transactions, Institute of British Geographers*, 38: 41–58.

BARRON, E. J. (1985), 'Explanations of the Tertiary global cooling trend', *Palaeogeography, Palaeoclimatology, Palaeoecology*, 50: 45–61.

BARRY, R. G. (1989), 'The present climate of the Arctic Ocean and possible past and future states', in Y. Herman, *The Arctic Seas* (New York): 1–46.

BARTLEIN, P. J., WEBB, T., and FLERI, E. (1984), 'Holocene climatic change in the northern Midwest: Pollen-derived estimates', *Quaternary Research*, 22: 361–74.

BATTARBEE, R. W. (1986), 'Diatom analysis', in B. E. Berglund (ed.), *Handbook of Holocene palaeoecology and palaeohydrology* (Chichester): 527–70.

BEADLE, L. C. (1974), *The inland waters of tropical Africa: An introduction to tropical limnology* (London).

BECKINSALE, R. P. (1965), 'Climatic change: a critique of modern theories', in J. B. Whittow and P. D. Wood (eds.), *Essays in Geography for Austin Miller* (Reading): 1–38.

BEHRE, K. E. (1989), 'Biostratigraphy of the last glacial period in Europe', *Quaternary Science Reviews*, 8: 25–44.

BENBOW, M. C. (1990), 'Tertiary coastal dunes of the Eucla Basin, Australia', *Geomorphology*, 3: 9–29.

BENEDICT, J. B. (1968), 'Recent glacial history of an Alpine area in the Colorado Front Range, U.S.A.', *Journal of Glaciology*, 7: 77–87.

BERGER, A., IMBRIE, J., HAYS, J., KULKA, G., and SALTZMAN, B. (eds.) (1984), *Milankovitch and climate*, 2 vols., (Dordrecht).

BERGGREN, W. A. (1969), 'Cainozoic stratigraphic, planktonic foraminiferal zonation and the radiometric time-scale', *Nature*, 224: 1072–5.

BERGHINZ, C. (1976), 'Venice is sinking into the sea', in R. Tank (ed.), *Focus on environmental geology*, 2nd edn. (New York): 512–18.

BERNABO, J. C. and WEBB, III T. (1977), 'Changing patterns in the Holocene pollen record of north-eastern North America: A mapped summary', *Quaternary Research*, 8: 64–96.

BETANCOURT, J. L., VAN DEVENSER, T. R., and MARTIN, P. S. (1990) (eds.), *Packrat middens. The last 40000 years of biotic change* (Tucson, Arizona).

BEVERTON, R. J. H. and LEE, A. J. (1965), 'Hydrographic fluctuations in the north Atlantic and some biological consequences', in C. G. Johnson and L. P. Smith (eds.), *The biological significance of climatic changes in Britain* (London): 79–107.

BINNS, R. E. (1972), 'Flandrian strandline chronology for the British Isles and correlation of some European Post-glacial strandlines', *Nature*, 235: 206–10.

BIRKELAND, P. W. (1972), 'Late Quaternary eustatic sea-level changes along the Malibu coast, Los Angeles County, California', *Journal of Geology*, 80: 432–44.

——(1973), 'Use of relative dating methods in a stratigraphic study of rock glacier deposits, Mt. Sopris, Colorado', *Arctic and Alpine Research*, 5: 401–16.

——(1984), *Soils and geomorphology* (New York).

BIRKS, H. J. B. (1986), 'Quaternary biotic changes in terrestial and lacustrine environments, with particular reference to north-west Europe', in B. E. Berglund (ed.), *Handbook of Holocene Palaeoecology and Palaeohydrology* (Chichester): 3–65.

——(1990), 'Changes in vegetation and climate during the Holocene of Europe', in M. M. Boer and R. S. de Groot (eds.), *Landscape-ecological impact of climatic change* (Amsterdam): 133–58.

——and BIRKS, H. H. (1980), *Quaternary Palaeoecology* (London).

BLOOM, A. L. (1967), 'Pleistocene shorelines: A new test of isostasy', *Bulletin, Geological Society of America*, 78: 1477–94.

——(1971), 'Glacial eustatic and isostatic controls of sea-level since the Last Glaciation', in K. K. Turekian (ed.), *The Late Cainozoic Glacial Ages* (New Haven, Conn.): 355–79.

——(1983), 'Sea-level and coastal morphology of the United States through the Late Wisconsin glacial maximum', in S. C. Porter (ed.), *Late Quaternary environments of the United States* (London), 215–29.

BOECK, W. L., SHAW, D. T., and VONNEGUT, B. (1975), 'Possible consequences of global dispersion of Krypton 85', *Bulletin of the American Meteorological Society*, 56: 527.

BONATTI, E. (1966), 'North Mediterranean climate during the last Würm Glaciation', *Nature*, 209: 984.

——and GARTNER, S. (1973), 'Caribbean climate during Pleistocene Ice Ages', *Nature*, 244: 563–5.

BORTENSCHLAGER, S. and PATZELT, G. (1969), 'Warmezeitliche Klima- und Gleischerswankungen im Pollin Profil Eines Hochgelegenen Moores (2270 m) Der Venediggergruppe', *Eiszeitalter und Gegenwart*, 20: 116–22.

BOWEN, A. J. (1972), 'The tidal regime of the River Thames; long-term trends and their possible causes', *Philosophical Transactions Royal Society of London A*, 272: 187–99.

BOWEN, D. Q. (1970), 'South-east and central South Wales', in C. A. Lewis (ed.), *The glaciations of Wales and adjoining regions* (London): 197–227.

BOWEN, D. Q. (1973), 'The Pleistocene history of Wales and the Borderland', *Geological Journal*, 8/2: 207–24.

——(1978), *Quaternary Geology* (Oxford).

——RICHMOND, G. M., FULLERTON, D. S., ŠIBRAVA, V., FULTON, R. J., and VELICHKO, A. A. (1986), 'Correlation of Quaternary Glaciations in the northern hemisphere', *Quaternary Science Reviews*, 5: 509–10.

——ROSE, J., MCCANE, A. M., and SUTHERLAND, D. G. (1986), 'Correlation of Quaternary glaciation in England, Ireland, Scotland and Wales', *Quaternary Science Reviews*, 5: 299–340.

BOWLER, J. M. (1976), 'Aridity in Australia: age, origins, and expression in aeolian landforms and sediments', *Earth Science Review*, 12: 279–310.

——HOPE, G. S., JENNINGS, J. N., SINGH, G., and WALKER, D. (1976), 'Late Quaternary climates of Australia and New Guinea', *Quaternary Research*, 6: 359–94.

BOWLES, F. A. (1976), 'Palaeoclimatic significance of quartz/illite variations in cores from the eastern equatorial North Atlantic', *Quaternary Research*, 5: 225–35.

BRADLEY, R. S. and MILLER, G. H. (1972), 'Recent climatic change and increased glacierization in the eastern Canadian Arctic', *Nature*, 237: 385–7.

——(1985), *Quaternary paleoclimatology* (Boston).

——DIAZ, H. F., EISCHEID, J. K., JONES, P. D., KELLY, P. M., and GOODESS, C. M. (1987), 'Precipitation fluctuations over northern hemisphere land areas since the mid-19th century', *Science*, 237: 171–5.

BRAKENRIDGE, G. R. (1978), 'Evidence for a cold, dry full-glacial climate in the American southwest', *Quaternary Research*, 9: 22–40.

BRAUDEL, F. (1972), *The Mediterranean and the Mediterranean world in the age of Philip II* (London) vol. 1: 642 ff.

BRAY, J. R. (1970), 'Temporal patterning of post-Pleistocene glaciation', *Nature*, 228: 353–4.

——(1974), 'Volcanism and glaciation during the past 40 millennia', *Nature*, 252: 679–80.

BRIFFA, K. R., BARTHOLIN, T. S., ECKSTEIN, D., JONES, P. D., KARLÉN, W., SCHWEINGRUBER, F. H., and ZETTERBERG, P. (1990), 'A 1400-year tree-ring record of summer temperatures in Fennoscandia', *Nature*, 346: 434–9.

BRINK, N. W. T. and WEIDICK, A. (1974), 'Greenland ice-sheet history since the Last Glaciation', *Quaternary Research*, 4: 429–40.

BROECKER, W. S. (1975), 'Climatic change: are we on the brink of a pronounced global warming?', *Science*, 189: 460–3.

——(1981), 'Glacial to interglacial changes in ocean and atmosphere chemistry', in A. Berger (ed.), *Climatic variations and variability: facts and theories* (Dordrecht): 111–21.

——and Denton, G. H. (1989), 'The role of ocean–atmosphere reorganizations in glacial cycles', *Geochimica et Cosmochimica Acta*, 53: 2465–501.

—— ——(1990), 'The role of ocean–atmosphere reorganizations in glacial cycles', *Quaternary Science Reviews*, 9: 305–41.

——and KAUFMANN, A. (1965), 'Radiocarbon chronology of Lake Lahontan and Lake Bonneville II, Great Basin', *Bulletin, Geological Society of America*, 76: 537–66.

——and THURBER, D. L. (1965), 'Uranium Series dating of corals and oolites from Bahaman and Florida Keys Limestones', *Science*, 149: 58–60.

——TAKAHASHI, T., SIMPSON, H. J., and PENG, T. H. (1979), 'Fate of fossil fuel carbon dioxide and the global carbon budget', *Science*, 206: 409–18.

—— —— GODDARD, J., KU, T. L., MATTHEWS, R. K., and MESOLLELA, K. J. (1968), 'Milankovitch hypothesis supported by precise dating of coral reefs and deep-sea sediments', *Science*, 159: 297–300.

BROOKES, I. A. (1989), 'Early Holocene basinal sediments of the Dakleh Oasis region, south central Egypt', *Quaternary Research*, 32: 139–52.

BROOKS, C. E. P. (1926), *Climate through the ages* (London).

BROWN, J. A. (1976), 'Shortening of growing season in the U.S. Corn Belt', *Nature*, 260: 420–1.

BROWN, K. S., SHEPPARD, P. M., and TURNER, J. R. G. (1974), 'Quaternary refugia in tropical America: evidence from race formation in Heliconius butterflies', *Proceedings, Royal Society* B: 369–78.

BROWN, P. R. (1953), 'Climatic fluctuation in the Greenland and Norwegian Seas', *Quarterly Journal of the Royal Meteorological Society*, 79: 272–81.

BRYAN, K. (1941), 'Pre-Columbian agriculture in the south-west as conditioned by periods of alluviation', *Annals Association of American Geographers*, 31: 219–42.

BRYSON, R. A. (1974), 'A perspective on climate change', *Science*, 184: 753–60.

——(1989), 'Late Quaternary volcanic modulation of Milankovitch climate forcing', *Theoretical and Applied Climatology*, 39: 115–25.

——and BARREIS, D. A. (1967), 'Possibility of major climatic modifications and their implications: North-west India, A case for study', *Bulletin, American Meteorological Society*, 48: 136–42.

BUCHARDT, B. (1978), 'Oxygen isotope palaeotemperatures from the Tertiary period in the North Sea area', *Nature*, 275: 121–3.

BUDYKO, M. A., DROZDOV, D. A., and YUDIN, M. I. (1971), 'The impact of economic activity on climate', *Soviet Geography*, 12: 666–79.

BUNTING, A. H., DENNETT, M. D., ELSTON, J., and MILFORD, J. R. (1976), 'Rainfall trends in the West African Sahel', *Quarterly Journal of the Royal Meteorological Society*, 102: 59–64.

BUTZER, K. W. (1961), 'Climatic change in arid regions since the Pliocene', *Arid Zone Research* (UNESCO), 17: 31–56.

——(1971), 'Recent history of an Ethiopian delta', *University of Chicago, Dept. of Geography, Research Paper* 136: 184.

——(1972) *Environment and archaeology: An ecological approach to prehistory* (London): 730 ff.

——(1975), 'Geological and ecological perspectives on the Middle Pleistocene', in K. W. Butzer and G. L. Isaac, *After the Australopithecines* (The Hague): 857–73.

Butzer, K. W. (1977) 'Environment, culture and human evolution', *American Scientist*, 65: 572–84.

—— and Hansen, C. L. (1968), *Desert and river in Nubia: geomorphology and prehistoric environments at the Aswan Reservoir* (University of Wisconsin Press).

—— Isaac, G. L., Richardson, J. L., and Washbourne-Kamau, C. (1972), 'Radiocarbon dating of East African lake levels', *Science*, 175: 1069–75.

Buuman, P. (1980), 'Palaeosols in the Reading Beds (Palaeocene) of Alum Bay, Isle of Wight, UK', *Sedimentology*, 27: 593–606.

Calder, N. (1974), *The weather machine* (London).

Callendar, G. S. (1961), 'Temperature fluctuations and trends over the Earth', *Quarterly Journal of the Royal Meteorological Society*, 87: 1–12.

Catt, J. A. (1986), *Soils and Quaternary geology* (Oxford).

—— (1988), *Quaternary geology for scientists and engineers* (Chichester).

Causse, C., Coque, R., Ch. Fontes, J., Gasse, F., Gibert, E., Ben Ouezdou, H., and Zouari, K. (1989), 'Two high levels of continental waters in the southern Tunisian chotts at about 90 and 159 ka', *Geology*, 17: 922–5.

Chaline, J. (1972), *Le Quaternaire* (Paris).

Chandler, T. J. and Gregory, S. (eds.) (1976), *The climate of the British Isles* (Harlow).

Chappell, J. and Thom, B. (1977), 'Sea levels and coasts', in J. Allen, J. Golson, and R. Jones (eds.), *Sunda and Sahul* (New York): 275–91.

—— and Veeh, H. H. (1978), 'Late Quaternary tectonic movements and sea-level changes at Timor and Atauro Island', *Bulletin, Geological Society of America*, 89: 356–68.

Charlesworth, J. K. (1957), *The Quaternary Era* (London): 1700 ff.

Charney, J., Stone, P. H., and Quirk, W. J. (1975), 'Drought in the Sahara: a biogeophysical feedback mechanism', *Science*, 187: 434–5.

Chepalyga, A. L. (1984), 'Inland sea basins', in A. A. Velichko (ed.), *Late Quaternary environments of the Soviet Union* (London): 229–47.

Chester, D. K. (1988), 'Volcanoes and climate: Recent volcanological perspectives', *Progress in Physical Geography*, 12/1: 1–35.

Childe, V. G. (1954), *New light on the most ancient east* (London).

Chu Ko-Chan (1973), 'A preliminary study on the climate fluctuations during the last 5000 years in China', *Scientia Sinica*, 16: 226–56.

Churchill, D. M. (1965), 'The displacement of deposits formed at sea-level 6500 years ago in southern Britain', *Quaternaria*, 7: 239–49.

Claiborne, R. (1973), *Climate, man and history* (London): 444 ff.

Clapperton, C. M. (1979), 'Glaciation in Bolivia before 3.27 Myr.', *Nature*, 277: 375–7.

—— (1990) (ed.), 'Quaternary glaciations in the southern hemisphere', *Quaternary Science Reviews*, 9: 121–304.

Clark, D. L. (1982), 'Origin, nature and world climate effect of Arctic Ocean ice cover', *Nature*, 300: 321–5.

CLARK, J. A. (1976), 'Greenland's rapid post-glacial emergence: a result of ice-water gravitational attraction', *Geology*, 4: 310–12.

——and LINGLE, C. S. (1979), 'Predicted relative sea-level changes (18 000 years BP to present) caused by late-glacial retreat of the Antarctic ice-sheet', *Quaternary Research*, 11: 279–98.

CLARK, J. D. (1975), 'Africa in prehistory: Peripheral or paramount?', *Man*, 10: 175–98.

CLARK, J. G. D. (1970), 'Mesolithic times', in *Cambridge Ancient History*, 3rd edn. vol. i: 90–121.

CLARKE, R. H. (1970), 'Quaternary sediments off south-east Devon', *Quarterly Journal of the Geological Society*, 125/3: 277–318.

CLIMAP PROJECT MEMBERS (1976), 'The surface of the Ice-Age earth', *Science*, 191: 1131–7.

COETZEE, J. A. (1964), 'Evidence for a considerable depression of the vegetation belts during the Upper Pleistocene on the East African mountains', *Nature*, 204: 564–6.

COHMAP members (1988), 'Climatic changes of the last 18 000 years: Observations and model simulations', *Science*, 241: 1043–52.

COLEMAN, J. M. and SMITH, W. G. (1964), 'Late recent rise of sea-level', *Bulletin, Geological Society of America*, 75: 833–40.

COLES, B. P. L. and FUNNELL, B. M. (1981), 'Holocene palaeoenvironments of Broadland, England', *Special Publs. Int. Ass. Sediment*, 5: 123–31.

COLHOUN, E. A. (1988), 'Recent morphostratigraphic studies of the Australian Quaternary', *Progress in Physical Geography*, 12/2: 264–81.

COLINVAUX, P. A. (1972), 'Climate and the Galapagos Islands', *Nature*, 240: 17–20.

——(1987), 'Amazon diversity in light of the paleoecological record', *Quaternary Science Reviews*, 6: 93–114.

COLLINS, B. W. and FRASER, R. (1971) (eds.), 'Recent crustal movements', *Bulletin, Royal Society of New Zealand*, No. 9.

COLLINSON, M. E. and HOOKER, J. J. (1987), 'Vegetational and mammalian formal changes in the early Tertiary of southern England', in E. M. Friis, W. G. Chaloner, and P. R. Crane (eds.), *The origins of angiosperms and their biological consequences* (Cambridge): 259–304.

——FOWLER, K., and BOULTER, M. C. (1981), 'Floristic changes indicate a cooling climate in the Eocene of southern England', *Nature*, 291: 315–17.

COLMAN, S. W. and PIERCE, K. L. (1981), 'Weathering rinds on andesitic and basaltic stones as a Quaternary age indicator, western United States', *United States Geological Survey, Professional Paper*, 1210.

CONOVER, J. H. (1967), 'Are New England winters getting milder?—11', *Weatherwise*, 20: 58–61.

COOK, P. J. and POLACH, H. A. (1973), 'A Chenier sequence at Broad Sound, Queensland, and evidence against a Holocene high sea-level', *Marine Geology*, 14: 253–68.

COOKE, H. B. S. (1973), 'Pleistocene chronology: long or short?', *Quaternary Research*, 3: 206–20.

COOKE, H. J. (1975), 'The Palaeoclimatic significance of caves and adjacent landforms in western Ngamiland, Botswana', *Geographical Journal*, 141: 430–44.

COOKE, R. U. and REEVES, R. W. (1976), *Arroyos and environmental change in the American south-west* (Oxford): 213 ff.

COOPE, G. R. (1975), 'Climatic fluctuations in north-west Europe since the Last Interglacial, indicated by fossil assemblages of Coleoptera', in A. E. Wright and F. Moseley (eds.), *Ice ages: ancient and modern* (Liverpool): 153–68.

—— (1975), 'Mid-Weichselian climatic changes in western Europe reinterpreted from Coleopteran assemblages', *Bulletin, Royal Society of New Zealand*, 13: 101–8.

—— MORGAN, A., and OSBORNE, P. J. (1971), 'Fossil Coleoptera as indicators of climatic fluctuations during the Last Glaciation in Britain', *Palaeogeography, Palaeoclimatology, Palaeoecology*, 10: 87–101.

COQUE-DELHUILLE, B. (1987), *Les massifs du sud-ouest Anglais et sa bordeure sedimentaire: étude géomorphologique* (Paris).

—— and VEYRET, Y. (1989), 'Limite d'englacement et évolution periglaciaire des Îles Scilly: l'interêt des arènes in situ et remainié', *Zeitschrift für Geomorphologie Supplementband*, 72: 79–96.

COUNCIL ON ENVIRONMENTAL QUALITY and the DEPARTMENT OF STATE (1982), *The Global 2000 Report to the President* (Harmondsworth).

COX, A., DOELL, R. R., and DALRYMPLE, G. B. (1968), 'Radiometric time-scale for geomagnetic reversals', *Quarterly Journal of the Geological Society*, 124: 53–66.

—— (1968), 'Polar wandering, continental drift and the onset of Quaternary glaciation', *Meteorological Monographs*, 8: 112–25.

CREBER, G. T. and CHALONER, W. G. (1985), 'Tree growth in the Mesozoic and early Tertiary and the reconstruction of palaeoclimates', *Palaeogeography, Palaeoclimatology, Palaeoecology*, 52: 35–60.

CRISP, D. J. (1959), 'The influence of climatic changes on animals and plants', *Geographical Journal*, 125: 1–19.

CRITTENDEN, M. D. (1963), 'New data on the isostatic deformation of Lake Bonneville', *United States Geological Survey Professional Paper*, 454-E: 31 ff.

CURRY, R. R. (1969), 'Holocene climatic and glacial history of the Central Sierra Nevada', *Geological Society of America Special Paper*, 123: 1–47.

CUSHING, D. (1976), 'The impact of climatic change on fish stocks in the North Atlantic', *Geographical Journal*, 142: 216–27.

DALEY, B. (1972), 'Some problems concerning the Early Tertiary climate of southern Britain', *Palaeogeography, Palaeoclimatology, Palaeoecology*, 11: 177–90.

DALY, R. A. (1926), *Our mobile Earth* (New York): 342 ff.

DAMUTH, J. E. and FAIRBRIDGE, R. W. (1970), 'Arkosic sands of the Last Glacial stage in the tropical Atlantic off Brazil', *Bulletin, Geological Society of America*, 81: 189–206.

DANSGAARD, W. (1969), 'One thousand centuries of climatic record from Camp Century on the Greenland ice-sheet', *Science*, 166: 377–80.

——JOHNSEN, S. J., and CLAUSEN, H. B. *et al.* (1970), 'Ice cores and paleoclimatology', in I. U. Olsson (ed.), *Radiocarbon variation and absolute chronology* (New York): 337–51.

——WHITE, J. W. C., and JOHNSEN, S. J. (1989), 'The abrupt termination of the Younger Dryas climate event', *Nature*, 339: 532–3.

—— —— ——and LANGWAY, C. C. (1971), 'Climatic record revealed by the Camp Century Ice Core', in K. K. Turekian (ed.), *The Late Cenozoic glacial ages* (New Haven, Conn.): 37–56.

—— ——REEH, N., OUNDERSTRUP, N., CLAUSEN, H. B., and HAMMER, C. U. (1975), 'Climatic changes, Norsemen and modern man', *Nature*, 255: 24.

DARWIN, C. (1936 edn.), *Origin of species* (London).

DAVEAU, S. (1965), 'Dunes ravinées et depôts du Quaternaire récent dans le Sahel mauretanien', *Revue de Géographie de l'Afrique Occidental*, 1–2: 7–48.

DAVIS, N. E. (1972), 'The variability of the onset of spring in Britain', *Quarterly Journal of the Royal Meteorological Society*, 98: 763–77.

DAVIS, P. T. and OSBORN, G. (1988) (eds.), 'Holocene glacier fluctuations', *Quaternary Science Reviews*, 7/2: 113–242.

DAVITAYA, F. F. (1969), 'Atmospheric dust content as a factor affecting glaciation and climatic change', *Annals, Association of American Geographers*, 59: 552–60.

DEACON, J. and LANCASTER, N. (1988), *Late Quaternary Palaeoenvironments of Southern Africa* (Oxford).

DEEVEY, E. S. (1949), 'Biogeography of the Pleistocene', *Bulletin, Geological Society of America*, 60: 1315–416.

——and FLINT, R. F. (1957), 'Post-Glacial hypsithermal interval', *Science*, 125: 182–4.

DEGENS, E. T. and HECKY, R. E. (1974), 'Palaeoclimatic reconstruction of Late Pleistocene and Holocene based on biogenic sediments from the Black Sea and a tropical African lake', *Colloques Internationaux du CNRS*, 219: 13–23.

DE JONG, J. D. (1967), 'The Quaternary of the Netherlands', in K. Rankama (ed.), *The Quaternary*, vol. 2: 301–426.

DELMAS, R. J., ASCENCIO, J. M., and LEGRAND, M. (1980), 'Polar ice evidence that atmospheric CO_2 20 000 yr. BP was 50% of present', *Nature*, 284: 155–7.

DEMOUGEOT, E. (1965), 'Variations climatiques et invasions', *Rev. Hist.*, 228: 1–22.

DENNETT, M. D., ELSTON, J., and RODGERS, J. A. (1985), 'A reappraisal of rainfall trends in the Sahel', *Journal of Climatology*, 5: 353–61.

DENTON, G. H. and KARLÉN, W. (1973), 'Holocene climatic variations: their pattern and possible cause', *Quaternary Research*, 3: 155–205.

DENTON, G. H. and HUGHES, T. J. (1981), *The last great ice-sheets* (New York).

——and PORTER, S. C. (1970), 'Neo-glaciation', *Scientific American*, 222/6: 101–10.

DE PLOEY, J. (1965), 'Position Géomorphologique, Genèse et Chronologie de Certains Depôts Superficiels au Congo Occidental', *Quaternaria*, 7: 131–54.

DEUSER, W. G., ROSS, E. H., and WATERMAN, L. S. (1976), 'Glacial and pluvial periods: their relationship revealed by Pleistocene sediments of the Red Sea and Gulf of Aden', *Science*, 191: 1168–70.

DEVOY, R. J. N. (1987) (ed.), *Sea surface studies* (London).

DIAMOND, M. (1958), 'Precipitation trends in Greenland during past 30 years', *Journal of Glaciology*, 3: 177–80.

DIESTER-HAASS, L. (1980), 'Upwelling and climate off north-west Africa during the later Quaternary', *Palaeoecology of Africa*, 12: 229–38.

——and SCHRADER, H. J. (1979), 'Neogene coastal upwelling history off north-west and south-west Africa', *Marine Geology*, 29: 39–53.

DODSON, J. (1977), 'Late Quaternary palaeoecology of Wyrie Swamp, south-eastern South Australia', *Quaternary Research*, 8: 97–114.

DONN, W. L., FARRAND, W. R., and EWING, M. (1962), 'Pleistocene ice volumes and sea-level changes', *Journal of Geology*, 70: 206–14.

——and SHAW, D. M. (1963), 'Sea-level and climate of the past century', *Science*, 142: 1166–7.

DONNER, J. J. (1970), 'Land/sea-level changes in Scotland', in D. Walker and R. G. West (eds.), *Studies in the vegetational history of the British Isles* (Cambridge): 23–9.

DREIMANIS, A., HUTT, G., RAUKAS, A., and WHIPPEY, P. W. (1978), 'Dating methods of Pleistocene deposits and their problems: I, thermoluminescence dating', *Geoscience Canada*, 5: 55–60.

——and RAUKAS, A. (1975), 'Did Middle Wisconsin, Middle Weichselian, and their equivalents represent an interglacial or an interstadial complex in the northern hemisphere?', *Bulletin, Royal Society of New Zealand*, 13: 109–20.

——TERASMAE, J., and McKENZIE, G. D. (1966), 'The Port Talbot Interstade of the Wisconsin Glaciation', *Canadian Journal of Earth Sciences*, 3: 305–25.

DREWRY, D. J. (1975), 'Initiation and growth of the east Antarctic ice-sheet', *Journal, Geological Society of London*, 131: 255–73.

DUNN, G. E. and MILLER, B. I. (1960), *Atlantic hurricanes* (Louisiana State University Press): 326 ff.

DUPLESSY, J. C., LABEYRIE, J., LALOU., C., and NGUYEN, H. V. (1970), 'Continental climatic variations between 130 000 and 90 000 Years BP', *Nature*, 226: 631–3.

DURHAM, J. W. (1950), 'Cenozoic marine climates of the Pacific coast', *Bulletin, Geological Society of America*, 61: 1243–64.

DURY, G. H. (1965), 'Theoretical implications of underfit streams', *United States Geological Survey Professional Paper* 452C, C1–C43.

—— (1967), 'Climatic change as a geographical backdrop', *Australian Geographer*, 10: 231–42.

—— (1973), 'Palaeo-hydrologic implications of some pluvial lakes in northwestern New South Wales, Australia', *Bulletin, Geological Society of America*, 84: 3663–76.

EARDLEY, A. J. and GVOSDETSKY, V. (1960), 'Analysis of Pleistocene core from Great Salt Lake, Utah', *Bulletin, Geological Society of America*, 71: 1323–44.

EAST, W. G. (1938), *The geography behind history* (London): 200 ff.

EDWARDS, K. J. and WARREN, W. P. (eds.) (1985), *Quaternary History of Ireland* (London).

ELDREDGE, S., EL SAYEED KHALIL, S., NICHOLDS, N., ALI ABDAUA, A., and RYDJESKI, D. (1988), 'Changing rainfall patterns in western Sudan', *Journal of Climatology*, 8: 45–53.

EMBELTON, C. and KING, C. A. M. (1967), *Glacial and periglacial geomorphology* (London).

EMERY, K. O. (1969), 'The continental shelves', *Scientific American*, 221: 107–22.

—— NIINO, H., and SULLIVAN, B. (1971), 'Post Pleistocene levels of the East China Sea', in K. K. Turekian (ed.), *The Late Cenozoic glacial ages* (New Haven, Conn.): 381–90.

EMILIANI, C. (1961), 'Cenozoic climatic changes as indicated by the stratigraphy and chronology of deep-sea cores of globigerina-ooze facies', *Annals, New York Academy of Science*, 95: 521–36.

—— (1966*a*), 'Isotopic paleotemperatures', *Science*, 154: 851–7.

—— (1966*b*), 'Palaeotemperature analysis of Caribbean cores P 6304–8 and P 6304–9 and a generalized temperature curve for the past 425 000 years', *Journal of Geology*, 74: 109–26.

—— (1968), 'The Pleistocene Epoch and the evolution of man', *Current Anthropology*, 9: 27–47.

—— and FLINT, R. F. (1963), 'The Pleistocene record', in M. N. Hill (ed.), *The Sea*, vol. 3 (New York): 888–927.

—— ROOTH C., and STIPP, J. J. (1978), 'The late Wisconsin flood into the Gulf of Mexico', *Earth and Planetary Science Letters*, 41: 159–62.

EPSTEIN, S., SHARP, R. P., and GOW, A. J. (1970), 'Antarctic ice-sheet: stable isotope analysis of Byrd Station cores and interhemisphere climatic implications', *Science*, 168: 1570–2.

ERICSON, D. B., EWING, M., WOLLIN, G., and HEEZEN, B. C. (1961), 'Atlantic deep-sea sediment cores', *Bulletin, Geological Society of America*, 72: 193–286.

—— and WOLLIN, G. (1968), 'Pleistocene climates and chronology in deep-sea sediments', *Science*, 162: 1227–34.

EVANS, G. (1979), 'Quaternary transgressions and regressions', *Journal of the Geological Society*, 136: 125–32.

EVANS, P. (1971), 'Towards a Pleistocene time-scale', in *The Phanerozoic time-scale: A supplement*, Part 2 (London): 123–356.

EWING, M. (1971), 'The Late Cenozoic history of the Atlantic Basin, and its bearing on the cause of the Ice Ages', in K. K. Turekian (ed.), *The Late Cenozoic glacial ages* (New Haven, Conn.): 565–73.

EWING, M. and DONN, W. L. (1956), 'A theory of ice ages', *Science*, 123: 1061–6.

—— —— (1958), 'A theory of ice ages', *Science*, 127: 1159–62.

EYLES, N., EYLES, C. H., and McCABE, A. M. (1989), 'Sedimentation in an ice-contact subaqueous setting: The mid-Pleistocene "North Sea Drifts" of Norfolk, UK', *Quaternary Science Review*, 8: 57–64.

FAIRBANKS, R. G. (1989), 'A 17 000-year glacio-eustatic sea level record: Influence of glacial melting rates on the Younger Dryas event and deep ocean circulation', *Nature*, 342: 637–42.

FAIRBRIDGE, R. W. (1958), 'Dating the latest movements of the Quaternary sea-level', *Transactions, New York Academy of Sciences*, 20: 471–82.

—— (1961), 'Eustatic changes in sea-level', *Physics and Chemistry of the Earth*, 4: 99–185.

—— (1981), 'The concept of neotectonics: An introduction', *Zeitschrift für Geomorphologie, Supplementband*, 40: VII–XII.

—— and KREBS, O. A. (1962), 'Sea-level and the southern oscillation', *Geophysics Journal*, 6: 532–45.

FARRAND, W. R. (1971), 'Late Quaternary paleoclimates of the eastern Mediterranean area', in K. K. Turekian (ed.), *The Late Cenozoic glacial ages* (New Haven, Conn.): 529–64.

FAURE, H. (1966), 'Evolution des Grands Lacs Sahariens à l'Holocène', *Quaternaria*, 8: 167–75.

FIELD, W. O. (1954), 'Notes on the advance of Taku Glacier', *Geographical Review*, 44: 236–9.

FLEMMING, N. C. (1969), 'Archaeological evidence for eustatic change of sea-level and earth movements in the western Mediterranean during the last 2000 years', *Geological Society of America Special Paper*, 109.

—— (1972), 'Relative chronology of submerged Pleistocene marine erosion features in the western Mediterranean', *Journal of Geology*, 80: 633–62.

FLENLEY, J. R. (1979), 'The late Quaternary vegetational history of the equatorial mountains', *Progress in Physical Geography*, 3: 488–509.

—— (1979) *Equatorial rain forest: a geological history* (Sevenoaks).

FLINT, R. F. (1971), *Glacial and Quaternary geology* (New York).

—— and BOND, G. (1968), 'Pleistocene sand ridges and pans in western Rhodesia', *Bulletin, Geological Society of America*, 79: 299–313.

FLORSCHUTZ, F., MENENDEZ AMOR, J., and WIJMSTRA, T. A. (1974), 'Palynology of a thick Quaternary succession in southern Spain', *Palaeogeography, Palaeoclimatology, Palaeoecology*, 10: 233–64.

FONTES, J., CH, and BORTOLAMI, G. (1973), 'Subsidence of the Venice area during the past 40 000 years', *Nature*, 244: 339–41.

—— and GASSE, F. (1989), 'On the ages of humid Holocene and Late Pleistocene phases in North Africa: Remarks on "Late Quaternary climatic reconstruction for the Maghreb (North Africa)" by P. Rognon', *Palaeogeography, Palaeoclimatology, Palaeoecology*, 70: 393–8.

FRAKES, L. A. (1978), 'Cenozoic climates: Antarctica and the Southern Ocean', in A. B. Pittock, L. A. Frakes, D. Jenssen, J. A. Peterson, and J. W. Zillman (eds.), *Climatic change and variability* (Cambridge): 53–69.

FRANKLIN, J. F., NOIR, W. H., DOUGLAS, G. W., and WIBERG, C. (1971), 'Invasion of sub-alpine meadows by trees in the Cascade Range, Washington and Oregon', *Arctic and Alpine Research*, 3, 215–24.

FRENZEL, B. (1973), *Climatic fluctuations of the Ice Age* (Cleveland, Ohio): 300 ff.

——and BLUDAU, W. (1987), 'On the duration of the interglacial to glacial transition at the end of the Eemian Interglacial (Deep Sea Stage 5E): Botanical and sedimentological evidence', in W. H. Berger and L. D. Labeyrie (eds.), *Abrupt climatic change*: 151–62.

FREY, D. G. (1986), 'Cladocera analysis', in B. E. Berglund (ed.), *Handbook of Holocene palaeoecology and palaeohydrology* (Chichester): 667–92.

FRITTS, M. C. (1976), *Tree rings and climate* (London).

FUJI, N. (1988), 'Palaeovegetation and palaeoclimate changes around Lake Biwa, Japan during the last *ca.* 3 million years', *Quaternary Science Reviews*, 7: 21–8.

FUJITA, T. T. (1973), 'Tornadoes round the world', *Weatherwise*, 26.

FUKUI, E. (1970), 'The recent rise of temperature in Japan', *Japanese Progress in Climatology*: 46–55.

FULTON, R. J. (1968), 'Olympia Interglaciation, Purcell Trench, British Columbia', *Bulletin, Geological Society of America*, 79: 1075–80.

FUNDER, S. (1972), 'Deglaciation of the Scoresby Sund Fjord region, north-east Greenland', in R. J. Price and D. E. Sugden (eds.), *Polar geomorphology*, IBG Special Publication 4: 33–41.

GALLOWAY, R. W. (1970), 'The full glacial climate in south-western U.S.A.', *Annals, Association of American Geographers*, 60: 245–56.

GAMBOLATI, G., GATTO, P., and FREEZE, R. A. (1974), 'Predictive simulation of the subsidence of Venice', *Science*, 183: 849–51.

GASSE, F. (1977), 'Evolution of Lake Abhé (Ethiopia and T.F.A.I.) from 70 000 BP', *Nature*, 265: 42–5.

GATES, W. L. (1976), 'Modeling the Ice-Age climate', *Science*, 191: 1138–44.

GAYLORD, D. R. (1990), 'Holocene palaeoclimatic fluctuations revealed from dune and interdune strata in Wyoming', *Journal of Arid Environments*, 18: 123–38.

GEITZENAUER, K. R., MARGOLIS, S. V., and EDWARDS, D. S. (1968), 'Evidence consistent with Eocene glaciation in a south Pacific deep-sea sedimentary core', *Earth and Planetary Science Letters*, 4: 173–7.

GENTILLI, J. (1961), 'Quaternary climates of the Australian region', *Annals, New York Academy of Sciences*, 95: 465–501.

——(1971), *Climates of Australia and New Guinea* (Amsterdam): 405 ff.

GERASIMOV, I. B. (1969), 'Degradation of the last European ice-sheet', in H. E. Wright (ed.), *Quaternary geology and climate* (Washington, DC): 72–8.

GEYH, M. A. and WIRTH, K. (1980), '¹⁴C ages of confined groundwater from the Gwandu aquifer, Sokoto basin, northern Nigeria', *Journal of Hydrology*.

GILL, E. D. (1961), 'Cainozoic climates of Australia', *Annals, New York*

Academy of Sciences, 95: 461–4.

GILLESPIE, R., HORTON, D. R., LADD, P., MACUMBER, P. G., RICH, T. M., THORNE, R., and WRIGHT, R. V. S. (1978), 'Lancefield Swamp and the extinction of the Australian megafauna', *Science*, 200: 1044–8.

GILLETTE, D. A. and HANSON, K. J. (1989), 'Spatial and temporal variability of dust production caused by wind erosion in the United States', *Journal of Geophysical Research*, 94D: 2197–206.

GJAEVEROLL, O. (1963), 'Survival of plants on nunataks in Norway during the Pleistocene glaciation', in A. Love and D. Love (eds.), *North Atlantic biota and their history* (Oxford): 261–83.

GLASS, B., ERICSON, D. B., HEEZEN, B. C., OPDYKE, N. D., and GLASS, J. A. (1967), 'Geomagnetic reversals and Pleistocene chronology', *Nature*, 216: 437–42.

GLASSFORD, D. K. and KILLIGREW, L. P. (1976), 'Evidence for Quaternary westward extension of the Australian Desert into south-western Australia', *Search*, 7: 394–6.

GODWIN, H. (1956), *The history of the British flora* (Cambridge).

——SUGGATE, R. P., and WILLIS, E. H. (1958), 'Radiocarbon dating of the eustatic rise in ocean level', *Nature*, 181: 1518–19.

GOLDTHWAIT, R. P., McKELLAR, J. C., and CRONK, C. (1963), 'Fluctuations of the Crillon Glacier system, south-east Alaska', *Bulletin, International Association of Scientific Hydrology*, 8: 62–74.

GORDON, J. E. (1980), 'Recent climatic trends and local glacier margin fluctuations in West Greenland', *Nature*, 284: 156–9.

GOREAU, T. (1980), 'Frequency sensitivity of the deep-sea climatic record', *Nature*, 287: 620–2.

GOUDIE, A. S. (1972), 'The concept of Post-Glacial progressive desiccation', *Research Paper*, No. 4, School of Geography, University of Oxford.

——(1990), *The human impact on the natural environment* (Oxford).

GOUDIE, A. S. (ed.) (1990), *Geomorphological Techniques* (London).

——ALLCHIN, B., and HEDGE, K. T. M. (1973), 'The former extensions of the Great Indian Sand Desert', *Geographical Journal*, 139: 243–57.

——JONES, D. K. C., and BRUNSDEN, D. (1983), 'Recent fluctuations in some glaciers of the western Karakoram Mountains, Hunza Pakistan', *Proceedings of the International Karakoram Project*.

——RENDELL, H., and BULL, P. A. (1983), 'The loess of Tajik SSR'. *Proceedings of the International Karakoram Project*.

GOW, A. J. and WILLIAMSON, T. (1971), 'Volcanic ash in the Antarctic ice-sheet and its possible climatic implications', *Earth and Planetary Science Letters*, 13: 210–18.

GOWLETT, J. A. J., HARRIS, J. W. K., WALTON, D. A., and WOOD, B. A. (1981), 'Early archaeological sites, hominid remains and traces of fire from Chesowanja, Kenya', *Nature*, 294: 125–9.

GRAAF VAN DER, W. J. E., CROWE, R. W. A., BUNTING, J. A., and JACKSON, M. J. (1977), 'Relict early Cainozoic drainages in arid Western Australia', *Zeitschrift für Geomorphologie*, 21: 379–400.

GRANT, P. J. (1981), 'Recently increased tropical cyclone activity and inferences concerning coastal erosion and inland hydrological regimes in New Zealand and Eastern Australia', *Climatic Change*, 3: 317–32.

GRASTY, R. L. (1967), 'Orogeny, a cause of world-wide regression of the seas', *Nature*, 216: 779.

GRAYSON, D. K. (1977), 'Pleistocene avifaunas and the overkill hypothesis', *Science*, 195: 691–3.

GREGORY, S. (1956), 'Regional variations in the trend of annual rainfall over the British Isles, *Geographical Journal*, 122: 346–53.

——(ed.) (1988), *Recent climatic change* (Lymington).

GREEN, C. P. (1973), 'Pleistocene river gravels and the Stonehenge problem', *Nature*, 243: 214–16.

——(1974), 'Pleistocene gravels of the river Axe in south-western England, and their bearing on the southern limit of glaciation in Britain', *Geological Magazine*, 111: 213–20.

GRIBBIN, J. (1975), 'Aerosol and climate: hotter or cooler', *Nature*, 253: 162.

GRIFFIN, J. B. (1967), 'Climatic change in American prehistory', in R. W. Fairbridge (ed.), *The encyclopedia of atmospheric sciences and astrogeology*: 169–71.

GROOTES, P. M. (1978), 'Carbon-14 time scale extended', *Science*, 200: 11–15.

GROVE, A. T. (1958), 'The ancient Erg of Hausaland, and similar formations on the south side of the Sahara', *Geographical Journal*, 124: 528–33.

——(1967), *Africa south of the Sahara* (London).

——(1969), 'Landforms and climatic change in the Kalahari and Ngamiland', *Geographical Journal*, 135: 192–212.

GROVE, A. T. and GOUDIE, A. S. (1971) 'Late Quaternary lake levels in the Rift Valley of southern Ethiopia and elsewhere in tropical Africa', *Nature*, 234: 403–5.

——and WARREN, A. (1968), 'Quaternary landforms and climate on the south side of the Sahara', *Geographical Journal*, 134: 194–208.

——STREET, F. A., and GOUDIE, A. S. (1975), 'Former lake levels and climatic change in the rift valley of southern Ethiopia', *Geographical Journal*, 141: 177–94.

GROVE, J. M. (1966), 'The Little Ice Age in the Massif of Mont Blanc', *Transactions, Institute of British Geographers*, 40: 129–43.

——(1972), 'The incidence of landslides, avalanches and floods in western Norway during the Little Ice Age', *Arctic and Alpine Research*, 4: 131–8.

——(1979), 'The glacial history of the Holocene', *Progress in Physical Geography*, 3: 1–54.

——(1988), *The Little Ice Age* (London).

GUILCHER, A. (1969), 'Pleistocene and Holocene sea-level changes', *Earth Science Reviews*, 5: 69–98.

GUILDAY, J. E. (1967), 'Differential extinction during late Pleistocene and Recent times', in P. S. Martin and H. E. Wright (eds.), *Pleistocene extinctions* (New Haven, Conn.): 121–40.

GUTENBERG, B. (1941), 'Changes in sea-level, post-glacial uplift, and mobility of the Earth's interior', *Bulletin, Geological Society of America*, 52: 721–72.

HACK, J. T. N. (1941), 'Dunes of the western Navajo Country', *Geographical Review*, 31: 240–63.

HAFFER, J. (1969), 'Speciation in Amazonian forest birds', *Science*, 165: 131–7.

HAMILTON, R. A. (1958) (ed.), *Venture to the Arctic* (Harmondsworth).

HAMMEN, T. VAN DER (1972), 'Changes in vegetation and climate in the Amazon basin and surrounding areas during the Pleistocene', *Geologie en Mijnbouw*, 51: 641–3.

—— (1974), 'The Pleistocene changes of vegetation and climate in tropical South America', *Journal of Biogeography*, 1: 3–26.

—— MAARLEVELD, G. C., VOGEL, J. C., and ZAGWIJN, W. H. (1967), 'Stratigraphy, climatic succession, and radiocarbon dating of the Last Glacial in the Netherlands', *Geologie en Mijnbouw*, 46: 79–95.

—— WIJMSTRA, T. A., and ZAGWIJN, W. H. (1971), 'The floral record of the Late Cenozoic of Europe', in K. K. Turekian (ed.), *The Late Cenozoic glacial ages* (New Haven, Conn.): 391–424.

—— BARELDS, J., DE JONG, H., and DE VEER, A. A. (1981), 'Glacial sequence and environmental history in the Sierra Nevada Del Cocuy (Colombia)', *Palaeogeography, Palaeoclimatology, Palaeoecology*, 32: 247–340.

HANSEN, J., JOHNSON, D., LACIS, A., LEBEDEFF, S., LEE, P., RIND, D., and RUSSELL, G. (1981), 'Climatic impact of increasing atmospheric carbon dioxide', *Science*, 213: 957–66.

HAQ, B. U., BERGGREN, W. A., and VAN COUVERING, J. A. (1977), 'Corrected age of the Pliocene/Pleistocene boundary', *Nature*, 269: 483–8.

HARRIS, G. (1964), 'Climatic changes since 1860 affecting European birds', *Weather*, 19: 70–9.

HARVEY, L. D. D. (1978), 'Solar variability as a contributing factor to Holocene climatic change', *Progress in Physical Geography*, 4: 487–530.

HASTENRATH, S., MING-CHIN, W., and PAO-SHIN, C. (1984), 'Toward the monitoring and prediction of north-east Brazil droughts', *Quarterly Journal of the Royal Meteorological Society*, 118: 411–25.

—— and KUTZBACH, J. (1985), 'Late Pleistocene climate and water budget of the South American altiplano', *Quaternary Research*, 24: 249–56.

HAWKINS, A. B. (1971), 'The Late Weichselian and Flandrian transgression of south-west Britain', *Quaternaria*, 14: 115–30.

—— and KELLAWAY, G. A. (1971), Report on field meeting at Bristol and Bath with special reference to new evidence of glaciation, *Proceedings of the Geologists' Association*, 82: 267–92.

HAYS, J. D., IMBRIE, J., and SHACKLETON, N. J. (1976), 'Variations in the Earth's orbit: Pacemaker of the Ice Ages', *Science*, 194: 1121–32.

HECHT, A. D. (1974), 'Quantitative micropaleontology and the amplitude of glacial interglacial temperature changes in the Caribbean Sea, Gulf of Mexico, and equatorial Atlantic', *Colloques Internationaux du CNRS*, 219: 213–20.

HEIKKINEN, O. (1984), 'Forest expansion in the subalpine zone during the past hundred years, Mount Baker, Washington, USA', *Erdkunde*, 38: 194–202.

HENDY, C. H. and WILSON, A. T. (1968), 'Palaeoclimate data from speleothems', *Nature*, 219: 48–51.

HERMAN, Y. (1970), 'Arctic palaeo-oceanography in Late Cenozoic time', *Science*, 169: 474–7.

HEUSSER, C. J. (1961), 'Some comparisons between climatic changes in north-western North America and Patagonia', *Annals, New York Academy of Sciences*, 95: 642–57.

——SCHUSTER, R. L., and GILKEY, A. K. (1954), 'Geobotanical studies on the Taku Glacier anomaly', *Geographical Review*, 44: 224–36.

——and MARCUS, M. G. (1964), 'Historical variations of Lemon Creek Glacier, Alaska, and their relationship to the climatic history', *Journal of Glaciology*, 5: 77–86.

HEY, R. W. (1963), 'Pleistocene screes in Cyrenaica (Libya)', *Eiszeitalter und Gegenwart*, 14: 77–84.

HIGGINS, C. G. (1965), 'Causes of relative sea-level changes', *American Scientist*, 53: 464–76.

——(1969), 'Isostatic effects of sea-level changes', in H. E. Wright (ed.), *Quaternary geology and climate* (Washington, DC): 141–5.

HOBBS, J. E. (1988), 'Recent climatic change in Australasia', in S. Gregory (ed.), *Recent Climatic Change* (London): 285–97.

HOFFMAN, R. S. and JONES, J. K. (1970), 'Influence of Late Glacial and Post-Glacial events on the distribution of recent mammals on the northern Great Plains', in W. Dort and J. K. Jones (eds.), *Pleistocene and recent environments of the central Great Plains* (Kansas University Press): 355–94.

HOINKES, H. C. (1968), 'Glacier variation and weather', *Journal of Glaciology*, 7: 3–20.

HOLLIDAY, V. T. (1989), 'The Blackwater Draw Formation (Quaternary): A 1.4-plus m.y. record of eolian sedimentation and soil formation on the Southern High Plains', *Bulletin, Geological Society of America*, 101: 1598–607.

HOLLIN, J. W. (1965), 'Wilson's theory of ice ages', *Nature*, 208: 12–16.

HOLMES, J. A. and STREET-PERROTT, F. A. (1989), 'The quaternary glacial history of Kashmir, north-west Himalaya: a revision of de Terra and Paterson's sequence', *Zeitschrift für Geomorphologie Supplementband*, 76: 195–212.

HOOGHIEMSTRA, H. (1989), 'Quaternary And Upper-Pliocene glaciation and forest development in the tropical Andes: evidence from a long high resolution pollen record from the sedimentary basin of Bogota, Colombia', *Palaeogeography, Palaeoclimatology, Palaeoecology*, 72: 11–26.

HOOGHIEMSTRA, H. and AGWU, C. O. C. (1988), 'Changes in the vegetation and trade winds in equatorial north-west Africa 140 000–70 000 yr. BP as deduced from the marine pollen records', *Palaeogeography, Palaeoclimatology, Palaeoecology*, 66: 173–213.

HORNER, R. W. (1972), 'Current proposals for the Thames Barrier and the organization of the investigations', *Philosophical Transactions, Royal Society of London* A, 272: 179–85.

HOROWITZ, A. (1989), 'Continuous pollen diagram for the last 3.5 m.y. from Israel: Vegetation, climate and correlation with the oxygen isotope record', *Palaeogeography, Palaeoclimatology, Palaeoecology*, 72: 63–78.

HOUGHTON, J. T., JENKINS, G. J., and EPHRAUMS, J. J. (eds.) (1990), *Climate change: the IPCC scientific assessment* (Cambridge).

HOWE, G. M., SLAYMAKER, H. O., and HARDING, D. M. (1966), 'Flood hazard in Mid-Wales', *Nature*, 212: 584–5.

HSIEH, CHIAO-MIN (1976), 'Chu K'O-Chen and China's climatic change', *Geographical Journal*, 142: 248–56.

HUNTINGTON, E. (1945), *Mainsprings of civilisation* (New York).

HUNTLEY, D. J., GODFREY-SMITH, D. I., and THEWALT, M. L. (1985), 'Optical dating of sediments', *Nature*, 313: 105–7.

IKEYA, M. (1985), 'Electron spin resonance', in N. W. Rutter (ed.), *Dating methods of Pleistocene deposits and their problems*, Geoscience Canada Reprint Series 2: 73–87.

IMBRIE, J. and IMBRIE, K. P. (1979), *Ice Ages: solving the mystery* (London).

INNES, J. L. (1985), 'Lichenometry', *Progress in Physical Geography*, 9: 187–254.

IRWIN-WILLIAMS, C. and HAYNES, C. V. (1970), 'Climatic change and early population dynamics in the south-western United States', *Quaternary Research*, 1: 59–71.

ISAAC, E. (1970), *Geography of domestication* (Englewood Cliffs, NY): 132 ff.

IVERSEN, J. (1958), 'The bearing of glacial and interglacial epochs on the formation and extinction of plant taxa', *Uppsala Universiteit Arssk*, 6: 210–15.

JARZEMBOWSKI, E. A. (1980), 'Fossil insects from the Bembridge marls, Palaeogene of the Isle of Wight, southern England', *Bulletin of the British Museum (Natural History), Geology Series*, 33: 237–93.

JELGERSMA, S. (1966), 'Sea-level changes in the last 10 000 years', in *Royal Meteorological Society International Symposium on World Climate from 8000–0 B.C.*: 54–69.

JETT, S. C. (1964), 'Pueblo Indian migrations: an evaluation of the possible physical and cultural determinants', *American Antiquity*, 29: 281–300.

JOHN, B. S. (1979) (ed.), *The Winters of the World* (Newton Abbot, Devon).

JOHNSON, S. J., DANSGAARD, W., CLAUSEN, H. B., and LANGWAY, C. C. (1970), 'Climatic oscillations 1200–2000 A.D.', *Nature*, 227: 482.

JOHNSON, W. H. (1990), 'Ice-wedge casts and relict patterned ground in Central Illinois and their environmental significance', *Quaternary Research*, 33: 51–72.

JONES, D. K. C. (1981), *South-east and Southern England* (London).

JONES, P. D., WIGLEY, T. M. L., FOLLAND, C. K., PARKER, D. E., ANGELL, J. K., LEBEDEFF, S., and HANSEN, J. E. (1988), 'Evidence for global warming in the past decade', *Nature*, 228: 790.

KAFRI, U. (1969), 'Recent crustal movements in northern Israel', *Journal of Geophysics Research*, 74: 4246–58.

KAISER, K. (1969), 'The climates of Europe during the Quaternary Ice Age', in H. E. Wright (ed.), *Quaternary Geology and Climate* (Washington, DC): 10–37.

KALELA, O. (1952), 'Changes in the geographic distribution of Finnish birds and mammals in relation to recent changes in climate', *Fennia*, 75: 38–51.

KALNICKY, R. A. (1974), 'Climatic change since 1950', *Annals, Association of American Geographers*, 64: 100–12.

KARLÉN, W. (1973), 'Holocene glacier and climate variations, Kabnekaise Mountains, Swedish Lapland', *Geografiska Annaler*, 55 (A): 29–63.

KAYE, C. A. and BARGHOORN, E. (1964), 'Late Quaternary sea-level rise at Boston, Mass., and notes on the autocompaction of peat', *Bulletin, Geological Society of America*, 75: 63–80.

KEANY, J., LEDBETTER, M., WATKINS, N., and TER CHIEN, H. (1976), 'Diachronous deposition of ice-rafted debris in sub-Antarctic deep-sea sediments', *Bulletin, Geological Society of America*, 87: 873–82.

KELLAWAY, G. A. (1971), 'Glaciation and the stones of Stonehenge', *Nature*, 233: 30–5.

——REDDING, J. H., SHEPHARD-THORN, E. R., and DESTOMBES, J. P. (1975), 'The Quaternary history of the English Channel', *Philosophical Transactions of the Royal Society of London*, 278A: 189–218.

KENDALL, R. L. (1969), 'An ecological history of the Lake Victoria Basin', *Ecological Monographs*, 39: 121–76.

KENNETT, J. P. (1970), 'Pleistocene paleoclimates and foraminiferal biostratigraphy in sub-Antarctic deep-sea cores', *Deep-Sea Research*, 17: 125–40.

——and SHACKLETON, N. J. (1975), 'Laurentide ice-sheet meltwater recorded in Gulf of Mexico deep-sea cores', *Science*, 188: 147–50.

——and THUNELL, R. C. (1975), 'Global increase in Quaternary explosive volcanism', *Science*, 187: 497–503.

KENT, D. P., OPDYKE, N. D., and EWING, M. (1971), 'Climate change in the North Pacific using ice-rafted detritus as a climatic indicator', *Bulletin, Geological Society of America*, 82: 2741–54.

KERANEN, J. (1952), 'On temperature changes in Finland during the last 100 years', *Fennia*, 75: 5–16.

KERR, R. A. (1987), 'Milankovitch climate cycles through the ages', *Science*, 235: 973–4.

KERSHAW, A. P. (1974), 'A long continuous pollen sequence from north-eastern Australia', *Nature*, 251: 222.

——(1976), 'A late-Pleistocene and Holocene pollen diagram from Lynch's crater, northeastern Queensland, Australia', *New Phytologist*, 77: 469–98.

KIDSON, C. (1977), 'The coast of south-west England', in C. Kidson and M. J. Tooley (eds.), *The Quaternary history of the Irish Sea*: 257–98.

——(1986), 'Sea-level changes in the Holocene', in O. van der Plassche (ed.), *Sea-level research: a manual for the collection and evaluation of data* (Norwich): 27–64.

KIND, N. V. (1972), 'Late Quaternary climatic changes and glacial events in the Old and New World—radiocarbon chronology', *24th International Geographical Congress*, Section 12: 55–61.

KING, J. W. (1973), 'Solar radiation changes and the weather', *Nature*, 245: 443–6.

KING, L. C. (1962), *Morphology of the Earth* (Edinburgh).

KING, P. B. (1965), 'Tectonics of Quaternary time in middle North America', in H. E. Wright and D. G. Frey (eds.), *The Quaternary of the U.S.A.*: 831–70.

KLEIN, C. (1965), 'On the fluctuations of the level of the Dead Sea in the beginning of the nineteenth century', *State of Israel, Hydrological Service, Hydrological Paper*, 7.

KOLLA, V., and BISCAYE, P. E. (1977), 'Distribution and origin of quartz in the sediments of the Indian Ocean', *Journal of Sedimentary Petrology*, 47: 642–9.

—— —— and HANLEY, A. F. (1979), 'Distribution of quartz in late Quaternary Atlantic sediments in relation to climate', *Quaternary Research*, 11: 261–7.

KOSIBA, A. (1963), 'Changes in the Werenskiold Glacier and Hans Glacier in SW Spitzbergen', *Bulletin, International Association of Scientific Hydrology*, 8: 24–35.

KRANTZ, G. S. (1970), 'Human activities and megafaunal extinctions', *American Scientist*, 58: 164–70.

KRAUS, E. B. (1954), 'Secular changes in the rainfall regime of south east Australia', *Quarterly Journal of the Royal Meteorological Society*, 80: 591–611.

——(1955a), 'Secular variations of east coast rainfall regimes', *Quarterly Journal of the Royal Meteorological Society*, 81: 430–9.

——(1955b), 'Secular changes of tropical rainfall regimes', *Quarterly Journal of the Royal Meteorological Society*, 81: 198–210.

——(1956), 'Secular changes of the standing circulation', *Quarterly Journal of the Royal Meteorological Society*, 82: 289–300.

KU, T. L. (1968), 'Protactinium 231 methods of dating coral from Barbados Island', *Journal of Geophysics Research*, 73: 2271–6.

KUDRASS, H. R., ERLENKEUSER, H., VOLLBRECHT, R., and WEISS, W. (1991), 'Global nature of the Younger Dryas cooling event inferred from oxygen isotope data from Sulu Sea cores', *Nature*, 349: 406–9.

KUHLE, M. (1987), 'Subtropical mountain- and highland-glaciation as ice-age triggers and the waning of the glacial periods in the Pleistocene', *GeoJournal*, 14: 393–421.

KUKLA, G. J. (1975), 'Loess stratigraphy of central Europe', in K. W. Butzer and G. L. Isaac (eds.), *After the Australopithecines* (The Hague): 99–188.

—— (1977), 'Pleistocene land-sea correlations I. Europe', *Earth Science Reviews*, 13: 307–74.

—— (1989), 'Long continental records of climate: An introduction', *Palaeogeography, Palaeoclimatology, Palaeoecology*, 72: 1–9.

KURTEN, B. (1972), *The Ice Age* (Stockholm).

KUTZBACH, J. E. (1981), 'Monsoon climate of the early Holocene: climate experiment with the earth's orbital parameters for 9000 years ago', *Science*, 214: 59–61.

—— and STREET-PERROTT, F. A. (1985), 'Milankovitch forcing of fluctuations in the level of tropical lakes from 18 to 0 Kyr BP', *Nature*, 317: 130–4.

KVASOV, D. D. and BLAZHCHISHIN, V. (1978), 'The key to sources of the Pliocene and Pleistocene glaciation is at the bottom of the Barents Sea', *Nature*, 273: 138–40.

KVITOVIC, J. and VANKO, J. (1971), 'Studium sucasnych pohybov zemskej Kory NA Slovensku', *Geograficky Casopis*, 23: 124–32.

LADURIE, E. LE R. (1972), *Times of feast, times of famine* (London).

LA MARCHE, V. C. Jr. (1974), 'Paleoclimate inferences from long tree-ring records', *Science*, 183: 1043–8.

LAMB, H. H. (1966a), *The changing climate* (London).

—— (1966b), 'Climate in the 1960s with special reference to East African lakes', *Geographical Journal*, 132: 183–212.

—— (1967), 'Britain's changing climate', *Geographical Journal*, 133: 445–68.

—— (1969a), 'Climate fluctuations', in H. Flohn (ed.), *World survey of climatology*, vol. 2 (Amsterdam): 173–249.

—— (1969b), 'The new look of climatology', *Nature*, 223: 1209–15.

—— (1970), 'Volcanic dust in the atmosphere: with a chronology and assessment of its meteorological significance', *Philosophical Transactions, Royal Society of London* A, 266: 425–533.

—— (1971), 'Volcanic activity and climate', *Palaeogeography, Palaeoclimatology, Palaeoecology*, 10: 203–30.

—— (1972), *Climate: present, past and future*, vol. 1 (London): 613 ff.

—— (1974), 'The current trend of world climate', *Climatic Research Unit Research Publications*, 3.

—— (1977), *Climate: Present, past and future*; ii, *Climatic History and the future* (London).

—— (1982), *Climate, history and the modern world* (London).

—— and MORTH, H. T. (1978), 'Arctic ice, atmosphere circulation, and world climate', *Geographical Journal*, 144: 1–22.

—— PROBERT-JONES, J. R., and SHEARD, J. W. (1962), 'A new advance of the Jan Mayen glaciers and a remarkable increase of precipitation', *Journal of Glaciology*, 4: 355–65.

LANCASTER, N. (1979), 'Quaternary environments in the arid zone of southern Africa', *Environmental Studies, University of the Witwatersrand, Occasional Paper*, 22.

LANCASTER, N. (1989), 'Late Quaternary palaeoenvironments in the south-western Kalahari', *Palaeogeography, Palaeoclimatology, Palaeoecology*, 70: 367–76.

LANDSBERG, H. E. (1976), 'Whence global climate: hot or cold? an essay review', *Bulletin of the American Meteorological Society*, 57: 441–3.

LARSEN, J. A. and BARRY, R. G. (1974), 'Palaeoclimatology', in J. D. Ives and R. G. Barry (eds.), *Arctic and alpine environments* (London).

LAWLER, D. M. (1987), 'Climatic change over the last millenium in Central Britain', in K. J. Gregory, J. Lewin and J. B. Thornes (eds.), *Palaeohydrology in Practice* (Chichester): 99–129.

LAWRENCE, D. B. (1950), 'Glacier fluctuation for six centuries in south-eastern Alaska and its relation to solar activity', *Geographical Review*, 40: 191–223.

——(1958), 'Glaciers and vegetation in south-east Alaska', *American Scientist*, 46: 89–122.

——and LAWRENCE, E. G. (1961), 'Response of enclosed lakes to current glacio-pluvial climatic conditions in middle-latitude western North America', *Annals, New York Academy of Sciences*, 95: 341–50.

LEAKEY, L. S. B. (1965), *Olduvai Gorge*, i. *A preliminary report on the geology and fauna* (Cambridge).

——and GOODALL, V. M. (1969), *Unveiling man's origins* (London).

LEHMER, D. J. (1970), 'Climate and culture history in the middle Missouri Valley', in W. Dort and J. K. Jones (eds.), *Pleistocene and recent environments of the central Great Plains* (Kansas University Press): 117–29.

LEINEN, M. and HEATH, G. R. (1981), 'Sedimentary indicators of atmospheric activity in the northern hemisphere during the Cenozoic', *Palaeogeography, Paleoclimatology, Palaeoecology*, 36: 1–21.

——and SARNTHEIN, M. (1989), '*Paleoclimatology and paleometeorology: Modern and past patterns of global atmospheric transport* (Dordrecht).

LEOPOLD, L. B. (1951), 'Rainfall frequency: an aspect of climatic variation', *Transactions, American Geophysics Union*, 32: 347–57.

——LEOPOLD, E. B., and WENDORF, F. (1963), 'Some climatic indicators in the period A.D. 1200–1400 in New Mexico', *Arid Zone Research*, 20: 265–70.

——WOLMAN, M. G., and MILLER, J. P. (1964), *Fluvial processes in geomorphology* (San Francisco): 522 ff.

LEROI-GOURHAN, A. (1974), 'Analyses Polliniques, Pre-Histoire et Variations Climatiques Quaternaires', *Colloques Internationaux du CRNS*, 219: 61–6.

LEVER, A. and McCAVE I. N. (1983), 'Eolian components in Cretaceous and Tertiary North Atlantic sediments', *Journal of Sedimentary Petrology*, 53: 811–32.

LEWIS, C. A. (ed.) (1970), *The glaciations of Wales and adjoining regions* (London).

LILJEQUIST, G. H. (1943), 'The severity of the winters at Stockholm, 1757–1942, *Geografiska Annaler*, 25: 81–97.

LISITZIN, E. (1964), 'Land uplift as sea-level problem', *Fennia*, 89: 7–10.

LIVINGSTONE, D. A. (1967), 'Post-glacial vegetation of the Ruwenzori Mountains in equatorial Africa', *Ecological Monographs*, 37: 25–52.

LLOYD, J. W. (1973), 'Climatic variations in north central Chile from 1866 to 1971', *Journal of Hydrology*, 19: 53–70.

LOFFLER, H. (1986), 'Ostracod analysis', in B. E. Berglund (ed.), *Handbook of Holocene Palaeoecology and palaeohydrology* (Chichester): 693–702.

LONGWELL, C. R. (1960), 'Interpretation of the levelling data', *United States Geological Survey Professional Paper*, 295: 33–8.

LORIUS, C., MERLIVAT, L., JOUZEL, J., and FOUCHET, M. (1979), 'A 30 000-yr isotope climatic record from Antarctic ice', *Nature*, 280: 644–8.

LOWE, J. J. and GRAY, M. J. (1980), 'The stratigraphic subdivision of the Late glacial of NW Europe: a discussion', in J. J. Lowe, M. J. Gray, and J. E. Robinson (eds.), *Studies in the late glacial of north-west Europe* (Oxford): 157–75.

——and WALKER, M. J. C. (1984), *Reconstructing Quaternary environments*, (London).

LU YANCHOU (1981), *Pleistocene climatic cycles and variations of CaCO₃ contents in a loess profile* (in Chinese).

LYALL, I. T. (1970), 'Recent trends in spring weather', *Weather*, 25: 163–5.

MABBUTT, J. A. (1971), 'The Australian arid zone as a prehistoric environment', in D. J. Mulvaney and J. Golson (eds.), *Aboriginal man and environment in Australia* (Canberra): 66–79.

——(1977), *Desert landforms* (Cambridge, Mass.).

MAHANEY, W. C. (ed.) (1976), *Quaternary Stratigraphy of North America* (London).

MAHERAS, P. (1988), 'Changes in precipitation conditions in the western Mediterranean over the last century', *Journal of Climatology*, 8: 179–89.

——and KOLYVA-MACHERA, F. (1990), 'Temporal and spatial characteristics of annual precipitation over the Balkans in the twentieth century', *International Journal of Climatology*, 10: 495–504.

MALDE, H. E. (1964), 'Environment and man in arid America', *Science*, 145: 123–9.

MALEY, J. (1989), 'Late Quaternary climatic changes in the African rainforest: Forest refugia and the major role of sea surface temperature variations', in M. Leinen and M. Sarnthein (eds.), *Paleoclimatology and paleometeorology: Modern and past patterns of global atmospheric transport* (Dordrecht): 585–616.

MANLEY, G. (1953), 'The mean temperature of central England 1698–1952', *Quarterly Journal of the Royal Meteorological Society*, 79: 242–61.

——(1964), 'The evolution of the climatic environment', in W. Watson and J. B. Sissons (eds.), *The British Isles: a systematic geography* (London).

——(1966), 'Problems of the climatic optimum: The contribution of glaciology', in *Royal Meteorological Society Symposium on world climate 8000–0 B.C.*: 34–9.

——(1971), 'Interpreting the meteorology of the Late and Post-Glacial', *Palaeogeography, Palaeoclimatology, Palaeoecology*, 10: 163–75.

——(1974), 'Central England temperatures: monthly means 1659–1973',

Quarterly Journal of the Royal Meteorological Society, 100: 389–405.

MARKGRAF, V. (1974), 'Palaeoclimatic evidence derived from timberline fluctuations', *Colloques Internationaux du CNRS*, 219: 67–76.

MARTIN, J. H. (1990), 'Glacial-Interglacial CO_2 change: The iron hypothesis', *Paleoceanography*, 5: 1–13.

MARTIN, P. S. (1963), 'Early man in Arizona: the pollen evidence', *American Antiquity*, 29: 67–73.

——(1966), 'Africa and Pleistocene overkill', *Nature*, 212: 339–42.

——(1967), 'Prehistoric overkill', in P. S. Martin and H. E. Wright (eds.), *Pleistocene extinctions* (New Haven, Conn.): 75–120.

MASON, B. J. (1976), 'The nature and prediction of climatic changes', *Endeavour*, 35: 51–7.

MASTERS, P. M. and FLEMMING, N. C. (eds.) (1983), *Quaternary coastlines and marine archaeology* (London).

MASURIER, W. E. LE (1972), 'Volcanic record of Antarctic glacial history: implications with regard to Cenozoic Sea-Levels', in R. J. Price and D. E. Sugden (eds.), *Polar geomorphology*, Institute of British Geographers Special Publication No. 4: 59–74.

MATHEWS, W. H. and CURTIS, G. H. (1966), 'Date of the Plio-Pleistocene boundary in New Zealand', *Nature*, 212: 979–80.

MATTHES, F. E. (1939), Report of Committee on Glaciers, *American Geophysical Union Transactions*, 20: 518–23.

MATTHEWS, J. A. (1974), 'Families of lichenometric dating curves from the Horbreen gletschervorfeld, Jotunheiman Norway', *Norsk Geogr. Tidsskr*, 28: 215–35.

MAYR, F. (1964), 'Untersuchungen uber Ausmass und Golgen der Klima- und Gletscherwankungen seit dem Beginn der post-glazialen Warmezeit', *Zeitschrift für Geomorphologie*, 8: 257–85.

MCBURNEY, C. B. M. and HEY, R. W. (1955), *Prehistory and Pleistocene geology in Cyrenaican Libya* (Cambridge).

MCCREA, W. H. (1975), 'Ice ages and the galaxy', *Nature*, 255: 607–9.

MCDOUGALL, I. and STIPP, J. J. (1968), 'Isotopic dating evidence for the age of climatic deterioration and the Pliocene–Pleistocene boundary', *Nature*, 219: 51–3.

——and WENSINK, H. (1966), 'Palaeomagnetism and geochronology of the Pliocene–Pleistocene lavas in Iceland', *Earth and Planetary Science Letters*, 1: 232–6.

MCINTYRE, A., RUDDIMAN, W. F., and JANTZEN, R. (1972), 'Southward penetrations of the North Atlantic Polar Front: faunal and floral evidence of large-scale surface water mass movements over the last 225 000 years', *Deep-Sea Research*, 19: 61–77.

MCLURE, H. A. (1976), 'Radiocarbon chronology of Late Quaternary lakes in the Arabian desert', *Nature*, 263: 755–6.

MCMASTER, L. E. and GARRISON, R. L. (1966), 'Morphology and sediments of the continental shelf of southern New England', *Marine Geology*, 4: 273–89.

McNutt, M. and Menard, H. W. (1978), 'Lithospheric flexure and uplifted atolls', *Journal of Geophysical Research*, 83 (B3): 1206–12.

Meadows, A. J. (1975), 'A hundred years of controversy over sunspots and weather', *Nature*, 256: 95–7.

Meggers, B. J. (1975), 'Application of the biological model of diversification to cultural distributions in tropical lowland South America', *Biotropica*, 7: 141–61.

Meier, M. F. (1965), 'Glaciers and climate', in H. E. Wright and D. G. Frey (eds.), *The Quaternary of the U.S.A.*: 795–805.

Mengel, R. M. (1970), 'The North American central plans as an isolating agent in bird speciation', in W. Dort and J. K. Jones (eds.), *Pleistocene and Recent environments of the central Great Plains* (Kansas University Press): 279–340.

Mercer, J. H. (1969), 'The Allerød oscillation: a European climatic anomaly', *Arctic and Alpine Research*, 1: 227–34.

——(1972), 'The lower boundary of the Holocene', *Quaternary Research*, 2: 15–24.

——(1978), 'West Antarctica ice-sheet and CO_2 greenhouse effect: A threat of disaster', *Nature*, 271: 321.

Mesollela, K. J., Matthews, R. K., Broecker, W. S., and Thurber, D. L. (1969), 'The astronomical theory of climatic change: Barbados data', *Journal of Geology*, 77: 250–74.

Michel, P. (1968), 'Genèse et Évolution de la Vallée du Sénégal de Bakel à l'Embouchure (Afrique Occidentale)', *Zeitschrift für Geomorphologie*, 12: 318–49.

Micklin, P. P. (1972), 'Dimensions of the Caspian Sea problem', *Soviet Geography*, 13: 589–603.

Miller, C. D. (1969), 'Chronology of Neo-glacial moraines in the Dome Peak area, North Cascade Range, Washington', *Arctic and Alpine Research*, 1: 49–66.

Miller, G. H., Hollin, J. T., and Andrews, J. T. (1979), 'Aminostratigraphy of U.K. Pleistocene deposits', *Nature*, 281: 539–43.

Milliman, J. D. and Emery, K. O. (1968), 'Sea-Levels during the past 35 000 years', *Science*, 162: 1121–3.

Milton, D. (1974), 'Some observations of global trends in tropical cyclone frequencies', *Weather*, 29: 267–70.

Mitchell, G. F. (1972), 'Soil deterioration associated with prehistoric agriculture in Ireland', *20th International Geological Congress Symposium*, 1: 59–68.

——and Orme, A. R. (1967), 'The Pleistocene deposits of the Isles of Scilly', *Quarterly Journal, Geological Society of London*, 123: 59–92.

——Penny, L. F., Shotton, F. S., and West R. G. (1973), *A correlation of Quaternary Deposits in the British Isles* (London).

Mitchell, J. M. (1963), 'On the world-wide pattern of secular temperature change', *Arid Zone Research*, 20: 161–81.

——(1965), 'Theoretical palaeoclimatology', in H. E. Wright and D. G. Frey (eds.), *The Quaternary of the U.S.A.*: 881–901.

——(1968) (ed.), *Causes of climatic change*, Meteorological Monographs, 8.

MOLNAR, P. and ENGLAND, P. (1990), 'Late Cenozoic uplift of mountain ranges and global climate change: chicken or egg?', *Nature*, 346: 29–34.

MONTFORD, H. M. (1970), 'The terrestrial environment during Upper Cretaceous and Tertiary Times', *Proceedings of the Geologists' Association*, 81: 181–204.

MOORE, P. D. (1975), 'Origin of blanket mires', *Nature*, 256: 267–9.

MOREAU, R. E. (1963), 'Vicissitudes of the African biomes in the Late Pleistocene', *Proceedings, Zoological Society of London*, 141: 395–421.

MÖRNER, N. A. (1969), 'The Late Quaternary history of the Kattegatt Sea and the Swedish west coast', *Sveriges Geologiska Undersokning Series C*, NR.640, Arsbok, 63, NR.3.

—— (1971*a*), 'Eustatic and climatic oscillations', *Arctic and Alpine Research*, 3: 167–71.

—— (1971*b*), 'The Plum Point Interstadial: age, climate and subdivision', *Canadian Journal of Earth Science*, 8: 1423–31.

—— (1972), 'Time-scale and ice accumulation during the last 125 000 years as indicated by the Greenland O^{18} curve', *Geological Magazine*, 109: 17–24.

—— (1976), 'Eustasy and geoid changes', *Journal of Geology*, 84: 123–51.

—— (1980) (ed.), *Earth rheology, isostasy and eustasy* (New York).

—— and WALLIN, B. (1977), 'A 10 000-year temperature record from Gotland, Sweden', *Palaeogeography, Palaeoclimatology, Palaeoecology*, 21: 113–38.

MORRISON, H. E. S. (1968), 'Pleistocene vegetation and climate in Uganda', *Journal of Ecology*, 56: 363–84.

MULVANEY, D. J. and GOLSON, J. (eds.) (1971), *Aboriginal man and environment in Australia* (Cambridge).

NAESER, C. W. and NAESER, N. D. (1988), 'Fission-track dating of Quaternary events', in D. J. Easterbrook (ed.), *Dating Quaternary sediments, Geological Society of America Special Paper*, No. 227.

NEALE, J. and FLENLEY, J. (1981) (eds.), *The Quaternary in Britain* (Oxford).

NEWELL, N. D. (1961), 'Recent terraces of tropical limestone shores', *Zeitschrift für Geomorphologie Supplementband*, 3: 87–106.

—— and BLOOM, A. L. (1970), 'The reef flat and "two-meter eustatic terrace" of some Pacific atolls', *Bulletin, Geological Society of America*, 81: 1881–93.

NEWMAN, W. S., FAIRBRIDGE, R. W., and MARCH, S. (1971), 'Marginal subsidence of glaciated areas: United States, Baltic and North Seas', *Quaternaria*, 14: 39–40.

NICHOL, J. E. (1991), 'The extent of desert dunes in northern Nigeria as shown by image enhancement', *Geographical Journal*, 157: 13–24.

NICHOLS, H. (1967), 'Central Canadian palynology and its relevance to north-western Europe in the late Quaternary period', *Review of Palaeobotany and Palynology*, 2: 231.

NICHOLSON, S. E. (1981), 'The historical climatology of Africa,' in J. H. L. Wigley, M. T. Ingram, and G. Farmer (eds.), *Climate and History* (Cambridge): 249–70.

—— and FLOHN, H. (1981), 'African climate changes in late Pleistocene and Holocene and the general atmospheric circulation', *IASH Publication*, 131:

295–301.

NISBET, E. G. (1990), 'The end of the ice age', *Canadian Journal of Earth Sciences*, 27: 148–57.

NUNN, P. D. (1986), 'Implications of migrating geoid anomalies for the interpretation of high-level fossil coral reefs', *Bulletin, Geological Society of America*, 97: 946–52.

—— (1990), 'Recent environmental changes on Pacific islands', *Geographical Journal*, 156: 125–40.

OESCHER, H. and LANGWAY, C. C. (1989) (eds.), *The environmental record in glaciers and ice-sheets* (Chichester).

OLAUSSON, E. and OLSSON, I. V. (1969), 'Varve stratigraphy in a core from the Gulf of Aden', *Palaeogeography, Palaeoclimatology, Palaeoecology*, 6: 87–103.

D'OLIER, B. (1975), 'Some aspects of the Late-Pleistocene-Holocene drainage of the river Thames in the eastern part of the London Basin', *Philosophical Transactions, Royal Society of London A*, 279: 269–77.

OLLIER, C. D. (1981), *Tectonics and landforms* (London).

OPDYKE, N. D., GLASS, B., HAYS, J. D., and FOSTER, J. (1966), 'Paleomagnetic study of Antarctic deep-sea cores', *Science*, 154: 349–57.

OPIK, E. J. (1958), 'Climate and the changing sun', *Scientific American*, 198: 85–92.

OSBORNE, P. J. (1974), 'An insect of early Flandrian Age from Lea Marston, Warwickshire, and its bearing on the contemporary climate and ecology', *Quaternary Research*, 4: 471–86.

OSMOND, J. K., CARPENTER, J. R., and WINDHOM, H. L. (1965), 'Th^{230}/U^{234} Age of the Pleistocene corals and oolites of Florida', *Journal of Geophysics Research*, 70: 1843–7.

OTTERMAN, J. (1974), 'Baring high albedo soils by overgrazing: a hypothesized desertification mechanism', *Science*, 186: 531–3.

PANT, G. B. and HINGANE, L. S. (1988), 'Climatic change in and around the Rajasthan Desert during the twentieth century', *Journal of Climatology*, 8: 391–401.

PARKER, D. E. and FOLLAND, C. K. (1988), 'The nature of climatic variability'. *Meteorological Magazine*, 117: 201–10.

PARKIN, D. W. and SHACKLETON, N. J. (1973), 'Trade winds and temperature correlations down a deep-sea core off the Saharan coast', *Nature*, 245: 455–7.

PARMENTER, C. and FOLGER, D. W. (1974), 'Eolian biogenic detritus in deep-sea sediments: a possible index of equatorial ice-age aridity', *Science*, 185: 695–8.

PARRISH, J. T. (1987), 'Global palaeogeography and palaeoclimate of the Late Creaceous and Early Tertiary', in E. M. Friis, W. G. Chaloner, and P. R. Crane (eds.), *The origins of angiosperms and their biological consequences* (Cambridge).

PARRY, M. L. (1975), 'Secular climatic change and marginal agriculture', *Transactions, Institute of British Geographers*, 64: 1–13.

—— (1978), *Climatic change, agriculture and settlement* (Folkestone).

PATZELT, G. (1974), 'Holocene variations of glaciers in the Alps', *Colloques Internationaux du CNRS*, 219: 51–9.

PEARSALL, W. H. (1964), 'After the ice retreated', *New Scientist*, 383: 757–9.

PENCK, A. and BRÜCKNER, E. (1909), *Die Alpen in Eiszeitalten* (Leipzig).

PENNINGTON, W. (1969), *The history of British vegetation* (London).

PETIT-MAIRE, N. (1989), 'Interglacial environments in presently hyperarid Sahara: Palaeoclimatic implications', in M. Leinen and M. Sarnthein (eds.), *Palaeoclimatology and palaeometeorology: Modern and past patterns of global atmospheric transport*, (Dordrecht): 637–61.

PHILLIPS, C. S. (1970) (ed.), *The Fenland in Roman times* (London).

PILBEAM, D. R. (1975), 'Middle Pleistocene hominids', in K. W. Butzer and G. L. Isaac (eds.), *After the Australopithecines* (The Hague), 809–56.

PIRAZZOLI, P. A. (1989), 'Present and near-future global sea-level changes', *Palaeogeography, Palaeoclimatology, Palaeoecology*, 75: 241–58.

PISIAS, N. G. and MOORE, T. C. (1981), 'The evolution of Pleistocene climate: A time series approach', *Earth and Planetary Science Letters*, 52: 450–8.

PLAFKER, G. (1965), 'Tectonic deformation associated with the 1964 Alaska earthquake', *Science*, 148: 1675–87.

PLASS, G. N. (1956), 'The Carbon Dioxide theory of climatic change', *Tellus*, 8: 140–54.

PLASSCHE, O. van der (ed.) (1986), *Sea-level research: A manual for the collection and evaluation of data* (Norwich).

PLINT, A. G. (1983), 'Sandy fluvial point-bar sediments from the Middle Eocene of Dorset, England', *Special Publications International Association of Sedimentologists*, 6: 355–68.

PORTER, S. C. (1975), 'Weathering rinds as a relative-age criterion: an application to sub-division of glacial deposits in the Cascade Range', *Geology*, 3: 909.

—— (1981), 'Recent glacier variations and volcanic eruptions', *Nature*, 291: 139–41.

—— (1986), 'Pattern and forcing of northern hemisphere glacier variations during the last millennium', *Quaternary Research*, 26: 27–48.

POTTER, G. L., ELLSAESSER, H. W., MACCRACKEN, M. C., and LUTHER, F. M. (1975), 'Possible climatic impact of tropical deforestation', *Nature*, 258: 697–8.

PRANCE, G. T. (1973), 'Phytogeographic support for the theory of Pleistocene forest refuges in the Amazon basin, based on evidence from distribution patterns in Caryocaraceae, Chrysobalanaceae, Dichapetalaceae and Lecythidaceae', *Acta Amazonica*, 3: 5–28.

PRELL, W. L., HUTSON, W. H., WILLIAMS, D. F., BE, A. W. H., GEITZENAUER, K., and MOLFINO, B. (1980), 'Surface circulation of the Indian Ocean during the Last Glacial Maximum, approximately 18 000 BP', *Quaternary Research*, 14: 309–36.

PRICE, W. A. (1958), 'Sedimentology and Quaternary geomorphology of South Texas', *Transactions, Gulf Coast Association of Geological Societies*, 8: 41–75.

PRICE-WILLIAMS, D., WATSON, A., and GOUDIE, A. S. (1982), 'Quaternary colluvial stratigraphy, archaeological sequences and palaeoenvironments in Swaziland, southern Africa', *Geographical Journal*, 148: 50–67.

PYE, K. (1987), *Aeolian dust and dust deposits* (London).

RACKHAM, O. (1980), *Ancient Woodland* (London).

RAIKES, R. (1967), *Water, weather and prehistory* (London): 208 ff.

RAPP, A. (1974), 'A Review of desertization in Africa—water vegetation, and man', *Secretariat for International Ecology* (Stockholm), Report No. 1.

RASMUSSON, E. M. (1987), 'Global climate change and variability: effects on drought and desertification in Africa', in M. H. Glantz (ed.), *Drought and Hunger in Africa* (Cambridge): 3–22.

RASOOL, S. I. and SCHNEIDER, S. H. (1971), 'Atmospheric Carbon Dioxide and aerosols: effects of large increases on global climate', *Science*, 173: 138–41.

RAYNAUD, D., CHAPPELLAR, J., BARNOLA, J. M., KOROTKEVICH, Y. S., and LORIUS, C. (1988), 'Climatic and CH_4 cycle implications of glacial-interglacial CH_4 change in the Vostok ice-core', *Nature*, 33: 655–9.

READING, A. J. (1990), 'Caribbean tropical storm activity over the past four centuries', *International Journal of Climatology*, 10: 365–76.

RECK, R. A. (1975), 'Aerosols and solar temperature changes', *Science*, 188: 728–30.

REED, C. A. (1970), 'Extinction of mammalian megafauna in the Old World Late Quaternary', *Bioscience*, 20: 284–8.

REEVES, C. C. (1966), 'Pluvial lake basins of west Texas', *Journal of Geology*, 74: 269–91.

REID, E. M. and CHANDLER, M. E. J. (1933) *The London clay flora* (London).

Report of the Study of Critical Environmental Problems (1970), *Man's impact on the global environment* (Cambridge, Mass.): 319 ff.

RHODES, E. J. (1988), 'Methodological considerations in the optical dating of quartz', *Quaternary Science Reviews*, 7: 395–400.

RICHMOND, G. M. (1970), 'Comparison of the Quaternary stratigraphy of the Alps and Rocky Mountains', *Quaternary Research*, 1: 3–28.

—— and FULLERTON, D. S. (1986), 'Summation of Quaternary glaciations in the United States of America', *Quaternary Science Reviews*, 5: 183–196.

RIEHL, H. (1956), 'Sea-suface temperature anomalies and hurricanes', *Bulletin, American Meteorological Society*, 37: 413–17.

RIPLEY, E. A. (1976), 'Drought in the Sahara: insufficient geophysical feedback?', *Science*, 191: 100.

RITCHIE, J. C., EYLES, C. H., and HAYNES, C. V. (1985), 'Sediment and pollen evidence for an early to mid Holocene humid period in the eastern Sahara', *Nature*, 314: 352–5.

—— and HAYNES, C. V. (1987), 'Holocene vegetation zonation in the eastern Sahara', *Nature*, 330: 645–7.

ROBERTS, N. (1983), 'Age, palaeoenvironments and climatic significance of Late Pleistocene Konya Lake, Turkey', *Quaternary Research*, 19: 154–71.

—— (1989), *The Holocene: an environmental history.* (Oxford).

—— EROL, O., DE MESSIER, T., and UERPMANN, H. P. (1979), 'Radiocarbon chronology of late Pleistocene Konya Lake, Turkey', *Nature*, 281: 662–4.

RODDA, J. C. (1970), 'Rainfall excesses in the United Kingdom', *Transactions, Institute of British Geographers*, 49: 49–60.

ROGNON, P. (1976) (ed.), 'Oscillations climatiques au Sahara depuis 40 000 ans', *Révue de géographie physique et de géologie dynamique*, 18: 147–282.

RONA, E. and EMILIANI, C. (1969), 'Absolute dating of Caribbean cores P6304–8 and P6304–9', *Science*, 163: 66–8.

RONAI, A. (1965), 'Neo-tectonic subsidence in the Hungarian basin', *Geological Society of American Special Paper*, 84: 219–32.

ROSE, J. (1981), 'Raised shorelines', in A. S. Goudie (ed.), *Geomorphological Techniques* (London): 327–41.

——(1987), 'Status of the Wolstonian Glaciation in the British Quaternary', *Quaternary Newsletter*, 53: 1–9.

——(1990), 'Raised shorelines', in A. S. Goudie (ed.), *Geomorphological Techniques* (London), 456–75.

——BOARDMAN, J., KEMP, R. A., and WHITEMAN, C. A. (1985), 'Palaeosols and the interpretation of the British Quaternary stratigraphy', in K. S. Richards, R. R. Arnett, and S. Ellis (eds.), *Geomorphology and Soils* (London): 348–75.

ROSEMAN, N. (1963), 'Climatic fluctuations in the Middle East during the period of instrumental record', *Arid Zone Research*, 20: 67–73.

ROSSIGNOL-STRICK, M. and DUZER, D. (1980), 'Late Quaternary West African climate inferred from palynology of Atlantic deep-sea cores', *Palaeoecology of Africa*, 12: 227–8.

RUDDIMAN, W. F. and MCINTYRE, A. (1981a), 'The North Atlantic ocean during the last deglaciation', *Palaeogeography, Palaeoclimatology, Palaeoecology*, 35: 145–214.

—— ——(1981b), 'Oceanic mechanisms for amplification of the 23 000-year ice-volume cycle', *Science*, 212: 617–27.

——and RAYMO, M. E. (1988), 'The past three million years: Evolution of climatic variability in the North Atlantic region', *Philosophical Transactions of the Royal Society of London* B, 318: 411–429.

——PRELL, W. L., and RAYMO, M. E. (1989), 'Late Cenozoic uplift in southern Asia and the American west: Rationale for general circulation modeling experiments', *Journal of Geophysical Research*, 94, D: 18379–91.

RUSSELL, F. S., SOUTHWARD, A. J., BOALCH, G. T., and BUTLER, E. I. (1971), 'Changes in biological conditions in the English Channel off Plymouth during the last half-century', *Nature*, 234: 468–70.

RUSSELL, R. J. (1967), 'Aspects of coastal morphology', *Geografiska Annaler*, 49: 299–309.

RUST, U. (1984), 'Geomorphic evidence of Quaternary climatic changes in Etosha, South-West Africa/Namibia', in J. C. Vogel (ed.), *Late Cainozoic palaeoclimates of the southern hemisphere* (Rotterdam): 279–86.

SALINGER, M. J. (1976), 'New Zealand temperatures since 1300 AD', *Nature*, 260: 310–11.

——and GUNN, J. M. (1975), 'Recent climatic warming around New Zealand', *Nature*, 256: 396–8.

SARMIENTO, J. L. and TOGGWEILER, J. R. (1984), 'A new model for the role of the oceans in determining atmospheric P_{CO_2}', *Nature*, 308: 621–4.

SARNTHEIN, M. (1972), 'Sediments and history of the Post-glacial transgression in the Persian Gulf and north-west Gulf of Oman', *Marine Geology*, 12: 245–66.

—— (1978), 'Sand deserts during Glacial Maximum and climatic optimum', *Nature*, 272: 43–6.

—— and DIESTER-HASS, L. (1977), 'Eolian sand turbidities', *Journal of Sedimentary Petrology*, 47: 868–90.

—— and KOOPMAN, B. (1980), 'Late Quaternary deep-sea record of north-west African dust supply and wind circulation', *Palaeoecology of Africa*, 12: 238–53.

—— WINN, K., DUPLESSY, J. C., and FONTUGNE, M. R. (1988), 'Global variations of surface ocean productivity in low and mid latitudes: Influence of CO_2 reservoirs on the deep ocean and atmosphere during the last 21 000 years', *Paleoceanography*, 3: 361–99.

SAUER, C. O. (1948), 'Environment and culture during the last deglaciation', *Proceedings, American Philosophical Society*, 92: 65–77.

—— (1968), *Northern mists* (University of California Press): 204 ff.

SAWYER, J. S. (1971), 'Possible effects of human activity on world climate', *Weather*, 26: 251–62.

SCHELL, I. I. (1961), 'Recent evidence about the nature of climatic changes and its implications', *Annals, New York Academy of Science*, 95: 251–70.

—— (1962), 'On the iceberg severity off Newfoundland and its prediction', *Journal of Glaciology*, 4: 161–72.

—— (1974), 'On the lag in the response of the ocean during a climatic change', *Climatic Research Unit Research Publication*, 2: 85–93.

SCHOFIELD, J. C. (1960), 'Sea-level fluctuations during the past four thousand years', *Nature*, 185: 836.

SCHOLL, D. W. (1964), 'Recent sedimentary record in mangrove swamps and rise in sea-level over the western coast of Florida', *Marine Geology*, 1: 344–66.

SCHUBERT, C. (1984), 'The Pleistocene and recent extent of the glaciers of the Sierra Nevada de Merida, Venezuela', *Erdwissenschaftliche Forschung*, 18: 269–78.

SCHUMM, S. A. (1963), 'The disparity between present rates of denudation and orogeny', *United States Geological Survey Professional Paper* 454–H: 13 ff.

SEDDON, B. (1971), *Introduction to biogeography* (London).

SEGOTA, T. (1966), 'Quaternary temperature changes in central Europe', *Erdkunde*, 20: 110–18.

—— (1967), 'Palaeotemperature changes in the Upper and Middle Pleistocene', *Eiszeitalte und Gegenwart*, 18: 127–41.

SELBY, M. J. (1973), 'Antarctica: the key to the Ice Age', *New Zealand Geographer*, 29: 134–50.

SELF, F. S. and SPARKS, R. J. S. (eds.) (1981), *Tephra studies*, (Dordrecht).

SEREBRYANNY, L. R. (1969), 'L'Apport de la Radiochronométrie et l'étude de l'histoire tardi-Quaternaire des Régions de Glaciation Ancienne de la Plaine Russe', *Révue de Geographie Physique et Geologie Dynamique*, 11: 293–307.

SERVANT-VILDARY S. (1973), 'Le Plio-Quaternaire ancien du Tchad: Évolution des associations de diatomées, stratigraphie, paléogéographie', *Cahiers ORSTOM*,

Sèrie Géologique, 5: 169–217.

SHACKLETON, N. J. (1967), 'Oxygen isotopic analyses and Pleistocene temperatures reassessed', *Nature*, 215: 15–17.

—— (1975), 'The stratigraphic record of deep-sea cores and its implications for the assessment of glacials, interglacials, stadials and interstadials in the mid-Pleistocene', in K. W. Butzer and G. L. Isaac (eds.), *After the Australopithecines* (The Hague): 1–24.

—— and 16 others (1984), 'Oxygen isotope calibration of the onset of ice-rafting and history of glaciation in the North Atlantic region', *Nature*, 307: 620–3.

—— (1987), 'Oxygen isotopes, ice volume and sea-level', *Quaternary Science Reviews*, 6: 183–90.

—— and OPDYKE, N. D. (1977), 'Oxygen isotope and palaeomagnetic evidence for early northern hemisphere glaciation', *Nature*, 270: 216–19.

SHANNAN, L., EVENARI, M., and TADMOR, N. H. (1967), 'Rainfall patterns in the central Negev Desert', *Israeli Exploration Journal*, 17: 163–84.

SHAW, P. A. and THOMAS, D. S. G. (1988), 'Lake Caprivi: A late Quaternary link between the Zambezi and middle Kalahari drainage system', *Zeitschrift für Geomorphologie* 32: 329–37.

SHENNAN, I. (1983), 'Flandrian and late Devensian sea-level changes and crustal movements in England and Wales', in D. E. Smith and A. G. Dawson (eds.), *Shorelines and isostasy* (London): 255–83.

—— (1986), 'Flandrian sea-level changes in the Fenland I: The geographical setting and evidence of relative sea-level changes', *Journal of Quaternary Science*, 1/2: 119–53.

—— (1986), 'Flandrian sea-level changes in the Fenland II: Tendencies of sea-level movement, altitudinal changes, and local and regional factors', *Journal of Quaternary Science*, 1: 155–79.

SHEPARD, F. P. (1963), 'Thirty-five thousand years of sea-level', in T. Clements (ed.), *Essays in marine geology in honor of K. O. Emery* (Los Angeles).

SHOTTON, F. W. (1966), 'Problems and contributions of methods of absolute dating within the Pleistocene Period', *Quarterly Journal of the Geological Society*, 122: 357–83.

—— (ed.) (1977), *British Quaternary Studies* (Oxford).

—— (1983), 'The Wolstonian stage of the British Pleistocene in and around its type area of the English Midlands', *Quaternary Science Reviews*, 2: 261–80.

SIEGENTHALER, U. and WENK, TH. (1984), 'Rapid atmospheric CO_2 variations and ocean circulation', *Nature*, 308: 624–6.

SIESSER, W. G. (1978), 'Aridification of the Namib Desert, evidence from oceanic cores', in E. M. van Zinderen Bakker (ed.), *Antarctic glacial history and world Palaeoenvironments* (Rotterdam): 105–13.

—— (1980), 'Late Miocene origin of the Benguela upwelling system off northern Namibia', *Science*, 208: 283–5.

SIMMONS, I. G. and TOOLEY, M. J. (eds.) (1981), *The environment in British prehistory* (London).

SINGH, G. (1971), 'The Indus valley culture seen in context of Post-Glacial climate and ecological Studies in north-west India', *Archaeology and Anthropology in Oceania*, 6: 177–89.

SISSONS, J. B. (1976), *Scotland* (London).

——(1980), 'The Loch Lomond Advance in the Lake District, northern England', *Transactions, Royal Society of Edinburgh, Earth Sciences*, 71: 13–27.

SLAUGHTER, B. H. (1967), 'Animal ranges as a clue to Late Pleistocene extinction', in P. S. Martin and H. E. Wright (eds.) *Pleistocene Extinctions*: 155–67.

SLAYMAKER, H. O., HOWE, G. M., and HARDING, D. M. (1967), 'Some aspects of the flood hydrology of the upper catchments of the Severn and Wye', *Transactions, Institute of British Geographers*, 41: 33–58.

SMALLEY, I. J. and VITA-FINZI, C. (1968), 'The formation of fine particles in sandy deserts and the nature of "desert" loess', *Journal of Sedimentary Petrology*, 38: 766–74.

SMITH, A. G. (1970), 'The influence of Mesolithic and Neolithic man on British vegetation: a discussion', in D. Walker and R. G. West (eds.), *The vegetational history of the British Isles* (Cambridge): 81–96.

——(1981) 'The neolithic', in I. G. Simmons and M. J. Tooley (eds.), *The environment in British prehistory* (London).

SMITH, C. G. (1967), 'Winters at Oxford since 1815', *Oxford Magazine*, March: 4 ff.

SMITH, D. E. and DAWSON, A. G. (eds.) (1983), *Shorelines and Isostasy* (London).

SMITH, G. I. (1968), 'Late-Quaternary geologic and climatic history of Searles Lake, Southeastern California', in R. B. Morrison and H. E. Wright (eds.), *Means of correlation of Quaternary Successions* (University of Utah Press).

——and STREET-PERROTT, F. A. (1983), 'Pluvial lakes of the western United States', in S. C. Porter (ed.), *Late-Quaternary environments of the United States* (London): 190–212.

SMITH, H. T. U. (1965), 'Dune morphology and chronology in central and western Nebraska', *Journal of Geology*, 73: 557–78.

SMITH, P. J. (1973), *Topics in geophysics* (Bletchley, Bucks.).

SNYDER, C. T. and LANGBEIN, W. B. (1962), 'The Pleistocene lake in Spring Valley, Nevada and its climatic implications', *Journal of Geophysical Research*, 67: 2385–94.

SOLECKI, R. (1963), 'Prehistory in Shanidar Valley, northern Iraq', *Science*, 139: 179–83.

SOMBROEK, A. G., MBUVI, J. P., and OKWARO, H. W. (1976), 'Soils of the semi-arid savanna zone of north-eastern Kenya', *Kenya Soil Survey, Miscellaneous Soil Paper*, No. M2.

SONNTAG, C., THORWEITE, RUDOLPH, J., LOHNERT, E. P., JUNGHANS, C., MUNNICK, K. O., KLITZSCH, E., EL SHAZLY, E. M., and SWAILEM, F. M. (1980), 'Isotopic indentification of Saharan groundwater: Groundwater formation in the past', *Palaeoecology of Africa*, 12: 159–71.

SPARKS, B. W. and WEST, R. G. (1972), *The Ice Age in Britain* (London): 302 ff.

SPENCER, T. and DOUGLAS, I. (1985), 'The significance of environmental change: diversity, disturbance and tropical ecosystems', in I. Douglas and T. Spencer (eds.), *Environmental change and tropical geomorphology* (London): 13–33.

STAGER, J. C. (1988), 'Environmental changes at Lake Cheshi, Zambia since 40 000 years BP', *Quaternary Research*, 29: 54–65.

STEARNS, C. E. and THURBER, D. L. (1967), 'Th^{230}/U^{234} Dates of Late Pleistocene marine fossils from the Mediterranean and Moroccan littorals', *Progress in Oceanography*, 4: 293–305.

STEENSBERG, A. (1951), 'Archaeological dating of the climatic change in north Europe about AD 1300', *Nature*, 168: 692–4.

STEPHENSON, R. L. (1965), 'Quaternary human evolution of the Plains', in H. E. Wright and D. G. Frey (eds.), *The Quaternary of the U.S.A.*: 685–707.

STOCKTON, E. W. (1990), 'Climatic variability on the scale of decades to centuries', *Climatic Change*, 16: 173–83.

STODDART, D. R. (1969), 'Sea-level change and the origin of sand cays: radiometric evidence', *Journal of the Marine Biological Association of India*, 11: 44–58.

—— (1971), 'Environment and history in Indian Ocean reef morphology', *Symposium, Zoological Society of London*, 28: 3–38.

—— (1973), 'Coral reefs: the last two million years', *Geography*, 58: 313–23.

STREET, F. A. (1977), 'Late Quaternary precipitation estimates for the Ziway–Shala basin, southern Ethiopia', *Palaeoecology of Africa*, 10/11: 135–43.

—— and GASSE, F. (1981), 'Recent developments in research into the Quaternary climatic history of the Sahara', in J. A. Allan (ed.), *The Sahara: Ecological change and early economic history* (London): 7–28.

—— and GROVE, A. T. (1976), 'Environmental and climatic implications of Late Quaternary lake level fluctuations in Africa', *Nature*, 261: 385–90.

—— —— (1979), 'Global maps of lake-level fluctuations since 30 000 yr. BP', *Quaternary Research*, 12: 83–118.

STREET-PERROTT, F. A., MITCHELL, J. F. B., MARCHAND, D. S., and BRUNNER, J. S. (1990), 'Milankovitch and albedo forcing of the tropical monsoons: a comparison of geological evidence and numerical simulations for 9000 y BP,' *Transactions, Royal Society of Edinburgh: Earth Sciences*, 81: 407–27.

STRIDE, A. H. (1959), 'On the origin of the Dogger Bank in the North Sea', *Geological Magazine*, 96: 33–44.

STRONGFELLOW, M. F. (1974), 'Lightning incidence in Britain and the solar cycle', *Nature*, 249: 332–3.

SUMBLER, M. G. (1983), 'A new look at the type Wolstonian glacial deposits of Central England', *Proceedings, Geologists' Association*, 94: 23–31.

SUTHERLAND, D. G. (1984), 'The Quaternary deposits and landforms of Scotland and the neighbouring shelves: a review', *Quaternary Science Reviews*, 3: 157–254.

TALBOT, M. R. (1981), 'Holocene changes in tropical wind intensity and rainfall: Evidence from southeast Ghana', *Quaternary Research*, 16: 201–20.

TALLIS, J. H. (1975), 'Tree remains in southern Pennine peats', *Nature*, 256: 483–4.

TANNER, W. F. (1968), 'Tertiary sea-level Symposium: Introduction', *Palaeogeography, Palaeoclimatology, Palaeoecology*, 5: 7–14.

TARLING, D. H. and TARLING, M. H. (1971), *Continental drift* (London).

TAUBER, H. (1970), 'The Scandinavian varve chronology and C14 dating', in I. U. Olsson (ed.), *Radiocarbon variation and absolute chronology* (New York): 173–96.

THEAKSTONE, W. H. (1965), 'Recent changes in the Glaciers of Svartisen', *Journal of Glaciology*, 5: 411–31.

THIEDE, J. (1979), 'Wind regimes over the late Quaternary southwest Pacific Ocean', *Geology*, 7: 259–62.

THOM, B. G. (1973), 'The dilemma of high interstadial sea-levels during the Last Glaciation', *Progress in Geography*, 5: 167–246.

——HAILS, J. R. and MARTIN, A. R. H. (1969), 'Radiocarbon evidence against Post-Glacial higher sea-levels in eastern Australia', *Marine Geology*, 7: 161–8.

THOMAS, D. S. G. (1984), 'Ancient ergs of the former arid zones of Zimbabwe, Zambia and Angola', *Transactions, Institute of British Geographers*, NS9: 75–88.

——and SHAW, P. A. (1991), *The Kalahari environment* (Cambridge).

THOMPSON, L. G., MOSLEY-THOMPSON, E., DANSGAARD, W., and GROOTES, P. M. (1986), 'The Little Ice Age as recorded in the stratigraphy of the tropical Quelccaya Ice Cap', *Science*, 234: 361–4.

THOMSON, J. and WALTON, A. (1972), 'Redetermination of chronology of Aldabra atoll by ^{230}Th/^{234}U Dating', *Nature*, 240: 145.

THORARINSSON, S. (1940), 'Recent glacier shrinkage and eustatic changes of sea-level', *Geografiska Annaler*, 22: 131–59.

THURBER, D. L. *et al.* (1965), 'Uranium-Series ages of Pacific atoll coral', *Science*, 149: 55–8.

TJIA, H. D. (1970), 'Rates of diastrophic movement during the Quaternary in Indonesia', *Geologie en Mijnbouw*, 49: 335–8.

TOLONEN, K. (1986), 'Charred particle analysis', in B. E. Berglund (ed.), *Handbook of Holocene palaeoecology and palaeohydrology* (Chichester): 485–96.

TOOLEY, M. J. (1978), *Sea-level changes: north-west England during the Flandrian stage* (Oxford).

TREMBOUR, F. and FRIEDMAN, I. (1984), 'The present status of obsidian hydration dating', in W. C. Mahanney (ed.), *Quaternary dating methods* (Amsterdam): 141–51.

TRICART, J. (1974), 'Existence de Periodes Seches au Quaternaire en Amazonie et dans les Regions Voisines', *Révue de Géomorphologie Dynamique*, 23: 145–58.

TRILSBACH, A. and HULME, M. (1984), 'Recent rainfall changes in central Sudan and their physical and human implications', *Transactions of the Institute of British Geographers*, 9: 280–98.

TROELS-SMITH, J. (1956), 'Neolithic period in Switzerland and Denmark', *Science*, 124: 876–9.

TUCKER, G. B. (1975), 'Climate: is Australia's changing?', *Search*, 6: 323–8.

TURNER, C. and WEST, R. G. (1968), 'The subdivision and zonation of interglacial periods', *Eiszeitalter und Gegenwart*, 19: 93–101.

TURNER, J. (1962), 'The Tilia decline: an anthropogenic interpretation', *New Phytologist*, 61: 328–41.

TWIDALE, C. R. (1971), *Structural Landforms* (Cambridge, Mass.): 247 ff.

——(1972), 'Landform development in the Lake Eyre region, Australia', *Geographical Review*, 62: 40–70.

TYSON, P. D. (1986), *Climatic change and variability in southern Africa* (Cape Town).

United States Committee for the Global Atmospheric Research Program (1975), *Understanding climatic changes: A program for action* (Washington, DC).

UTTERSTRÖM, G. (1955), 'Climatic fluctuations and population problems in early modern history', *Scandinavian History Review*, 3: 1–47.

VALENTINE, J. W. and VEEH, H. H. (1969), 'Radiometric ages of Pleistocene terraces from San Nicolas Island, California', *Bulletin Geological Society of America*, 80: 1415–18.

VAN ANDEL, T. H. (1989), 'Late Quaternary sea-level changes and archaeology', *Antiquity*, 63: 733–45.

VAN CAMPO, E., DUPLESSY, J. C., PRELL, W. L., BARRATT, N., and SABATIER, R. (1990), 'Comparison of terrestrial and marine temperature estimates for the past 135 kyr off south-west Africa: a test for GCM simulations of palaeoclimate', *Nature*, 348: 209–12.

VAN VEEN, J. (1954), 'Tide-gauges, subsidence-gauges and flood-stones in the Netherlands', *Geologie en Mijnbouw*, 33: 214–19.

VAN ZINDEREN BAKKER, E. M. (1962), 'A Late Glacial and Post-Glacial correlation between East Africa and Europe', *Nature*, 194: 201.

——(1984), 'Elements for the chronology of late Cainozoic African climate', in W. C. Mahoney (ed.), *Correlation of Quaternary chronologies* (Norwich): 23–37.

VEEH, H. H., and CHAPPELL, J. (1970), 'Astronomical theory of climatic change: Support from New Guinea', *Science*, 167: 862–5.

——and GIEGENGACK, R. (1970), 'Uranium-Series ages of corals from the Red Sea', *Nature*, 226: 155–6.

——and VALENTINE, J. W. (1967), 'Radiometric ages of Pleistocene fossils from Cayucos, California', *Bulletin, Geological Society of America*, 78: 547–9.

——and VEEVERS, J. J. (1970), 'Sea-level at—175 M off the Great Barrier Reef, 13 600 to 17 000 years ago', *Nature*, 226: 526–7.

VEENSTRA, H. J. (1970), 'Quaternary North Sea coasts', *Quaternaria*, 12: 169–79.

VELICHKO, A. A. *et al.* (eds.) (1984), *Late Quaternary environments of the Soviet Union* (Ann Arbor, Mich.).

—— and FAUSTOVA, M. A. (1986), 'Glaciations in the East European region of the USSR', *Quaternary Science Reviews*, 5: 447–61.

VERSTAPPEN, H. TH. (1970), 'Aeolian geomorphology of the Thar Desert and palaeoclimates', *Zeitschrift für Geomorphologie, Supplementband*, 10: 104–20.

—— (1975), 'On palaeo climates and landform development in Malesia', in G. J. Bartstra and W. A. Casparie (eds.), *Modern Quaternary research in Southeast Asia* (Rotterdam): 3–35.

VIBE, C. (1967), 'Arctic animals in relation to climatic fluctuations', *Meddelelser om Grønland*, 170: 227 ff.

VITA-FINZI, C. (1969), *The Mediterranean valleys* (Cambridge): 140 ff.

—— (1973), *Recent earth history* (London): 130 ff.

—— (1986), *Recent earth movements: an introduction to neotectonics* (London).

VIVIAN, R. (1975), *Les Glaciers des Alpes Occidentales* (Grenoble).

VOGEL, J. C. (ed.) (1984), *Late Cainozoic palaeoclimates of the southern hemisphere* (Rotterdam).

VÖLKEL, J. and GRUNERT, J. (1990), 'To the problem of dune formation and dune weathering during the Late Pleistocene and Holocene in the southern Sahara and the Sahel', *Zeitschrift für Geomorphologie*, 34: 1–17.

VUILLEUMIER, B. S. (1971), 'Pleistocene changes in the fauna and flora of South America', *Science*, 173: 771–80.

WALSH, R. P., HUDSON, R. N., and HOWELLS, K. A. (1982), 'Changes in the magnitude–frequency of flooding and heavy rainfalls in the Swansea Valley since 1875', *Cambria*, 9: 36–60.

—— HULME, M., and CAMPBELL, M. D. (1988), 'Recent rainfall changes and their impact on hydrology and water supply in the semi-arid zone of the Sudan', *Geographical Journal*, 154: 181–98.

WALTON, K. (1966), 'Vertical movements of shorelines in highland Britain: an introduction', *Transactions Institute of British Geographers*, 39: 1–8.

WASSON, R. J. (1984), 'Late Quaternary palaeoenvironments in the desert dunefields of Australia', in J. C. Vogel (ed.) (1984), 419–32.

—— FITCHETT, K., MACKEY, B. and HYDE, R. (1988), 'Large-scale patterns of dune type, spacing and orientation in the Australian continental dunefield', *Australian Geographer*. 19: 80–104.

WEARE, B. C., TEMKIN, R. L., and SNELL, F. M. (1974), 'Aerosols and climate: Some further considerations', *Science*, 186: 827–8.

WEGMANN, E. (1969), 'Changing ideas about moving shorelines', in C. J. Scheer (ed.), *Toward a history of geology* (Cambridge, Mass.): 386–414.

WELLS, G. L. (1983), 'Late glacial circulation over North America revealed by aeolian features', in F. A. Street-Perrott, M. Bevan, and R. Ratcliffe (eds.), *Variation in the global Water Budget* (Dordrecht), 317–30.

WELLS, P. V. (1976), 'Macrofossil analysis of wood rat (*Neotoma*) middens as a key to the Quaternary vegetational history of arid America', *Quaternary Research*, 6: 223–48.

WENDORF, F., SCHILD, R., SAID, R., HAYNES, C. V., GAUTIER, A., and KOBUSIEWICZ, P. (1976), 'The prehistory of the Egyptian Sahara', *Science*, 193: 103–16.

WEST, R. G. (1972), *Pleistocene geology and biology* (London).

WESTERN, D. and PRAET, C. V. (1973), 'Cyclical changes in the habitat and climate of an East African ecosystem', *Nature*, 241: 104–6.

WEXLER, H. (1961), 'Additional comments on the warming trend at Little America, Antarctica', *Weather*, 16: 56.

WEYL, P. K. (1968), 'The role of the oceans in climatic change: a theory of the ice ages', *Meteorological Monographs*, 8: 37–62.

WHITTOW, J. B. (1973), 'Shoreline evolution and the eastern coast of the Irish Sea', *Quaternaria*, 12: 185–96.

—— SHEPHERD, A., GOLDTHORPE, J. E., and TEMPLE, P. H. (1963), 'Observations on the glaciers of the Ruwenzori', *Journal of Glaciology*, 4: 581–616.

—— and WOOD, P. D. (eds.) (1965), *Essays in Geography for Austin Miller* (Reading).

WHYTE, R. O. (1963), 'The significance of climatic change for natural vegetation and agriculture', *Arid Zone Research*, 20: 381–93.

WICKENS, G. E. (1975), 'Changes in the climate and vegetation of the Sudan since 20000 BP', *Boissiera*, 24: 43–65.

WIGLEY, T. M. L. and ATKINSON, T. C. (1977), 'Dry years in south-east England since 1698', *Nature*, 265: 431–4.

—— and JONES, P. D. (1987), 'England and Wales precipitation: a discussion of recent changes in variability and an update to 1985', *Journal of Climatology*, 7: 231–46.

—— INGRAM, M. J., and FARMER, G. (1981), *Climate and History* (Cambridge).

WILLIAMS, G. E. and POLACH, H. A. (1971), 'Radiocarbon dating of arid zone calcareous paleosols', *Bulletin, Geological Society of America*, 82: 3069–86.

WILLIAMS, J. (1975), 'The influence of snowcover on the atmospheric circulation and its role in climatic change: an analysis based on results from the near global circulation model', *Journal of Applied Meteorology*, 14: 137–52.

WILLIAMS, M. A. J. (1989), 'Pleistocene aridity in tropical Africa, Australia and Asia', in I. Douglas and T. Spencer (eds.), *Environmental change and tropical geomorphology* (London): 219–33.

—— and FAURE, H. (eds.) (1980), *The Sahara and the Nile* (Rotterdam).

WILLIAMS, R. B. G. (1969), 'Permafrost and temperature conditions in England during the last glacial period', in T. L. Péwé (ed.), *The periglacial environment* (Montreal), 339–410.

—— (1975), 'The British climate during the Last Glaciation: an interpretation based on periglacial phenomena', in A. E. Wright and F. Moseley (eds.), *Ice ages: ancient and modern* (Liverpool): 95–127.

WILLIS, E. H. (1961), 'Marine transgression sequences in the English Fenlands',

Annals, New York Academy of Sciences, 95: 368–76.

WILSON, A. T. (1964), 'Origin of ice ages: An ice shelf theory for Pleistocene glaciation', *Nature*, 201: 147–9.

——(1978), 'Pioneer agriculture explosion and CO_2 levels in the atmosphere', *Nature*, 273: 40–1.

——HENDY, C. H., and REYNOLDS, C. P. (1979), 'Short-term climate change and New Zealand temperatures during the last millennium', *Nature*, 279: 315–17. 279: 315–17.

WINSTANLEY, D. (1973), 'Rainfall patterns and general atmospheric circulation', *Nature*, 245: 190–4.

WOILLARD, G. (1978), 'Grande Pile peat bog: a continuous pollen record for the last 140 000 years', *Quaternary Research*, 9: 1–21.

——(1979), 'Abrupt end of the last interglacial s.s. in North-East France', *Nature*, 281: 558–62.

WOLFE, J. A. (1978), 'A palaeobotanical interpretation of Tertiary climates in the northern hemisphere', *American Scientist*, 66: 694–703.

WOLLIN, G., ERICSON, D. B., and RYAN, W. B. F. (1971), 'Variations in magnetic intensity and climate changes', *Nature*, 232: 549–51.

—— ——and WOLLIN, J. (1974), 'Geomagnetic variations and climatic change 2 000 000 BC–1970 AD', *Colloques Internationaux du CNRS*, 219: 273–86.

——KUKLA, G. J., ERICSON, D. B., RYAN, W. B. F., and WOLLIN, J. (1973), 'Magnetic intensity and climatic changes 1925–1970', *Nature*, 242: 34–6.

WOOD, C. A. and LOVETT, R. R. (1974), 'Rainfall, drought and the solar cycle', *Nature*, 252: 594–6.

WOODBURY, R. B. (1961), 'Climatic changes and prehistoric agriculture in the south-western United States', *Annals, New York Academy of Sciences*, 95: 705–9.

WOODWELL, G. M. (1978), 'The Carbon Dioxide question', *Scientific American*, 238: 34–43.

WORSLEY, P. (1990), 'Lichenometry', in A. S. Goudie (ed.), *Geomorphological techniques* (London): 422–8.

WORSLEY, T. R. and HERMAN, Y. (1980), 'Episodic ice-free Arctic Ocean in Pliocene and Pleistocene time: calcareous nannofossil evidence', *Science*, 210: 323–5.

WRIGHT, H. E. (1968), 'Climatic change in the eastern Mediterranean region: The natural environment of early food production in the mountains north of Mesopotamia', *Final report, University of Minnesota, Contract NONR–710* (33), Task No. 380–129.

——(ed.) (1983), *Quaternary environments of the United States* (London).

——(ed.) (1984), *Late Quaternary environments of the United States*, ii. *The Holocene* (London).

WRIGHT, H. E. and FREY, D. G. (eds.) (1965), *The Quaternary of the United States* (Princeton University Press).

WRIGHT, W. B. (1937), *The Quaternary ice age* (London).

WYRWOLL, K. H. and MILTON, D. (1976), 'Widespread late Quaternary aridity in western Australia', *Nature*, 264: 429–30.

ZAGWIJN, W. H. (1974), 'The Pliocene–Pleistocene boundary in western and southern Europe', *Boreas*, 3: 75–97.

——(1975), 'Variations in climate as shown by pollen analysis especially in the Lower Pleistocene of Europe', in A. F. Wright and F. Moseley (eds.), *Ice ages: ancient and modern* (Liverpool): 137–52.

ZEUNER, F. E. (1959), *The Pleistocene period: its climate, chronology and faunal successions*, 2nd edn. (London).

ZUBAKOV, V. A. (1969), 'La Chronologie des Variations Climatiques au Cours du Pleistocene en Siberie Occidentale', *Révue de Géographie Physique et Géologie Dynamique*, 11: 315–24.

Index

Index compiled by Frank Pert